Progress in Scale Modeling, Volume II

Kozo Saito • Akihiko Ito • Yuji Nakamura
Kazunori Kuwana
Editors

Progress in Scale Modeling, Volume II

Selections from the International Symposia on Scale Modeling, ISSM VI (2009) and ISSM VII (2013)

Springer

Editors
Kozo Saito
Department of Mechanical Engineering
University of Kentucky
Lexington, KY, USA

Yuji Nakamura
Toyohashi University of Technology
Toyohashi, Japan

Akihiko Ito
Hirosaki University
Hirosaki-shi, Aomori, Japan

Kazunori Kuwana
Yamagata University
Yonezawa-shi, Yamagata, Japan

ISBN 978-3-319-10307-5 ISBN 978-3-319-10308-2 (eBook)
DOI 10.1007/978-3-319-10308-2
Springer Cham Heidelberg New York Dordrecht London

Library of Congress Control Number: 2008932278

Printed on acid-free paper

Springer is part of Springer Science+Business Media (www.springer.com)

Preface

This volume stands as a remedy for the lack of available material on scale modeling and as an essential resource for many university courses, both graduate and undergraduate. *Progress in Scale Modeling, Volume 2* builds on the success of *Volume 1* as the definitive source of information on this powerful simplifying and clarifying tool used by scientists and engineers across many disciplines. The book's diverse slate of authors helps readers "see the forest, when there are many, many trees" by elucidating techniques used when it would be too expensive, or too difficult, to test a system of interest in the field.

Topics addressed in the current edition include scale modeling to study natural disasters and its impacts on environment and structures, developing and testing energy generation systems and methods, and others. This book also

- Enables readers to evaluate essential and salient aspects of profoundly complex systems, mechanisms, and phenomena at scale.
- Offers engineers and designers a new point of view, liberating creative and innovative ideas and solutions.
- Serves the widest range of readers across the engineering disciplines and in science and medicine.

Volume 1 was a selection of seminal papers given at the first 5 sessions of the International Symposium on Scale Modeling (ISSM), which were held between 1988 and 2006. During the intervening years ISSM VI and ISSM VII were held in 2009 and 2013, respectively. The papers in this volume represent those two later symposia. The International Scale Modeling Committee consists of worldwide researchers in scale modeling, and celebrated its 25th anniversary in 2013. The committee is interdisciplinary, attracting researchers from a broad range of scientific and engineering disciplines.

Lexington, KY, USA — Kozo Saito
Hirosaki-shi, Aomori, Japan — Akihiko Ito
Toyohashi, Japan — Yuji Nakamura
Yonezawa-shi, Yamagata, Japan — Kazunori Kuwana

Contents

Part II Engineering Design Performance Evaluation and Fundamental Understanding Using Scale Models

Scale Modeling in the Age of High-Speed Computation

K. Saito and F.A. Williams

Introduction

The first volume on scale modeling, "Progress in Scale Modeling," was published in 2008, collecting a total of 44 articles, selected from the first through fifth international symposia on the topic [1], covering fire and explosion; combustion; materials processing, manufacturing, and environment; and medical applications. This volume, the second on scale modeling, collects a total of 24 papers, selected from 68 papers presented at both the sixth and seventh international symposia. The focus of volume 2 is the use of scale modeling in natural disasters and structural failure, in engineering design and performance evaluation, and in the development of fundamental understanding. Some papers are directed to finding the direct root cause of the events, while others address scaling laws for these problems. In contrast to the first volume, in which only four of the contributions introduced or employed numerical methods, a larger fraction of the papers in the present volume are oriented toward computation. This reflects the exponential growth in capabilities of high-speed computing and thereby calls for consideration of the role of numerical simulations in developing and understanding the phenomena investigated. The computational papers in this volume importantly address the validation of their results through comparison with results of either scale model experiments or the full-scale phenomena.

During the Seventh International Symposium on Scale Modeling in Hirosaki, Japan, the publication committee asked us to write a brief introduction on the basics of scale modeling for the benefit of rather broad audiences, including both experts

K. Saito (✉)
IR4TD, College of Engineering, University of Kentucky, Lexington, KY 40506-0503, USA
e-mail: ksaito@uky.edu

F.A. Williams
Department of Mechanical and Aerospace Engineering, USCD, La Jolla, CA 92093, USA

K. Saito et al. (eds.), *Progress in Scale Modeling, Volume II*,
DOI 10.1007/978-3-319-10308-2_1

and beginners in scale modeling. In thinking about the best way to respond to this request, we reviewed two authoritative textbooks: Emori, Saito, and Sekimoto's *Scale Modeling in Engineering: Theory and Applications* (Mokei Jikken no Riron to Ohyou, in Japanese) [2] and Emori and Schuring's *Scale Modeling in Engineering: Fundamentals and Applications* [3]. The first version of the text [2] was originally published by Gihodo Publishing Co. in 1973, with Emori and Schuring as co-authors, but that version is difficult to find. The second book [3], the English version of the first book, was published in 1977 by Pergamon Press. After the death of R.I. Emori in 1996, Gihodo (a Japanese publisher) asked Saito and Sekimoto to revise the first version, which was written by Emori himself, to address broader audiences, including college students, general engineers, and policy makers. Saito and Sekimoto decided to expand the scope of the presentation on scale modeling, to include computer simulation, Kufu principles (to be discussed more fully below) for obtaining assumptions and for solving problems, and a new method to assess the applicable limits of scale models and of scaling laws.

These two comprehensive textbooks [2, 3] fully cover the basics of scale modeling and provide applications of scale modeling to various fields of engineering and science. Both consist of two different parts: part 1, basics of scale modeling, and part 2, applications. The basics of scale modeling encompass the concepts of scaling and simulation, the definitions of pi numbers, scaling laws and methods for deriving them, the relaxation technique to achieve scaling requirements by relaxing conflicting scaling laws, and case studies which include multidisciplinary engineering and scientific problems. Large portions of the two books are dedicated to case studies because Emori and Schuring [3] believed that most engineering problems are so different that there seldom are general rules that can be employed to obtain scaling relations that are achievable in practice. More often than not, complete sets of scaling laws result in conflicting scaling requirements, as was demonstrated, for example, in a treatment of scaling mass fires by Williams [4]. All three authors, Emori, Schuring, and Williams, thus see the necessity for engineers and scientists to learn how to reduce strict scaling requirements in order to obtain achievable scaling laws. In effect, these authors declare that all successful scale modeling really represents partial scaling (i.e., it is based on various approximations), so that scale modeling is more of an art than a strict science.

This partial scaling nature of successful scale modeling has a direct counterpart in the numerical-simulation techniques that make use of advances in high-speed computation. There are corresponding approximations in the models employed in numerical simulations. To emphasize that in a colorful way, it has been stated [5] that "all models are wrong, but some are useful." Thus, irrespective of whether computers or experiments are considered, there is a common essential underlying type of starting point in the approaches, namely, determining how to make assumptions that can help to obtain useful results. If the assumptions are wrong, then regardless of whether the selected technique is numerical simulation or scale modeling, the final results will be wrong. What is needed (and not easily taught) are ways to make helpful assumptions.

As a result of the above thinking, we chose to address here the unique important basic features of scale modeling that can lead to helpful assumptions. These aspects were only vaguely touched upon in the above texts, or they were not included at all. Thus, this chapter addresses, in the relationship between scale modeling and numerical simulation, the question of how to make reasonably good assumptions (RGA) that contain useful information and valuable new ideas. RGA differs from not good assumptions (NGA), which lead to a dead end without offering new ideas or help. Making assumptions is the first step in deductive scientific methods, including both numerical and scale modeling. Inductive rather than deductive approaches are needed here, since descriptions of successful methods are difficult to find.

An inductive approach to the development of assumptions is Kufu [6], which, in a sense, amounts to a highly developed professional intuition. This chapter attempts to explain the role of Kufu and its extension to scale modeling, Kufu eyes, in obtaining RGA rather than NGA. Reference [2] briefly introduces Kufu as the fourth method in relation to the current three scientific methods, experiment, theory, and numerical computation, but details are not offered in that reference. Readers who are further interested in Kufu may consult References [6–8]. A western version of Kufu may be considered to be "Gut Feelings," as described by Gigerenzer [9]. The University of Kentucky offers a semester-long course (ME 565, Scale Modeling in Engineering) every fall semester using the above texts, which covers additional case study examples and a 6-week-long scale-modeling project, where students can engage in actual Kufu practice. The contents of ME 565 are available through the IR4TD Web site [10].

The following list provides some scale-modeling problems from ME 565 as examples of applications. Interested readers are encouraged to accept the challenges of these problems, through which they will directly experience what we are discussing in this chapter: Kufu and the art of scale modeling.

1. If a six-pound roast of beef requires three hours of cooking time, how many hours would a three-pound roast require at the same temperature?
2. There is a rough trend among different species of mammals that the heavier the weight, the longer the life expectancy. Introduce RGA to obtain the scaling laws between the body weight and the life expectancy among different mammals.
3. Estimate the life expectancy of dinosaurs using the same RGA obtained for mammals and compare the results with literature values to check whether or not the RGA for mammals can be applied to dinosaurs. If not, find the reason why.
4. As a scale-modeling application to automobile safety, develop scaling laws for the collision speed and vehicle mass, which are needed to reconstruct a two-vehicle collision process by designing two different 1/16th-scale model vehicles. Describe the assumptions that are required for the obtained scaling laws.
5. The Tacoma Narrows (suspension-type) Bridge was completed on July 1, 1940 and collapsed on November 7, 1940 under 50 MPH gusty wind conditions. Develop scaling laws to reconstruct this event using a 1/50th-scale model;

determine the ratio of the wind speed between the full scale and the scale model and the ratio of the full-scale bridge weight to that of the scale model. An expanded discussion of this problem is provided later in the section on relaxation and partial scaling.

A student can add examples to this list and can test results by conducting scale model experiments. One reason to introduce these diversified examples is to help students develop a thinking process for solving problems using Kufu and to help them to capture the general principles of scale modeling and apply them to solve other problems.

Throughout this article, in speaking of numerical methods, it will be understood that the use of high-speed computation is intended in its broadest sense. Computers require algorithms for computation, all of which may be considered to be types of mathematical descriptions. For the computational papers in the present volume, the underlying mathematics generally relies on partial differential equations, finite-difference approximations to which are programmed to enable the computations to be performed. There are, however, other kinds of mathematical descriptions that can underlie the algorithms equally well. They may, for example, be based, instead, on ordinary differential equations, on algebraic equations, on integers rather than continuous variables, or on inequalities. Our comments are intended to apply to any such high-speed computation, although specific example will mainly be drawn from partial differential equations. Numerical simulations for problems 1 and 5 of the preceding list, for example, may be based on partial differential equations, while problem 4 might be approached through algebraic equations and problems 2 and 3 through ordinary differential equations.

Scale Modeling and the Law Approach

Scale modeling involves the use of physical models, whose size is either smaller or larger than that of the full-scale system (which is called the prototype in the technical literature), to conduct experiments for the purpose of testing the scaling laws, attempting to validate them (confirmation tests), or to use validated scale models in various applications (production tests). Figure 1 shows a flow diagram that describes the concepts of scale modeling in relationship to those of numerical modeling. Both scale modeling and numerical modeling begin, in the first step, with assumptions, as was indicated in the introduction. There is, however, no well-established logical prescription available for obtaining RGA, and an individual researcher must rely on her/his unique thinking or intuition to arrive at RGA. Through this struggle, some may be able to develop Kufu eyes to obtain RGA rather than NGA. We offer further explanation of Kufu eyes later, because, in a major respect, it is the most important element in Fig. 1 concerning how to obtain RGA. This step is followed by the deductive processes indicated in the figure, which are different, depending on whether the approach to be applied is scale

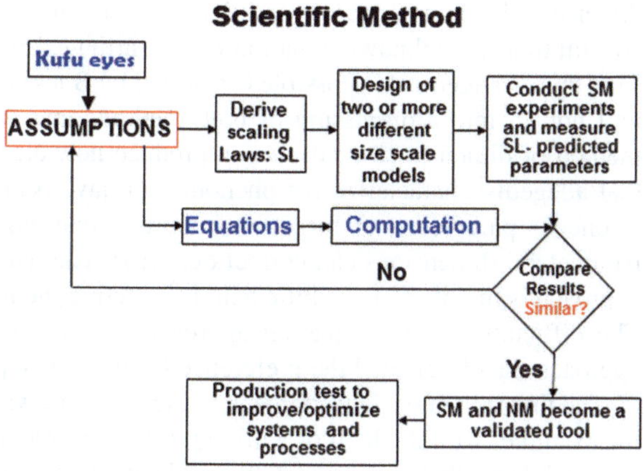

Fig. 1 Basic concept of scale modeling and numerical simulation

modeling or numerical simulation. In other words, if we come up with RGA, then the final result can be right, independent of the technique, while if we start with NGA, then the final result must be wrong. Moreover, the two approaches can be complementary, as discussed in a later section on the relationships between scale modeling and numerical modeling.

We have indicated that scale modeling involves obtaining pi numbers, for which three different kinds of approaches are available: the parameter approach (identifying parameters involved in the full-scale phenomena and then using Buckingham's pi theorem), the equation approach (ascertaining the governing equations and then converting them into non-dimensional form), and the law approach (determining the governing physical laws and then describing each of them with representative characteristic parameters). The above texts [2, 3] detail all three approaches and recommend the use of the law approach in most cases, since it can help researchers to better understand the governing mechanisms and hence provides a better chance of reaching RGA. Let us first address the relationships of the equation approach and the parameter approach to the law approach.

Contrary to the law approach, the equation approach requires much more detailed information about the phenomenon because it rests upon the development of governing equations by identifying specific interactions among the various parameters involved. This can be quite difficult, and also it often is unnecessary for successful scale modeling. When fully developed equations are available, the equation approach offers the possibility of numerical simulation by directly solving the governing equations. It therefore also underlies addressing problems by high-speed computation. Nevertheless, it is less universally available in that the law approach can give pi numbers even when such detailed parametric relationships are not known.

Concerning the parameter approach, the authors of the above texts [2, 3] provide the following explanation: "it assumes that the governing equations are dimensionally homogeneous, and that, because of this, the governing parameters constituting the unknown equation can be arranged in dimensionless pi numbers. That this arrangement is possible is asserted by Buckingham's pi theorem. Here we will not go into a discussion of both methods, but we must emphasize, that no method of dimensional analysis can produce new arguments after the parameters that allegedly characterize the phenomenon have been chosen. If we should pick erroneous parameters or forget an important one, no subsequent analysis of the parameters' dimensions can correct our error. The real weakness of the parameter approach is that it offers so little help in selecting the important parameters [2, 3]." The difficulties of the equation approach and the weaknesses of the parameter approach thus motivated the preference for the law approach.

The first step of the law approach in Fig. 1 begins with the identification of RGA, for example, with the help of Kufu eyes. Here researchers are required to identify physical laws that govern the full-scale phenomena. Physical laws in the law approach are often written in terms of forces and energy, since numerous engineering problems deal with them, as has been explained [2, 3]. For example, if researchers have initially identified as three possible forces in a full-scale phenomenon an inertia force, a viscous force, and a gravity force, then, with further thinking and careful observations, they feel justified in making the assumption that the major effects arise from inertia force and gravity forces, while the viscous force is secondary, they end up with one pi number (the presumed controlling dimensionless number), the ratio of the inertia and gravity forces, which is the Froude number. This then becomes the scaling law that appears in the second box along the scale-modeling path in Fig. 1. Such scaling laws give us guidelines and criteria on how to design scale model experiments.

The third box along the scale-modeling path in the figure recommends designing two, three, or possibly more experiments of different sizes. More experiments provide better tests of confirmation but also are more expensive, necessitating trade-offs. In selecting scale models, relatively large-scale ratios are desirable, because normally the larger the scale ratio, the easier it is to test the accuracy of the assumptions. But, on the other hand, all scaling laws apply only over limited ranges of conditions, and therefore, if the scale ratio is too large, there is a higher chance of not satisfying the assumptions of the relationships between the full scale and the scale model that are being tested, and the scaling law becomes invalid. Hence, in this respect as well, trade-offs are needed.

When full-scale data are not available, measurements for at least two different size scale models must be performed to test any scaling law. The fourth box along the scale-modeling path in Fig. 1 represents conducting these scale model tests and measuring the parameters whose values are predicted by the scaling laws. Comparing the results of the measurements with the predictions completes the confirmation test. If the results are deemed to be sufficiently similar, then the scaling law can be considered to be sufficiently well validated for the tests performed; otherwise, the assumptions must be revised to obtain different scaling laws that are then tested by repeating the same procedures. This five-step process is called confirmation testing in scale modeling, and, if successful, it can be followed by production

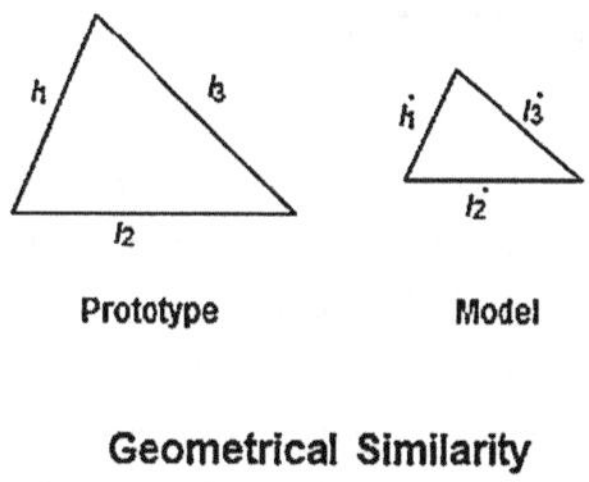

$$\frac{l_1'}{l_1} = \frac{l_2'}{l_2} = \frac{l_3'}{l_3} = \text{const.}$$

This can be modified into :

$$\frac{l_1'}{l_2'} = \frac{l_1}{l_2} \text{ and } \frac{l_2'}{l_3'} = \frac{l_2}{l_3}$$

$$\pi_i' = \pi_i$$

Fig. 2 Geometrical similarity and two different mathematical expressions

Similarity

Geometrical Similarity
- Different scale lengths

Dynamical Similarity
- Momentum (velocity, pressure, mass)
- Energy (temperature, heat, work)
- Chemical Reaction (chemical species concentration, temperature, reaction rate)
- Electro-Magnetic field (current, voltage, power)

Fig. 3 Geometrical similarity and dynamic similarity

tests, in which the scaling laws are applied at different scales of practical interest. It must be kept in mind, however, that the confirmation applies only over the scale range tested and may not represent well the full-scale phenomenon if its scale is too different from those tested.

As a simple example, Fig. 2 illustrates geometrical similarity with two different mathematical expressions. The first formula, $l_1'/l_l = l_2'/l_2 = l_3'/l_3 = \text{const.}$, is the normal definition of geometrical similarity, where the prime stands for the scale model and the absence of the prime the full scale. When the first expression is converted to the second formula, $l_1/l_2 = l_1'/l_2'$ *and* $l_1/l_3 = l_1'/l_3'$, it expresses a different view of the physical meaning. The left-hand side of the second formula relates different length-scale ratios of full-scale figure, while the right-hand sides represent the corresponding length-scale ratios of the scale model. When lengths are replaced by the governing physical laws, e.g., $l_1 =$ inertia force, $l_2 =$ gravity force, and $l_3 =$ viscous force, then $\text{Fr} = \text{Fr}'$ and $\text{Re} = \text{Re}'$ will be obtained, from the first two, and first and last, respectively, for the Froude number Fr and the Reynolds number Re. In this way the second geometrical similarity formula can be extended to dynamical similarity, as indicated in Fig. 3. The terms shown in parentheses in that figure, velocity, pressure, temperature, heat, etc., are parameters for which researchers may be interested in obtaining scaling correlations. These particular parameters can be expressed as combination of five primary dimensional parameters, namely, length, time, force, temperature, and electric current, which encompass a very wide range of problems of engineering interest.

Relaxation and Partial Scaling

It was stated above that all scale modeling is partial scaling in nature, which is consistent with scale modeling seeking approximate solutions to practical engineering problems. It is normal for researchers or engineers to be faced with conflicting scaling requirements when they first develop scaling laws. It is important, therefore, to find a way to relax the conflicting requirements to obtain achievable scaling. Emori and Schuring [3] use the term relaxation rather than reduction to emphasize that the process includes creative thinking to overcome conflicting scaling requirements, not only reducing the number of pi numbers. Phenomena are often called "scalable" if they naturally satisfy scaling requirements, but that rarely occurs. Instead, practitioners generally must seek approximate scaling laws by selecting major governing physical laws and ignoring marginal and nonessential laws. Relaxation is the process of doing this. Emori and Schuring [3] suggest several different ways to achieve relaxation, among which two worthy concepts are disregarding weak laws and circumventing strong laws.

In an extreme example of disregarding weak laws, Williams [4, 11] derived a total of 28 pi numbers for mass fires and then suggested keeping only one or two to achieve useful scaling, primarily retaining Fr, since inertial and buoyant forces are the major forces in fires. This is consistent with the first step in the guidelines of Emori and Schuring [3], who state "As a first step in relaxation, it is helpful to determine whether the laws causing scaling conflicts are governing the given system with equal or unequal strength. If they are governing it with unequal strength, the weakest laws can perhaps be disregarded within segments of the investigation if not throughout its entire range. If their influences are equally strong, none can be neglected outright, but correct results may still be attained by skillful circumvention of the most disturbing law."

Circumventing strong laws are to be applied for problems in which none of the conflicting laws can be disregarded. In this approach, the phenomenon is broken into a number of special cases, each governed by fewer laws than the entire phenomenon. If a large enough number of special cases is investigated, and if they are all relatively independent of each other, then an approximation to the total phenomenon can be obtained by superimposing results of the special cases.

As a specific example of disregarding weak laws and circumventing strong laws, consider the collapse of the Tacoma Narrows Bridge, example 5 in the introduction. An interesting discussion of the circumstances surrounding this disaster was published not long ago [12]. This event has been looked at from different points of view, including resonance between the frequency of Karman vortex shedding and the natural frequency of the bridge structure, as well as aeroelastic fluttering, independent of vortex shedding. Prior to its collapse, the bridge experienced sustained swinging motions, which may have accelerated the fatigue of its materials. It would be interesting to search for the root cause of this dramatic event by scale modeling, and the following suggestions are offered as a possible approach along these lines to help to understand the situation.

The wind creates both inertial (through pressure) and viscous forces on the structure and the roadway, while the gravity force acting on the bridge, the inertial force of the bridge itself, the force associated with the material strength of the bridge's elements, and the bridge system's dynamic response force all may be relevant. By combining these effects, it may be concluded that there are three independent pi numbers, which may be arranged as a Froude number, Fr = (inertia force)/(gravity force); a Strouhal number, St = (inertia force)/(dynamic response force); and a Cauchy number, Ca = (inertia force)/(material-strength force). The first of these arises by comparing the pressure and gravity forces on the bridge (even though buoyancy is negligible), and the second implicitly invokes a Reynolds number (ratio of inertial to viscous forces), since the frequency of shedding of Karman vortices depends on the Reynolds number.

Scaling requirements are next extracted from these parameters, with the idea of constructing and testing a model of the bridge at reduced scale in a wind tunnel in which the wind velocity v can be imposed externally. From the constancy of Fr, scaling would require $v \sim \ell^{1/2}$; from St, $M\omega^2/\kappa$ must remain constant; and from Ca, $Mv^2/\sigma\ell^3$ must remain constant, where M denotes the mass of the bridge, ω is the shedding frequency of Karman vortices, κ represents the dynamic constant of the bridge system for harmonic vibrations, and σ stands for the strength constant of the material of the bridge. Since St depends on the Reynolds number Re, constancy of which would require $v \sim \ell^{-1}$, the incompatibility of the three requirements is immediately apparent in the absence of adjustments of the mass and material properties of the model. For example, the normal dependence $M \sim \ell^3$ would not maintain constancy of both Fr and St, nor could it preserve constancy of both Fr and Ca, since constancy of Ca would then necessitate constancy of v. It could therefore be of interest to consider investigating two different special cases to circumvent laws that are incompatible.

From a close look at the swinging motion of the bridge, it may be inferred that the wind blows mainly in the horizontal direction, causing the bridge's twisting and swinging motion, while when the bridge collapses, the motion is mainly in the vertical direction only. Therefore, we can separate this entire event into two different events, namely, the swinging motion caused by gusty wind and the collapse of the bridge. The pre-bridge-collapse event is mainly restricted to the horizontal direction of motion, while the bridge collapse is a vertically oriented event. If these two events can be dealt with separately, we can derive two different scaling laws.

Before the bridge collapses, the inertial force of the wind on the bridge and the bridge's dynamic response force to the imposed wind are important, while the bridge's material-strength force, the inertial force of the bridge, and the gravity force acting on the bridge become important only later, when the bridge collapses. Therefore, St may be considered to be dominant for the horizontal motion, while Fr and Ca become important for the vertical motion. The first scaling law then would be based solely on St, the wind velocity v being increased so that the shedding frequency ω divided by the natural frequency $(\kappa/M)^{1/2}$ of the model equals $\mathrm{St}^{-1/2}$ of the prototype. In the second scaling law, that for the vertical motion, the incompatibility of Fr and Ca scaling must somehow be circumvented. This can be accomplished by adding weights to the scale model, distributed evenly, to enforce the condition $M \sim \ell^2$ under the assumption that $\sigma = \sigma'$ (i.e., using the same material for both the prototype and the scale model). An interesting example of the use of such distributed dummy weights (small lead discs) can be found [13]. This results in $v \sim \ell^{1/2}$ for both Fr and Ca scaling, thereby producing the desired compatibility. So far as we know, no one has reported such scale model experiments, and therefore, it is still unknown whether or not these proposed scaling laws can work.

Kufu and Kufu Eyes

There is an interesting old saying in Japan that stresses the importance of action (which in this case can be interpreted as experiment) over thinking. It can be translated as: one good seeing is better than hearing 100 times; one action is better than observing 100 times [7]. One good experiment can hold a rich deposit of information, which can be extracted by Kufu eyes, while Kufu can help develop intuition, new perspectives, and imagination by carefully observing the experiment and successfully speculating about what is controlling the phenomena that the experiment is showing. The Kufu eyes-based approach to successfully achieving RGA involves highly subjective activities. This is illustrated in Fig. 4, which shows a four-step cyclic process that can be followed in seeking RGA. Here Kufu eyes play an important role in the second step as the source of ideas and creativity for RGA, preceding deductive scientific methods. In this Kufu eyes step, science and art will meet and interact to create a stage from which innovative scale modeling may begin. For this reason, the authors of the text [2] emphasized the importance of Kufu in the best kind of RGA-making process, prior to developing scaling laws.

Students endeavoring to learn both art and science may best employ two different types of methods. Developing proficiency in art may be attained by simple, well-structured, repetitive training, which, through a highly nonlinear process, at some stage may rapidly convert ordinary eyes into Kufu eyes. On the other hand, development of excellence in science uses logical thinking, which is a

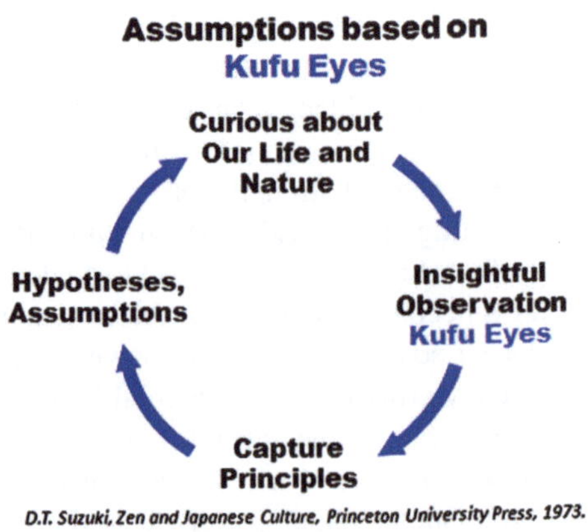

Fig. 4 The concept of Kufu eyes

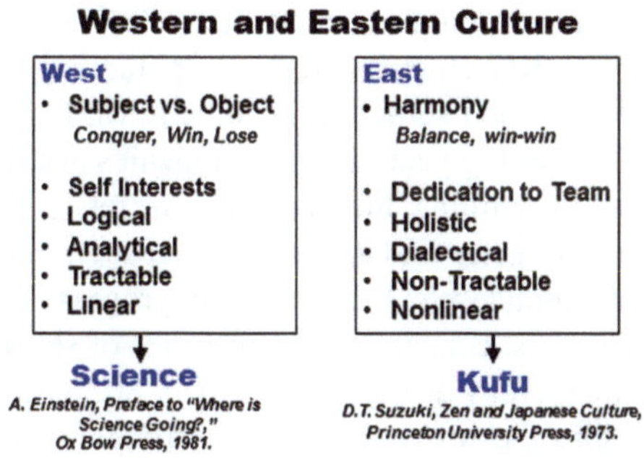

Fig. 5 Science based on the western culture and Kufu based on the eastern culture

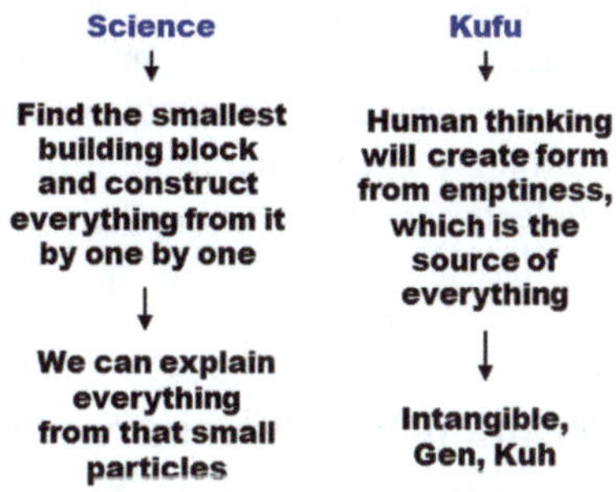

Fig. 6 Scientific method and Kufu

much more linear, step-by-step process to achieving understanding. This difference is, in a sense, the difference between the right-hand and left-hand paths in Fig. 5, where the eastern culture-based Kufu eyes and the western culture-based science are contrasted, the latter, on the left, following a reference to Einstein. The former process, on the right, is similar to learning Zen [8], the concepts of which, including Kufu, have exerted a significant influence on Japanese scientific research and technological development. One example of this latter influence is the Toyota Production System (TPS) [7], a highly successful manufacturing system designed to produce the highest quality products in the least expensive way, which has gained worldwide attention, not only from business and industry but also from organizations specializing in R&D, education, and services. The four-step process shown in Fig. 4 is one of the basic tools of TPS, in which Kufu eyes are related to a similar process, termed Kaizen eyes, Kaizen translating as continuous improvement, finding ways to identify waste and to institute better methods for making products. Figure 6 expands on Fig. 5 by further contrasting the characteristics of science and Kufu. Readers who are interested in further discussions of these differences may consult available references [6–9, 14].

Also relevant to Kufu eyes is an induction method of TPS, which emphasizes action first, then learning from that action [7]. This induction method can be helpful when the full-scale phenomenon is very poorly understood. In this approach, first a simple design of a just-try-and-see-how-it-works type of experiment is implemented, followed by careful observations of the results, aimed at capturing whatever unique aspects of the full-scale phenomenon can be gleaned therefrom, using Kaizen eyes. Although potentially unreliable in unfavorable circumstances, this method sometimes can be rewarding.

Over the years, many companies and organizations have exhibited strong interest in learning the supposed secrets of TPS, attempting to copy it, to improve performance and achieve similar high efficiency and effectiveness. The University of Kentucky's Lean System program [10], however, compiled statistics, over 15 years, concerning the histories of several hundred companies and organizations and thereby discovered that over 80 % of them failed and less than 10 % succeeded, while the rest continued to struggle. Interestingly enough, those that succeeded were found to have developed their own unique Kaizen culture, which was different from the original TPS [7]. This outcome emphasizes that TPS and the Kaizen system essentially are an art and that they do not hold any particular secret; their essence cannot be learned or copied easily because of the requisite nonlinear learning process. The same difficulty exists in teaching Kufu and Kufu eyes to students and in explaining their essence.

Relationships Among Scale Modeling, Numerical Modeling, and Full-Scale Experiments

Four aspects of scale modeling that have been listed in texts [2, 3] are (1) imagine the full-scale phenomena, (2) understand the mechanisms that control this full-scale phenomenon, (3) validate the numerical predictions of the scale model results, and (4) develop new products or improve the performance of existing products, on the basis of the scale model results. In numerical modeling, the analog of aspects (1) and (2) is the activity that results in the determination of the numerical algorithm. The scale-modeling confirmation tests, involved in aspect (3), correspond to comparison of numerical predictions with experiment for numerical modeling, while the production test aspect (4) translates into numerous current applications of numerical codes in engineering design. These Moore's law mushrooming applications of computer simulations, destined to end at some time in the not-too-distant future unless large-scale quantum computing becomes a reality, should be viewed with an element of skepticism, since they live in a virtual world, not real nature, regardless of how impressive their multidimensional images and color movie videos are [15]. By way of contrast, scale-modeling approaches, unfortunately experiencing markedly decreasing use in production tests and associated engineering applications in recent years, have the distinct advantage of offering direct contact with nature, thereby grounding them firmly in physical reality. Development of the correct understanding of the mechanisms of nature is the key objective, and that cannot be attained by a simple increase in computational power! Both scale modeling and validated numerical modeling are needed, each with its unique strengths and methodologies.

After aspects (1) and (2) have been completed, resulting in RGA, scale modeling can help numerical modeling in the important validation aspect (3), where numerical modeling has little to offer scale modeling. Numerical modeling may achieve reasonable agreement with full-scale experiments but fail to predict results of scale model experiments well. Since the scale model experiments are in the real world, such an observation would show that something is missing from the numerical model, thereby enabling that model to be studied further to achieve improvements. If, on the other hand, the agreement is reasonable, then confidence in both the numerical and scale models is increased. Since computational methods offer unique capabilities to produce extensive numerical results quickly and relatively inexpensively, they are very valuable for use in production aspects (4), provided that they perform well in the validation aspects (3). Winning teams thus would best consist of numerical simulators and scale-modeling experimenters, contributing their complementary inputs.

The role of full-scale experiments still remains, of course, very important. It would, for example, be quite beneficial if it were possible to conduct full-scale experiments to reconstruct natural disasters or to identify mechanisms of failure in large structures such as buildings, bridges, and towers, but difficulties concerning safety, cost, and time generally preclude that. It is for this reason that scale modeling and numerical modeling are called for. Numerical simulation may give us more detailed information than scale modeling, because results can be extracted from numerics that would be very difficult or impossible to measure experimentally, but scale modeling can help us validate assumptions, better than the numerical modeling. The combination of both approaches therefore not only helps us understand the phenomena but also can increase the chances of solving practical problems. Since one of this volume's focuses is the use of scale modeling in natural disasters and structural failure, we shall discuss in the following sections, in closing, two kinds of examples, different from those in the present volume, of how the combinations of scale modeling, numerical modeling, and full-scale measurements have been applied to such problems.

Investigation of the 1966 Breakup of a British Jetliner in the Air Near Mt. Fuji [16]

In 1966, a British jetliner suddenly began to disintegrate in the air, about 85 km from the summit of Mt. Fuji, at 4,900 m above sea level, under fine weather conditions, yet with strong winds, killing all 124 passengers plus crew members onboard. After the accident, a special investigation committee was formed, consisting of an interdisciplinary group of experts, including physicists, mechanical and aerospace engineers, metallurgists, and meteorologists [16]. Seiji Soma, a physical meteorologist, was involved in two different aspects of the investigation of this accident, namely, collecting the wind data relevant to the jetliner breakup and reconstructing the wind patterns created by Mt. Fuji on the basis of a scale model. He received a 5-year grant from the Japanese Ministry of Science and Technology for his modeling and for his Meteorology Institute to collect full-scale wind data as validation tests for the scale model. His final report, written in Japanese [16], details his findings and his conclusion that the maximum wind speed when the jetliner broke up in the air could be as high as 76 m/s, with a strong turbulent component.

He applied two different methods to obtain three-dimensional wind profiles. One employed 80 cm-diameter polyether balloons, coated with a thin aluminum layer and filled with helium, their weight adjusted by a counterweight to make them neutrally buoyant. The other employed a flashing cylinder, suspended by a parachute, emitting a high-intensity light pulse every 15 s. All devices were released from the top of Mt. Fuji under strong wind conditions, intended to be similar to those that prevailed on the day of the jetliner accident. Later, Soma found that the wind velocities for these tests likely were lower than those at the time of the accident, but they nevertheless helped him to obtain important data. A total of 25 trails were made over a 4-year period, from 1968 through 1971. The trajectories of the neutrally buoyant balloons and of the parachute-carried cylinders were recorded by video cameras at two different positions near ground level, close to Mt. Fuji, to obtain the flow pattern data. Figure 7 shows three different vertical trajectories of balloons under relatively strong wind conditions, between 30 and 35 m/s. For weaker wind conditions, the balloon trajectories were quite smooth and similar to each other in three corresponding tests, contrary to the large difference exhibited in the figure. The results implied strong turbulence at the higher wind velocities.

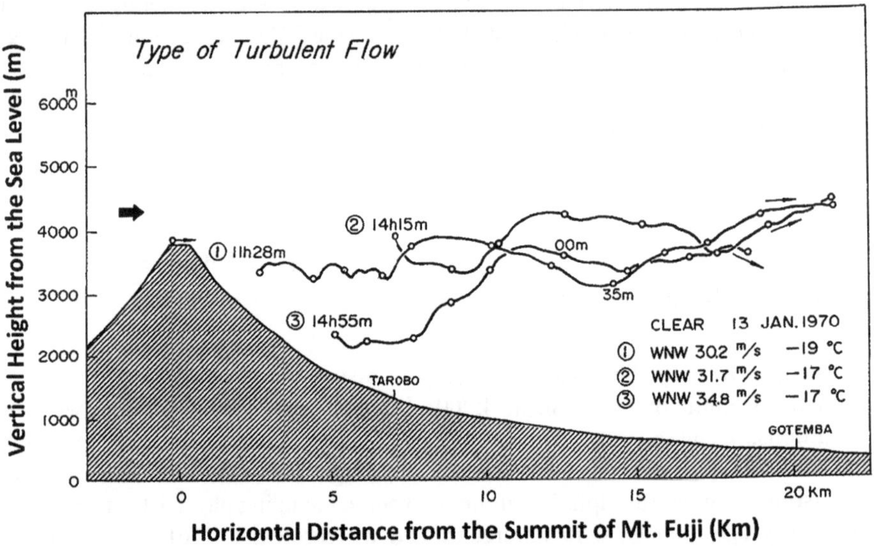

Fig. 7 Wind velocity profiles obtained by a non-gravity balloon released from the top of Mt. Fuji. Three trials were made on January 13, 1970 whose results are marked as 1–3

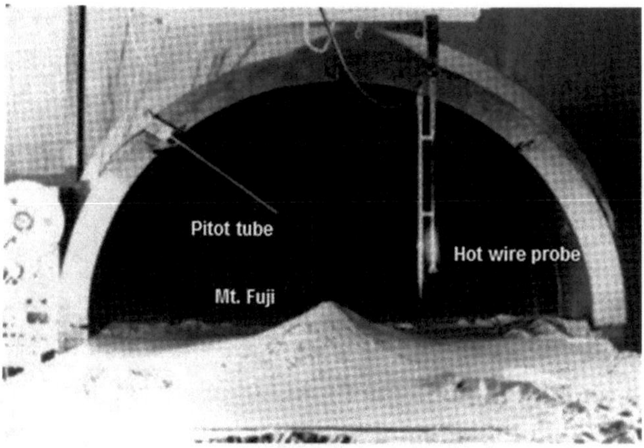

Fig. 8 A photograph of a 25,000th scale model of Mt. Fuji placed at the exit of an open loop wind tunnel where a pitot tube and a hot anemometer were shown

A 25,000th reduced-scale model of Mt. Fuji was designed and tested in a wind tunnel. An open-exit wind tunnel was used for these tests with the model at its exit, and the flow pattern near the model was visualized by a smoke-tracing method, with velocity profiles measured by a three-dimensional hot-wire anemometer. The wind velocities approaching the mountain were recorded by a fine pitot tube positioned at the exit of the wind tunnel. Figure 8 shows a photograph of the scale model mountain located at the exit of the open wind tunnel.

Figure 9 shows a comparison between the full-scale cloud pattern created by Mt. Fuji with a 35 m/s wind speed at the summit and the scale model wind pattern based on Fr scaling, $(76\ \mathrm{m/s})/(25{,}000)^{1/2} = 0.48\ \mathrm{m/s}$, corresponding to the maximum wind speed, 76 m/s, estimated by the investigation report [16], which was about twice as fast as that of the full-scale tests. Therefore, the cloud shape during the full-scale test extended in a somewhat more upward direction than the smoke-visualized scale model wind pattern, which bends in a more horizontal direction. But both pictures seem to exhibit a remarkable similarity in the general flow pattern, considering the very large ratio of scales, and they both suggest strong turbulence.

This example involved a combination of both scale-modeling and full-scale experiments, since reliable numerical simulation techniques were not available at that time. If this accident had taken place more recently, numerical simulation definitely should have been included as a third tool (but not to replace either the scale-modeling or the full-scale measurements). Even without numerical simulation, however, the investigation team completed a thorough investigation to identify the root cause of the accident to be the strong turbulent flow associated with the high wind speed. Strict regulations therefore were imposed on all flights near Mt. Fuji during conditions of strong winds. Because of these regulations, no similar accidents have occurred since 1966.

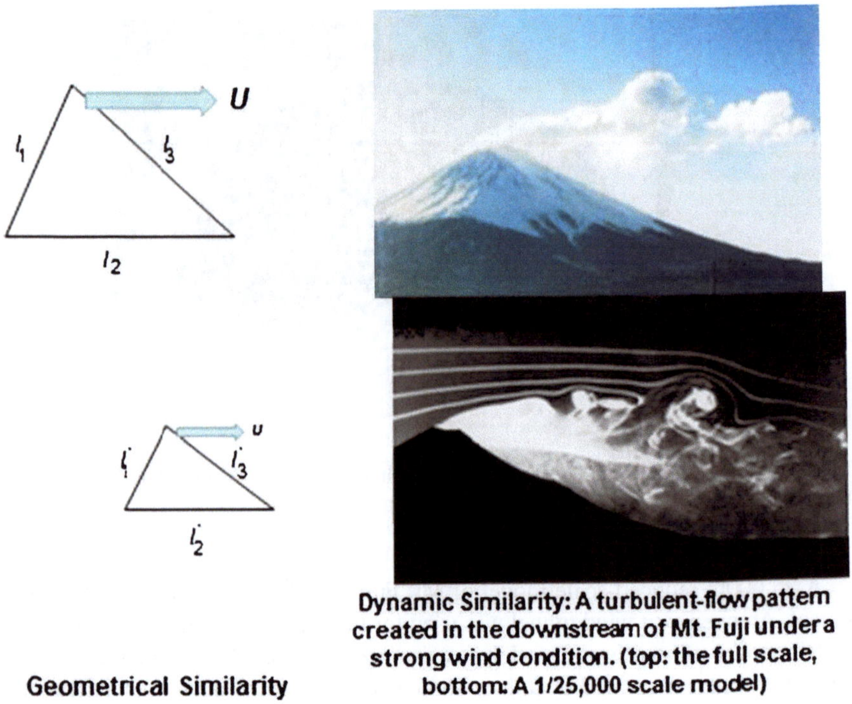

Fig. 9 A wind tunnel scale model experiment was conducted to investigate the 1966 British Jetliner breakup in the air near Mt. Fuji. The reconstructed flow pattern in the downstream side of the 25,000th scale model of Mt. Fuji (the *bottom photograph*) was favorably compared with the actual downstream wind flow pattern created by Mt. Fuji under the 100 m wind gust condition (the *top photograph*). Based on this validation of scaling similarity between the model and the full scale, detailed scale model experiments were conducted to reveal the root cause to be a large-scale downward turbulence that was strong enough to clash the jet liner

Investigation of the Collapse of The World Trade Center in a Terrorist Attack [17, 18]

Contrary to the investigation of the British airliner accident, the investigation of the collapse of the Twin Towers of the World Trade Center on September 11, 2001, after terrorists flew airplanes into them, was pursued mainly by a single modeling technique and numerical simulation, with experiments performed only on full-scale models of rooms in the towers that experienced fires [17, 18]. Some experts in the community believe that scale modeling of fires on complete floors of the towers could have provided helpful information for testing the numerical simulations. It could have been more revealing if the investigation team had applied all three techniques, thorough study of the full-scale phenomena, numerical simulation, and scale modeling, to increase the chances of finding the root cause of the tragic event.

Had this extensive investigation been even more thorough, possible causes not identified in the reported findings may have been indicated [19].

Ichiro Emori, the pioneer of traffic accident reconstruction using scale modeling [2, 20, 21], who developed the unique scale-modeling technique known as the Law Approach [2, 3], spent more than 30 years investigating over 800 accident cases [20, 21] and believed that it is the investigator's responsibility to carefully reconstruct each accident based on science and to learn from that process and then to suggest to policy makers how to improve the traffic safety system. He strongly felt that the investigator must not become involved in questions of right or wrong judgment, which is the duty of the courts and judicial system, but instead must focus on seeking factual causes. This outlook should continue to prevail today.

A Concluding Comment

Contrary to experimenters, many recent computational experts are often remote from the experimental site and do not experience enough interactions with experimenters to develop Kufu eyes. Such limited interaction does not help computational experts to attain Kufu eyes, and this is becoming even more problematic for students and future engineers and researchers who want to fully develop their capacity to solve engineering problems using the computational method. Increased interaction between scale modelers and numerical modelers needs to be encouraged.

References

1. Saito, K. (ed.): Progress in Scale Modeling. Springer (2008)
2. Emori, R.I., Saito, K., Sekimoto, K.: Scale Models in Engineering (Mokei Jikken no Riron to Ohyou), Third Edition (ISBN 4-7655-3252-6 C3053), Gihodo Publishing Co., Tokyo, Japan, Second Print 2008 in Japanese
3. Emori, R.I., Schuring, D.J.: Scale Models in Engineering: Fundamentals and Applications. Pergamon, New York, NY (1977)
4. Williams, F.A.: Scaling mass fires. Fire Res. Abst. Rev. **11**, 1–23 (1969)
5. Box, G.E.P., Draper, N.R.: Empirical Model-Building and Response Surfaces. Wiley, New York (1987)
6. Saito, K.: Kufu: Foundations of employee empowerment and kaizen. In: Saito, K. (ed.) Principles of Continuous Learning Systems. McGraw-Hill, New York (1995)
7. Saito, A., Saito, K., edited: Seeds of Collaboration: Seeking the Essence of the Toyota Production System. Larkspur Press, Monterey, KY (2012)
8. Suzuki, D.T.: Zen and Japanese Culture. Princeton University Press, Princeton (1973)
9. Gigerrenzer, G.: Gut Feeling, The Intelligence of the Unconscious. Viking Penguin, New York (2007)
10. IR4TD. http://www.ir4td.uky.edu/
11. Williams, F.A.: Modeling combustion phenomena. In: Saito, K. (ed.) Progress in Scale Modeling, pp. 179–196. Springer, New York (2008)
12. Petroski, H.: Tacoma narrows bridge. Am. Sci. **97**, 103–107 (2009)
13. Matsudaira, T.: Japan Society of Mechanical Engineers 82 (No. 733): 39–44 (1979) in Japanese

14. Saito, K., Futamura, B.: Face-to-face discussions on "Seeking Truth: Expanding Perception through the Harmony of the Western and Eastern Thought," Part 1 and 2, *Nippon Steel Monthly*, January–February, Nippon Steel Corp., 2-6-3 Otemachi, Chiyoda-ku, Tokyo, Japan (2009). http://www.nssmc.com/en/company/publications/monthly-nsc/pdf/2009062515075221917.pd
15. Williams, F.A.: Significance of scale modeling in engineering science. In: Saito, K. (ed.) Forward to Progress in Scale Modeling. Springer, New York (2008)
16. Soma, S.: A study on large scale turbulences created by strong wind in the downstream section of Mt. Fuji, A special report to The Ministry of Science and Technology and Meteorology Research Institute of Japan, March (1988) in Japanese
17. Gann, R.G., Hamins, A., McGrattan, K., Nelson, H.E., Ohlemiller, T.J., Prasad, K.R., Pitts, W. M.: Reconstruction of the fires and thermal environment in World Trade Center buildings 1, 2, and 7. Fire. Technol. **49**, 679–707 (2013)
18. Gann, R.G. (ed.): Federal building and fire safety investigation of the World Trade Center disaster, Final report on the collapse of the World Trade Center towers, NIST NCSTAR 1, Gaithersburg (2005)
19. Quintiere, J.G., WIlliams, F.A.: Comments on the National Institute of Standards and Technology Investigation of the 2001 World Trade Center Fires, FORUM contribution to the Journal of Fire Sciences, to appear (2014)
20. Emori, R.I.: Engineering spirits on the study of traffic safety. People Cars **7**, 4–17 (1996). in Japanese
21. Emori, R.I.: Mystery of traffic accidents, Asahi Bunko, ISBN4-02-261440-4, The Asahi Shinbun, Tokyo, Japan (2004), in Japanese

Part I
Natural Disasters and Structural Failures

Summary of Part I

Natural Disasters and Structural Failures

Natural disasters and structural failures are examples of what are hardly "predictable" and "controllable" issues that deliver severe damages/losses in various ways. Hence estimation of their potential damage (effect), understanding the possible causes/scenarios, and technology development for their prevention are enormous challenging yet indispensable tasks for engineers. Since these disasters and failures are generally extremely large scale (~km) and hard to reproduce intentionally, adopting the scale modeling concept to perform the small-scale tests would give one powerful/attractive approach to understand the key issues concerned. Editorial committees carefully surveyed all articles submitted to ISSM6 and ISSM7 on these issues and selected 14 most suitable articles from the three sessions contributing to this area. From the "Earthquake" session, applicability of the scale modeling concept to geotechnical engineering field is reviewed and how to utilize supercomputing technology to help to derive the effective scaling law for the effect of flotsam mixed Tsunami is presented. These articles are devoted to learning from the tragedy of the 2011 Great East Coast Earthquake in Japan and are excellent evidence of how scale model approach can be adopted to predict/estimate natural disaster.

From the "Fire and Explosion" session, papers give interesting scaling laws to represent fire behavior, explosive growth, and extinction character. For example, the growth of forest fire is modeled as the flame spread over the cardboard fuelbeds, and the similarity is ensured by a scaling law developed by authors. Selected articles effectively apply the various types of ambient conditions, e.g., various pressures and gravities, to access the same physics appeared in the actual fire and explosion based on the scaling law. This enables us to perform small-scale experiments to reproduce the large-scale phenomena, which is often hard to perform, and to have access to investigate them. Not only are there articles dealing with the typical large fire/explosion, likely pool fire, fire whirl, and accidental explosion, but also one adapted to a new type of fire caused by hydrogen generation via bacterial activity is selected to reveal the wide range of applicability of the scale modeling concept/approach in the various types of practical fire accidents.

From the "Structure" session, excellent work on utilizing centrifuge tests to examine the large scale construction is selected. This work was done by a general constructor, revealing that the scale modeling concept/approach can handle real engineering problems. Interesting work to estimate the fatigue load of a windmill by natural wind is selected in this volume since this is a very important original work to assist to constructing a better energy harvesting system exposed to various wind conditions.

Section A
Earthquake - Scale Modeling in the Geotechnical Engineering Field

Tadashi Kawai

Abstract In the geotechnical engineering field, scale modeling plays a very important role. Downscaled models are used because the targets of geotechnical engineering are usually huge, e.g., a dam, reclaimed land, a slope, or the base ground of various kinds of structures. However, since the characteristics of soils depend on the confining pressure, special treatment needs to be given to soils in these geotechnical models. In this paper, the types of problems which may be encountered in geotechnical engineering and the characteristics of soils are introduced. Then a special use of the scale model in geotechnical engineering is explained using problems involving a seawall, an underground structure, and a slope as examples of its potential application.

Introduction

In the field of geotechnical engineering, since the materials are not artificial and because they differ from one site to the next, model tests should be done for every construction. However, the reality is that before construction begins, geotechnical engineers rarely have the budget or the time to do model tests. Geotechnical engineers usually have to make a design based on the simple method described in the proper design guideline composed of empirical equations or based on a simulation created by a numerical method (e.g., FEM). As such, geotechnical model tests clearly have two major purposes; one is to find a common mechanism governing the design of all similar constructions, and the other is to provide a set of verification data for numerical simulation methods. For both of these, it is necessary to keep in mind that soil is a rather unique material comprised of numerous types and sizes of particles and to have a sound understanding of the soil mechanics that describes their material characteristics.

T. Kawai (✉)
Civil Engineering, Tohoku University, 6-6-06 Aoba, Aoba-ku, Sendai, Miyagi 980-8579, Japan
e-mail: t-kawai@civil.tohoku.ac.jp

K. Saito et al. (eds.), *Progress in Scale Modeling, Volume II*,
DOI 10.1007/978-3-319-10308-2_2

In a standard textbook of soil mechanics used in colleges worldwide, many subjects are introduced, such as the kinds and nature of soil particles, permeability, shear strength, consolidation, bearing capacity, earth pressures, and slope stability. The four latter categories are clearly problems that depend on the initial and boundary conditions. However, it should be noted that since a soil is a merely collection of particles, permeability and shear strength are qualities that also strongly depend on the initial and boundary conditions. Because of the various characteristics of soils which need to be taken into consideration, scale modeling in the geotechnical field is complicated. In this paper the author introduces a special model test method, which will be referred to simply as the centrifuge test. A detailed explanation of the soil characteristics is given in this paper after introducing a variety of different geotechnical engineering structures; then, the centrifuge test method is described with its scaling law, and finally a few examples of the geotechnical scale modeling are provided.

Targets of Geotechnical Engineering

As can be seen in Fig. 1, geotechnical engineering is relevant to each and every construction on earth. In our societies, seawalls need to be erected in order to prevent coastal erosion, and sufficient lifeline systems, such as sewers, waterworks, electricity, and gas, need to be put into place to make our lives comfortable. As Japan has limited plains, we need to cut and fill slopes to level the ground. In coastal areas, large landfills have been constructed for various purposes. In the construction of all of these structures, it is essential to take into account the influences of natural

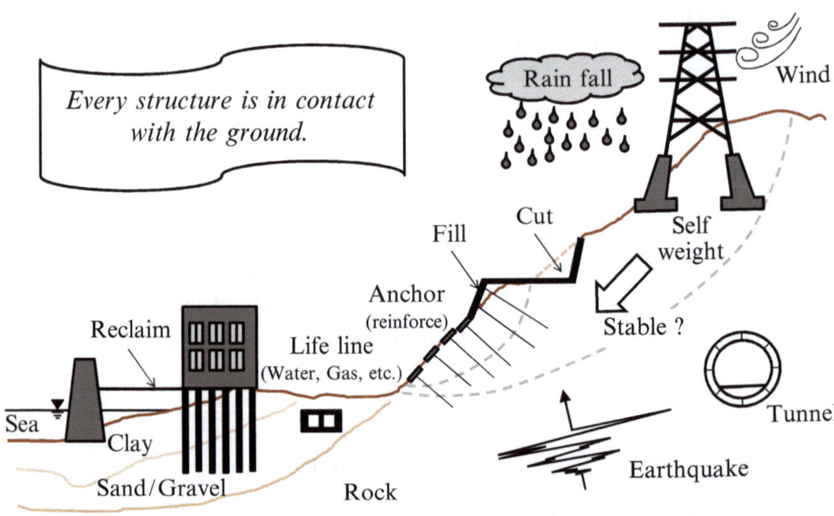

Fig. 1 The challenges in geotechnical engineering

hazards, like earthquakes, with their shaking and associated tsunamis, and typhoons, with their heavy rainfall and strong winds. In addition, because the settlement of landfills is the result of their own weight and continues for years, geotechnical engineers need to estimate the total amount of the settlement prior to their construction.

Damage related to the ground can be classified into several types as shown in Fig. 2. If a large mass is placed on the ground, the type of damage shown in Fig. 2a and/or b might occur. Though the footing of the foreside (on the left) of the tower in Fig. 2c is similar to Fig. 2a, the pullout resistance of the footing at the back (on the right) has to be considered as well. Two major problems are depicted in Fig. 2d; the one is a landslide and the other is the lateral displacement of level ground. Landslides result in serious damage to structures both on the slope and below the slope. Even if no apparent landslide occurs on the slope, deviatoric stress derived from the height difference can cause the lateral displacement of the ground in the direction away from the foot of the slope. Lateral displacement results in damage to building foundations, such as piles. Unlike Fig. 2a–d, all of which are concerned with the equation of motion, Fig. 2e is related to the law of conservation of mass. Near the broken parts of the buried conduit caused by an earthquake or decay, soil particles may flow out. Consequently, the cavity that appears in the ground reaches the surface, resulting in the cave-in of the ground surface. Examples of Fig. 2a, which demonstrate the basics of geotechnical engineering, are shown in Figs. 3 and 4. This damage was incurred as a result of the 1995 Hyogo-ken Nanbu earthquake.

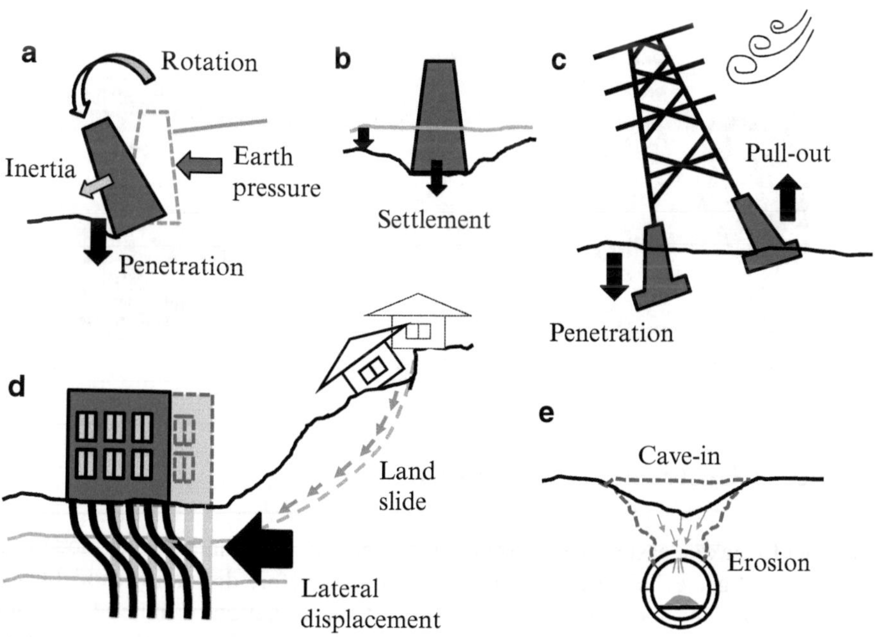

Fig. 2 Examples of problems in the field of geotechnical engineering

Fig. 3 A damaged caisson-type seawall after the 1995 Hyogo-ken Nanbu earthquake

Fig. 4 Slope failure incurred during the 1995 Hyogo-ken Nanbu earthquake

Basics of Geotechnical Engineering

In order to make it easier to understand the essence of geotechnical scale modeling, the basics of geotechnical engineering are described before discussing model tests.

Soil is any uncemented or weakly cemented accumulation of mineral particles formed by the weathering of rock, the void space between the particles containing water and/or air. The products are usually transported by gravity, wind, water, etc.

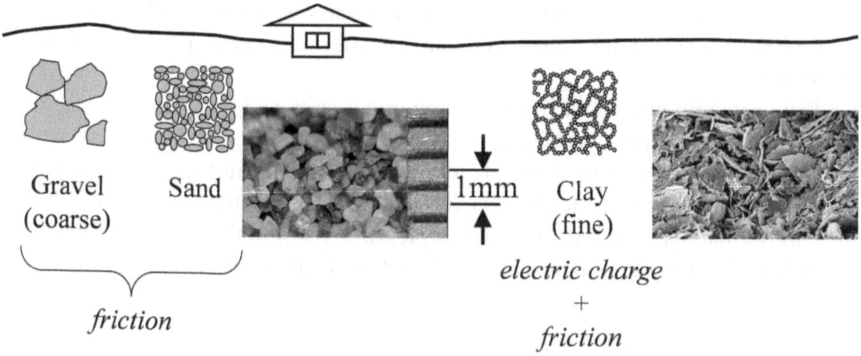

Fig. 5 Different types of soil particle

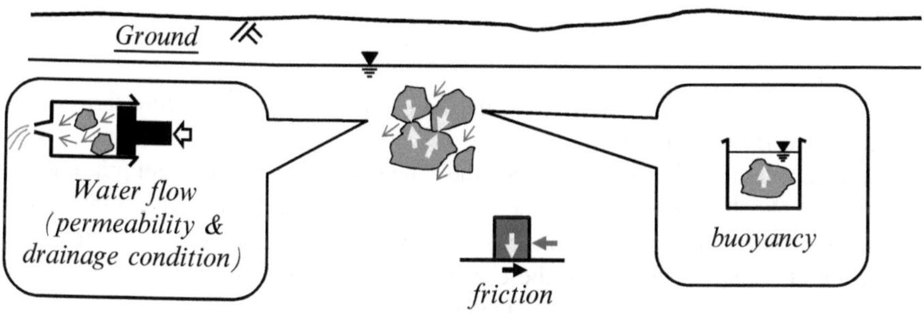

Fig. 6 The factors which affect soil mechanics when soil is below the water table

and are deposited in different locations. During this transportation process, the size and shape of particles change. The particles can be sorted into types according to their size ranges, as shown in Fig. 5. Those particles are referred to as gravel, sand, silt (omitted), and clay respectively according to the size of the particles. The shear strength of both gravel and sand is mainly provided by particle interlocking and friction, whereas that of clay is largely affected by ions. The surfaces of clay mineral particles tend to be negatively charged and the cations bond the clay mineral particles to each other. This means that the shear strength of the clay masses is strongly affected by these bondings.

Under the water table in the ground, the voids between the particles are filled with water. This water plays an important role in the resistance of any given soil. At least the permeability, the drainage condition, the friction, and the buoyancy, all of which are shown in Fig. 6, should be taken into consideration before any construction work begins.

Since shear stress can be resisted only by the skeleton of solid particles, by means of the forces developed at the interparticle contacts, a reduction of normal forces at every point of contact of the individual particles leads to a reduction of shear strength of the soil mass. This means that if buoyancy decreases the weight of the soil particles, the shear strength of the soil mass is also reduced. Therefore, in geotechnical engineering, the principle of effective stress [1] is of primary importance.

This principle applies only to fully saturated soils and consists of three stress components. Consider stresses at a depth where there is a soil mass with a horizontal surface with the water level at the top of the surface, where the total stress [σ (N/m^2)] is equal to the weight of all material (solids and water) per unit area above that depth, and the pore water pressure [p (N/m^2)] of the continuous void space filled with water is hydrostatic. Then the effective stress [σ' (N/m^2)] is

$$\sigma' = \sigma - p. \tag{1}$$

The shear strength of a soil mass is appropriately expressed by Coulomb in the following equation:

$$\tau_f = c' + \sigma' \tan \varphi \tag{2}$$

where τ_f (N/m^2) is shear strength of the soil mass, c' (N/m^2) is cohesion and φ (degree) is a friction angle.

An example of this relation between the effective stress and the strength of a soil mass is shown in Fig. 7, where the figure on the left indicates the relationship between the shear strain and shear stress in individual shearing tests, and the figure on the right shows the peaks of the curves plotted versus the respective effective stresses, at which each test was conducted.

Another effect of pore water should also be taken into consideration: its relation to the change in effective stress. For the sake of convenience, consider a piston filled with soil particles and water, as shown in Fig. 6. If the piston is pushed in so quickly that there is no sufficient time for the water to escape from the piston, the water pressure inside the piston will increase and shear strength of the soil mass will decrease, whereas if the piston is pushed in slowly enough that the water is allowed to drain, the water pressure will remain constant and the shear strength of the soil will not change. This means that shear strength of a soil mass depends not only on the effective stress but also on the drainage conditions. That is, the shear strength of

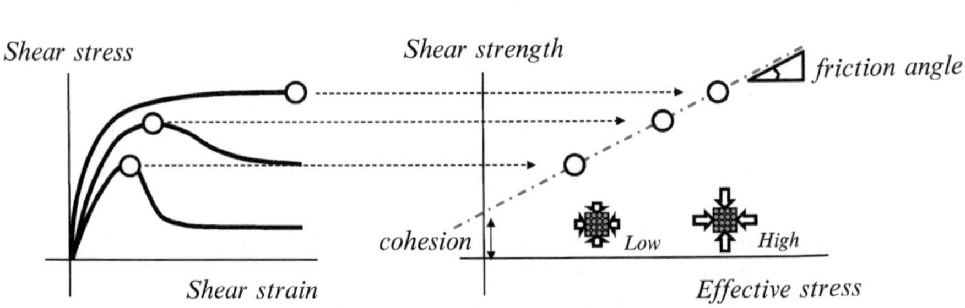

Fig. 7 The relationship between the strength of soil and the effective stress

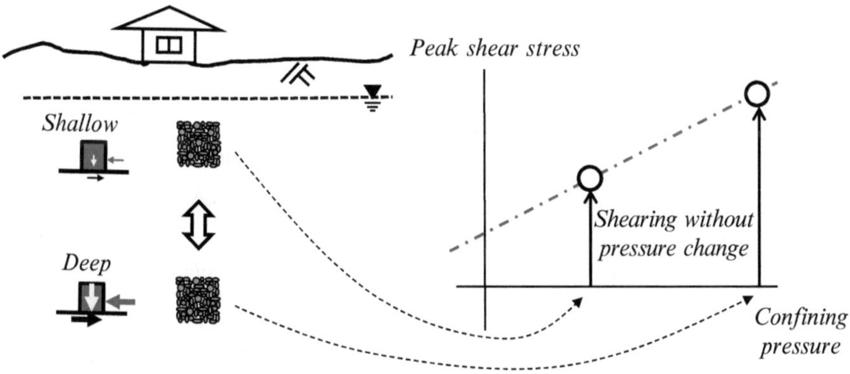

Fig. 8 The confining pressure dependency of soil strength with sufficient drainage

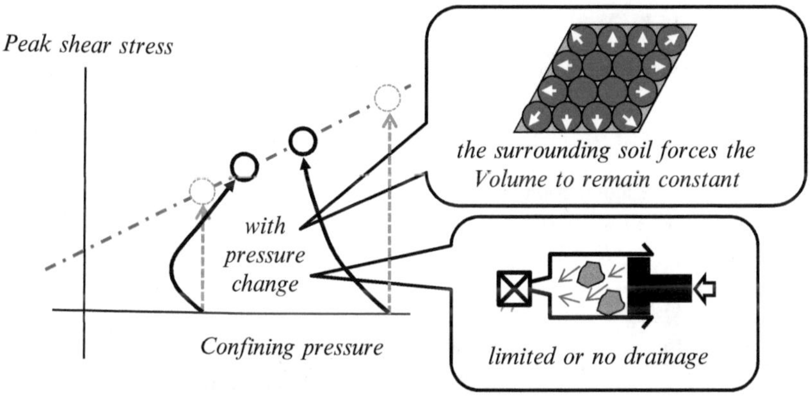

Fig. 9 The effect of the drainage condition on soil strength

a soil mass is not a material property but a variable which depends on the initial and boundary conditions, as shown in Figs. 8 and 9.

In the actual ground, the piston is the soil itself, because a soil mass is comprised of particles and is capable of expanding or shrinking in response to changes in stress, whether that stress is normal stress or shear stress. A positive volume change (expansion) caused by shear stress is called "dilatancy:" it is caused by each individual particle climbing up the neighbor particles due to the shear deformation of the whole mass. Dilatancy tends to occur when a rather dense soil, with short distances between the individual particles, is sheared. The shearing of a loose soil leads to a contraction, which is referred to as "negative dilatancy." For a medium dense soil, negative dilatancy is observed in the early stages of shearing, and in the later stage (positive), dilatancy will occur, as shown in Fig. 10.

This figure shows a typical result of a shearing test of medium dense sand. The solid line indicates the relationship between shear strain and shear stress, whereas the dashed line expresses the relationship between the shear strain and volumetric

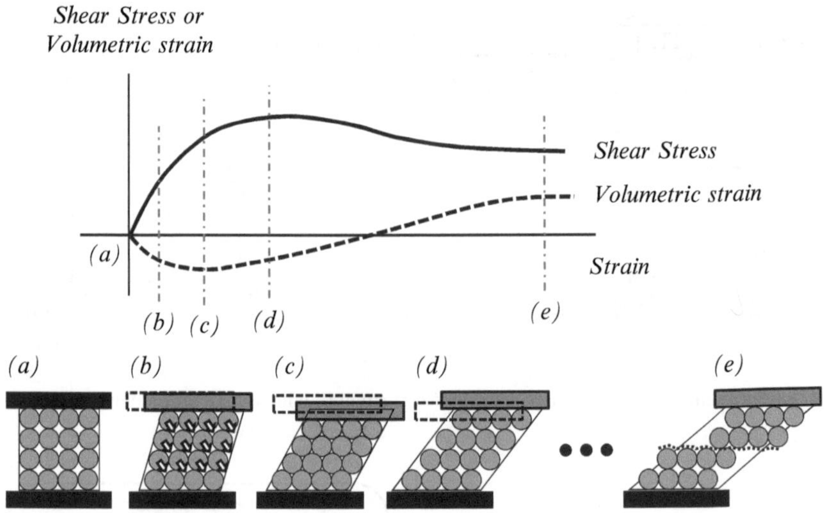

Fig. 10 Shear test results for sand and a schematic image of how sand particles move

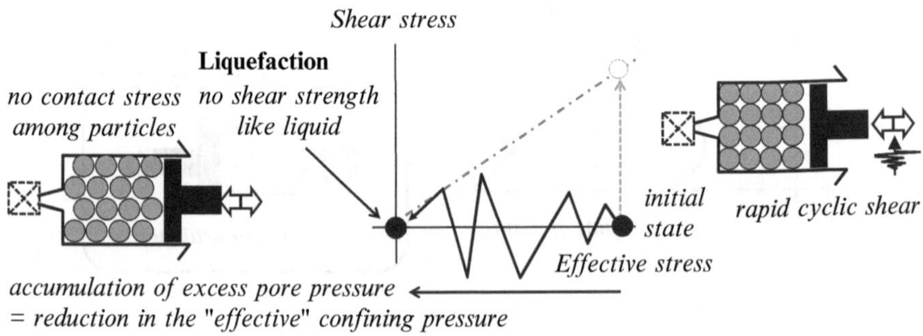

Fig. 11 Effect of the drainage condition on soil strength

strain of the same test. As shown by the schematic images in Fig. 10, a negative dilatancy occurs during (a)–(c), and further shearing causes a (positive) dilatancy during (c)–(e). Since the soil mass is merely gathering particles, localization may end up occurring at around (e). In the other shearing condition, "liquefaction" may occur instead of a localized shear band. Consider an earthquake repeatedly shaking the ground within the soil positions represented by (a)–(c) shown in Fig. 10. In that case, each cycle of shearing tends toward negative dilatancy (the contraction of the soil mass), but because there is insufficient time for the water to escape, the soil mass does not actually contract. Instead, the water pressure increases and is gradually accumulated, which we call "excess pore water pressure" (since the water pressure is higher than hydrostatic pressure), reaches the same value of the effective stress, as shown in Fig. 11.

Fig. 12 Damage due to ground liquefaction

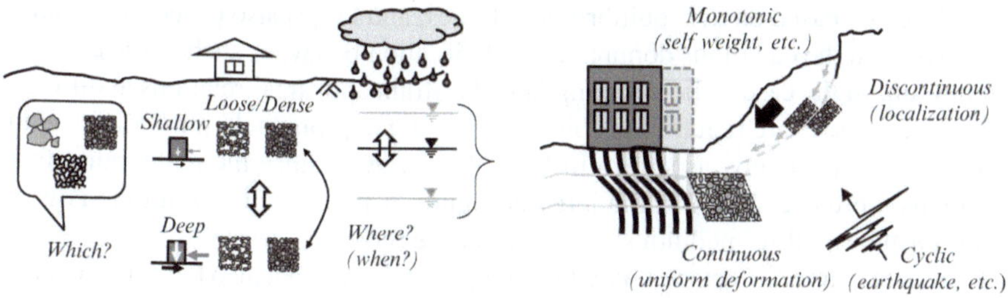

Fig. 13 The many factors which need to be taken into consideration

In that state, no effective stress means no normal contact stresses among the individual soil particles. It means that any slight shear force acting on the surface of the soil mass can deform it easily like shearing liquid. This state is very dangerous for all the structures constructed on it. Figure 12 shows typical phenomena caused by liquefaction. The picture on the left shows boiled sand through a crack on the ground surface, and the one in the middle shows the ground subsidence after an earthquake. The picture on the right is the most impressive: because of liquefaction, the ground became like a liquid, allowing structures to move around easily to an extent considered impossible before the earthquake.

As mentioned above, to solve the problems facing geotechnical engineers, several points must be taken into consideration to ensure soil strength is sufficient. Figure 13 summarizes the things which must be considered.

Since the strength of soils, i.e., gravel, sand, silt, and clay, depends on the initial and drainage conditions, the locations of both the soil itself and the water table should be taken into consideration. If the soil is a sand sedimentation in a loose state, liquefaction is a very real possibility during an earthquake. The failure patterns should also be taken into consideration, since the relations between strain and stress are nonlinear and the stress level tends to peak before localized deformation begins. The strength for uniform deformation is different from that of localized deformation. As such, the soil strength must be evaluated and the appropriate soil strength must be ensured in order to avoid the various geotechnical engineering problems outlined above.

Scale Modeling in Geotechnical Engineering

Scaling Law

In geotechnical engineering, model experiments have long been used. Because of the large scale of the target problem in question, a downscaled model is used in most cases. However, as the characteristics of soils are strongly influenced by the effective stress, it is difficult to reproduce the prototype by the downscaled model.

The similarity rules between any given model and the prototype have been studied since the 1950s in geotechnical engineering [2, 3]. Kagawa [4] proposed a scaling law based on the equilibrium of forces, and Iai [5] also proposed a scaling law based on the equations dominating soil vibration behavior, such as the equation of motion. In the case of Iai's scaling law, the strain and stress relations at different effective stresses are required to coincide by using appropriately normalized strain and stress respectively, as shown in Fig. 14. However, since the real strain–stress relations are those shown in the left side figure of Fig. 15, the similitude is valid only for the small-to-medium strain range at the most.

Volume change affects not only the strength of the soil but also the subsidence of the ground surface, and as such, it should be mimicked in a model. As shown in the right side figure of Fig. 15, the relation at low effective stress exhibits dilatancy

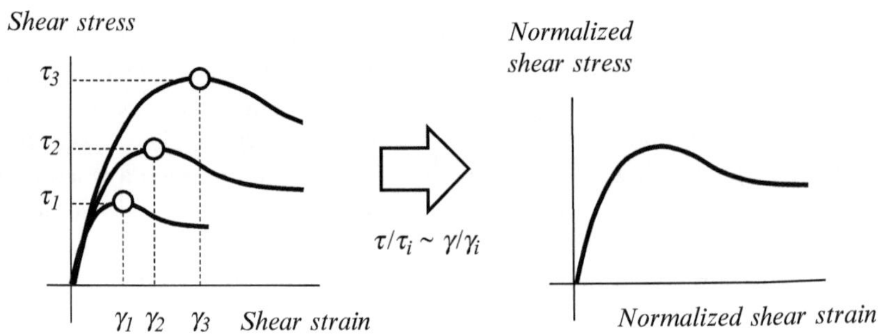

Fig. 14 Assumption needed to make Iai's similitude [5] valid

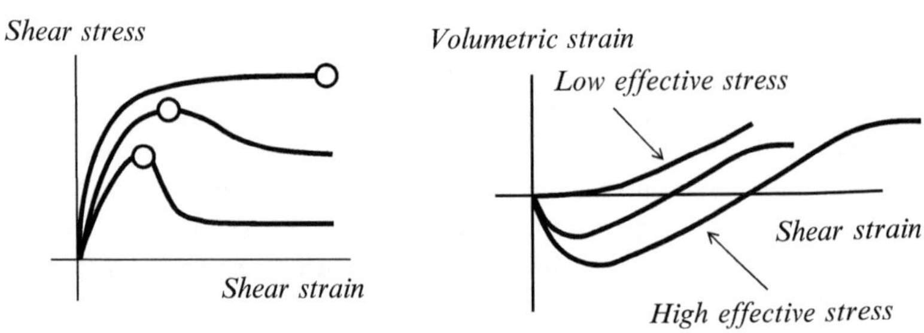

Fig. 15 Schematic image of characteristics of real soil during shear

from the origin, whereas the relation at high effective stress exhibits negative dilatancy at first and then dilates. This example of dense-sand behavior indicates that at least for dense sand, no adequate strain value to normalize the relations between shear strains so that they coincide can be found at different effective stresses.

Centrifuge Test

Because of the complicated nonlinearity of soil, it is difficult to realize a prototype using a downscaled model. The downscaled model is usually regarded as a small prototype itself at ordinary gravity. Pokrovsky and Fedorov [6] gave a solution to the difficulty by giving a centrifugal acceleration as extra gravity to a model. The history of the centrifuge tests has been summarized by Kimura and Kusakabe [7].

In the centrifuge test, in order to increase the effective stress of a model up to that of the prototype of interest, a large amount of centrifugal acceleration is given by conducting the experiment on a stage attached to the end of a rotating arm, as shown in Fig. 16, or perhaps on the inner surface of a rotating drum. A solution to model slope failure comprised of cohesive soil, which means the strength is not proportional to the effective stress, is also explained in Fig. 16 as an example.

In the centrifugal model experiment, a similarity rule is, of course, necessary. Shibata et al. [8] derived a scaling law based on the governing equations. In geotechnical engineering field, the equation of motion is described as follows:

$$-\frac{\partial \tau_{ij}}{\partial x_j} + \rho g \delta_{i3} = \rho \frac{\partial^2 u_i}{\partial t^2} \delta_{i3} \quad \text{: Kronecker delta } [i = 1, 2, 3(\text{vertical})] \qquad (3)$$

in which τ_{ij} (N/m^2) is the stress tensor, x_j (m) is the spatial coordinate, ρ (kg/m^3) is the soil density, g (m/s^2) is the gravitational acceleration, u_i (m) is the displacement

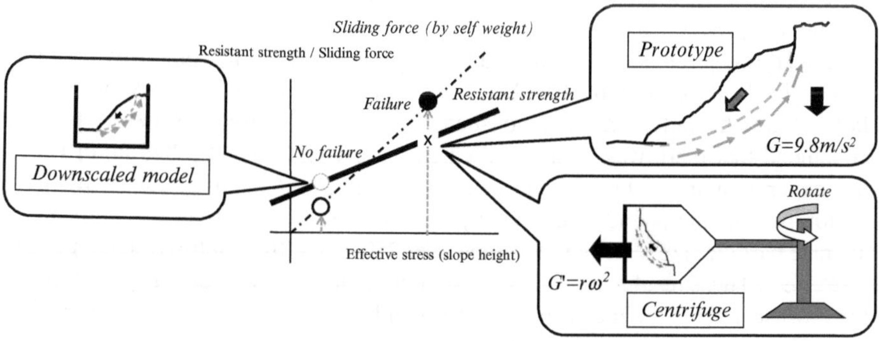

Fig. 16 A solution for the slope stability problem using a centrifuge

of i direction, and t (s) is time. The equation is firstly transformed into the relations among the dimensionless quantities as shown in Eq. (4). And then by keeping proportion between each member of Eq. (4), namely, by satisfying Eq. (5) between a model and a prototype, the similitude is derived.

$$-\frac{\partial \tau^*_{ij}}{\partial x^*_j} + \frac{\rho g L}{S}\delta_{i3} = \frac{\rho U L}{S T^2}\frac{\partial^2 u^*_i}{\partial t^{*2}} \tag{4}$$

$$\frac{\rho_p g_p L_p}{S_p} = \frac{\rho_m g_m L_m}{S_m} \quad \frac{\rho_p U_p L_p}{S_p {T_p}^2} = \frac{\rho_m U_m L_m}{S_m {T_m}^2} \tag{5}$$

The values with an asterisk are the dimensionless values in Eq. (4), and the subscripts "p" or "m" respectively designate prototype or model in Eq. (5). The L (m), S (N/m^2), U (m), and T (s) in Eq. (4) are the representative values of length, stress, displacement, and time used to make dimensionless values. These procedures to derive the similitudes are described in detail by Iai et al. [5] and Shibata et al. [8].

The validity of these scaling laws was confirmed by performing a centrifuge test with dry sand in a large-scale model test at 1G ordinal gravity as the prototype. Sato et al. [9] also confirmed that a centrifuge model comprised of pile, upper structure, and dry sand well reproduced the responses of a large-scale 1G test with the same proportion to the centrifuge model multiplied by the similitude. These were both examined using dry sand.

In the case of saturated ground, the similitude in terms of permeability and the similitude in terms of vibration are different since the former is derived from the equation of motion and the latter is derived from Darcy's law. Allard and Schenkeveld [10] studied the suitable type of fluid for satisfying both the similitudes at once. Hayashi et al. [11] confirmed the validity of the similitude by conducting an experiment using saturated sand with the fluid having similar characteristics.

As the same materials are used in centrifuge tests, the particle size multiplied by the similitude becomes significantly larger and may affect localization observed in the failure process. Kutter et al. [12], White et al. [13], and Corte et al. [14] examined this influence and reported the opposite: that is, the former two demonstrated in their experiment that there was a particle size dependency on the soil failure and the latter reported no such dependency. Further, Renzi et al. [15] concluded that there was a range in which the particle diameter had no effect on the experimental results and the same material could be used for modeling a prototype. Goodings and Gillette [16] also showed that keeping the ratio of the average particle size to height of a slope over 250 was sufficient to avoid the particle size effect. These results indicate that using the same soil to model the prototype for the downscaled model in centrifuge test is valid at least in the adequate range of the ratio of particle size to a representative length.

Examples of Scale Modeling in Geotechnical Engineering

In this section, three examples of scale modeling in geotechnical engineering, two of which are centrifuge tests and the third a shaking table test at 1G, are shown. Those are with reference to a seawall surrounding a man-made island, underground structures buried in dry sand or saturated sand, and slopes comprised of a stiff base layer and a weak sand layer which was expected to slide. All models were examined to learn the seismic performance using sinusoidal waves and/or synthetic earthquake waves. Table 1 shows the similitude adopted in those centrifuge tests. The slope shaking tests were conducted at 1G to observe the progress of failure in the weak layer, because the time similitude of a centrifuge test is λ and the duration of a shaking event was too short to take sufficient photos. In the 1G test, no scaling law was adapted, since the model was regarded as a small prototype.

Seismic Performance of a Seawall Model

A detailed experimental study examining the seismic performance of a seawall model was conducted using the large servo-hydraulic shaker on the large centrifuge [17–21]. As shown in Fig. 17, the seawall models consisted of a concrete caisson

Table 1 Similitude for the centrifugal field at λ G

	Similitude (model/prototype)
Centrifugal acceleration	$1/\lambda$
Length	λ
Period of vibration	λ
Stress/strain/mass density	1
Displacement	λ
Acceleration of vibration	$1/\lambda$
Water flow (permeability)	λ^2

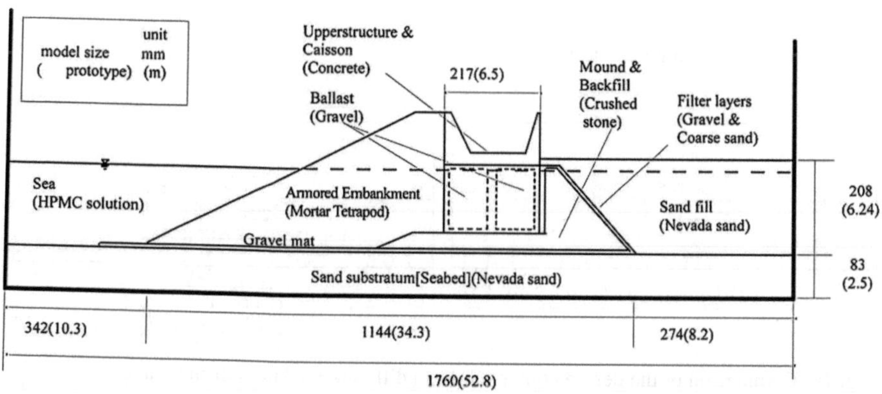

Fig. 17 Cross section of the seawall model

and superstructure retaining a sand fill that formed the island surface, with an embankment comprised of concrete armor units on the seaward side. A crushed stone mound and a crushed stone backfill were also placed under and behind the caisson respectively. These were all constructed on a sand seabed stratum except for the test modeling the seawall, which was directly placed on rock. The conditions of the sand seabed were chosen as the parameters to be tested.

A series of rather large base shaking events were commenced to record the response of the model mainly at a centrifugal acceleration of 30G, which means that the scaling factor of the model was 30. Typical time histories from a sinusoidal event are shown in Fig. 18. During strong shaking, the caisson was observed to displace laterally away from the backfill, as well as settle downwards. These movements of the caisson were mainly caused by the deformation of the sand seabed indicated in Fig. 19. These figures also indicate that the thickness of the sand seabed affected not only the amount of displacement but also the deformation patterns of the sand seabed.

In order to protect the land from the attacks of the ocean waves including a *tsunami*, those seawalls must keep the required height even directly after the experience of an earthquake. The displacements of the caisson after the 0.3G sinusoidal wave input in the various tests were summarized in Fig. 20. The results indicate that a thinner sand seabed led to smaller displacements and even the narrowest improved area of the sand seabed had a significant effect on reducing the amount of displacement of the caisson. This kind of knowledge about the seismic performance of the seawall is useful when designing a real seawall.

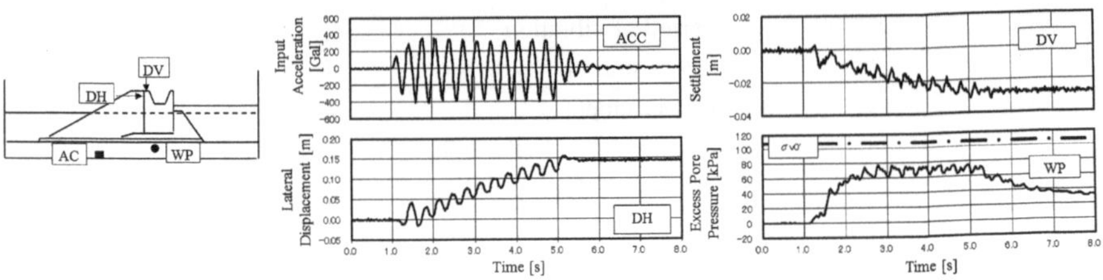

Fig. 18 Typical shaking table test results

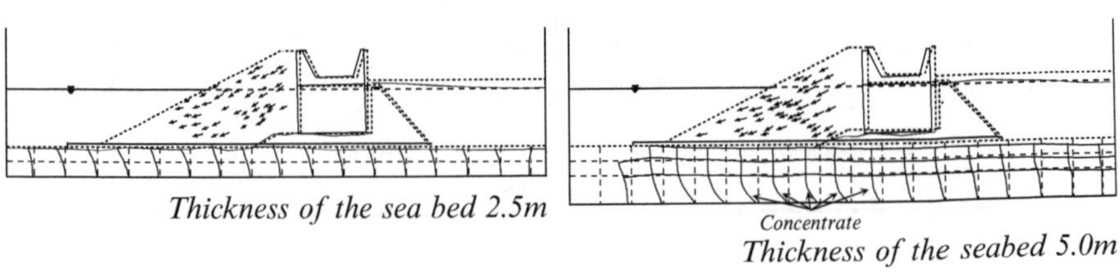

Fig. 19 Comparison of the deformation patterns in different thickness of the seabed

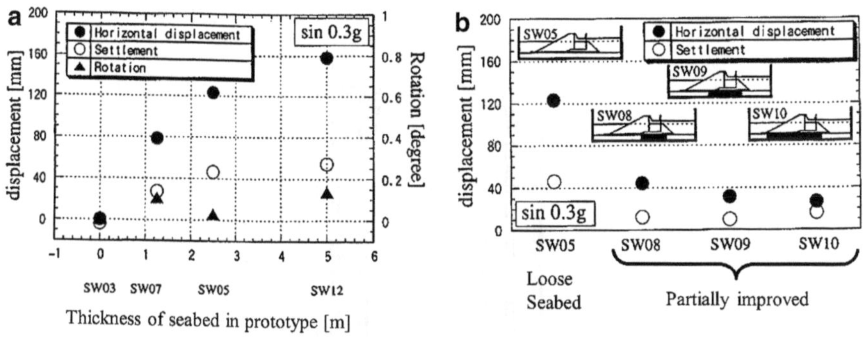

Fig. 20 Influences of the seabed conditions on the displacement of the caisson. (**a**) Dependency of thickness. (**b**) Effects of partial improvement

Seismic Performance of Underground Structures

In order to obtain basic knowledge about the deformation mechanisms of box culvert structures buried in saturated sand during an earthquake, a series of centrifuge shaking table tests were conducted using a metal structure and a reinforced concrete (RC) structure [22, 23]. Several two-way load cells were attached to each sidewall and upper slab of the metal structure to measure the acting normal and tangential force separately. A typical cross section of the whole model was tested and the cross sections of the metal structure and the RC structure are shown in Fig. 21.

A synthetic earthquake and sinusoidal waves of different frequency were input to the model. Those are shown in Fig. 22.

It was observed that horizontal earth pressure gradually increased up to the same value of the vertical earth pressure due to liquefaction of the surrounding saturated ground. This earth pressure increase led to the occurrence of many cracks at middle height of the inside surface of the sidewalls only of the RC structures buried in the saturated sand as shown in Fig. 23. Because no liquefaction occurs in dry sand, cracks in the sidewalls of the RC structure buried in the dry sand were observed only at the corners of the structure, and these were caused by the relative displacement between the top slab and the bottom slab, namely, the shear deformation of the structure as a whole.

As was the case with the structure buried in dry sand, corner cracks were observed for the structure in saturated sand. Such deformation was mainly caused by the force acting on the top slab as shown in Fig. 24, which was acquired from the metal structure buried in the saturated sand. This figure also shows that the increase of the horizontal earth pressure mentioned above has little influence on the shear deformation of the structure. The horizontal earth pressure was acting on both sides, in effect almost canceling each other.

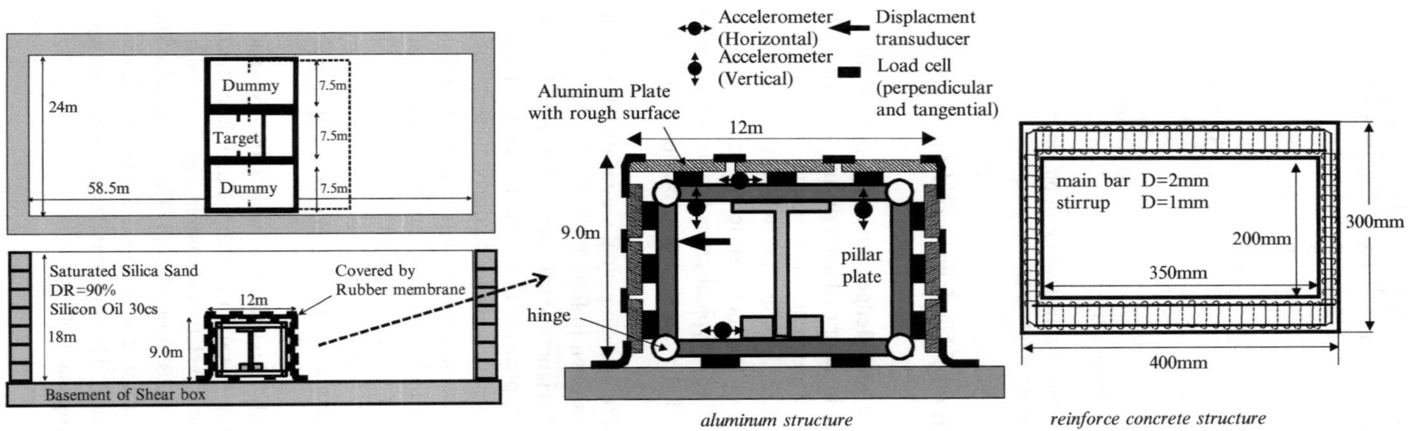

Fig. 21 Cross sections of the culvert structure models

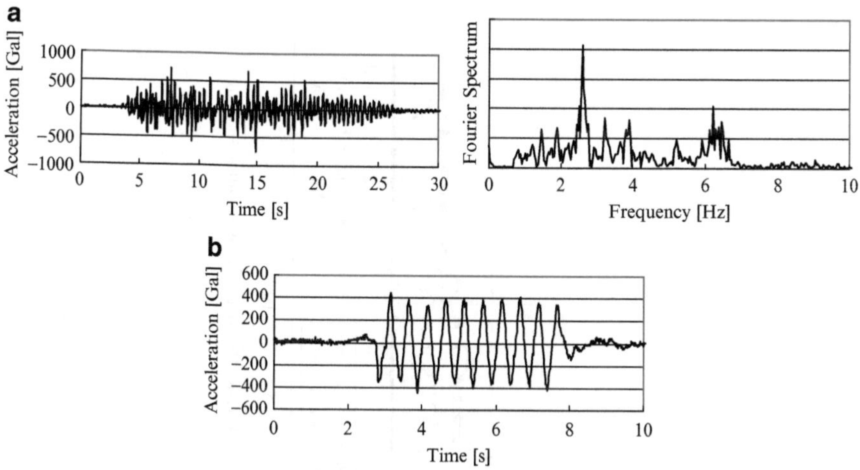

Fig. 22 Input waves for the shaking table tests. (**a**) Time history and Fourier spectrum of synthetic earthquake wave. (**b**) Sinusoidal wave

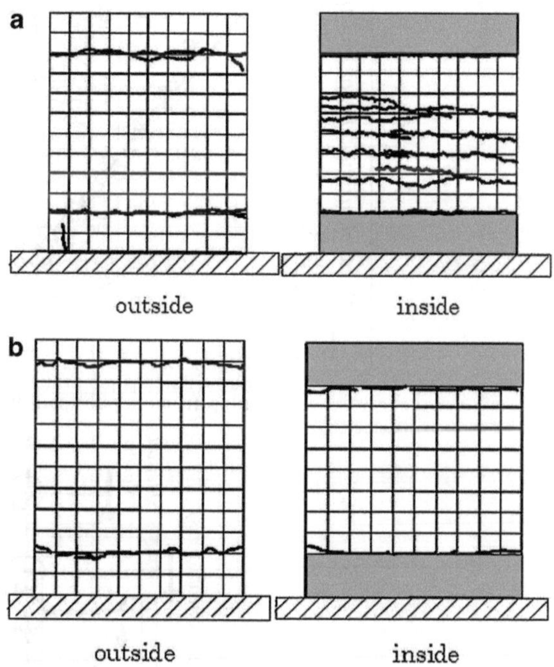

Fig. 23 The sketches of the crack on the surface of the sidewall. (**a**) Buried in the fully saturated sand. (**b**) Buried in the dry sand

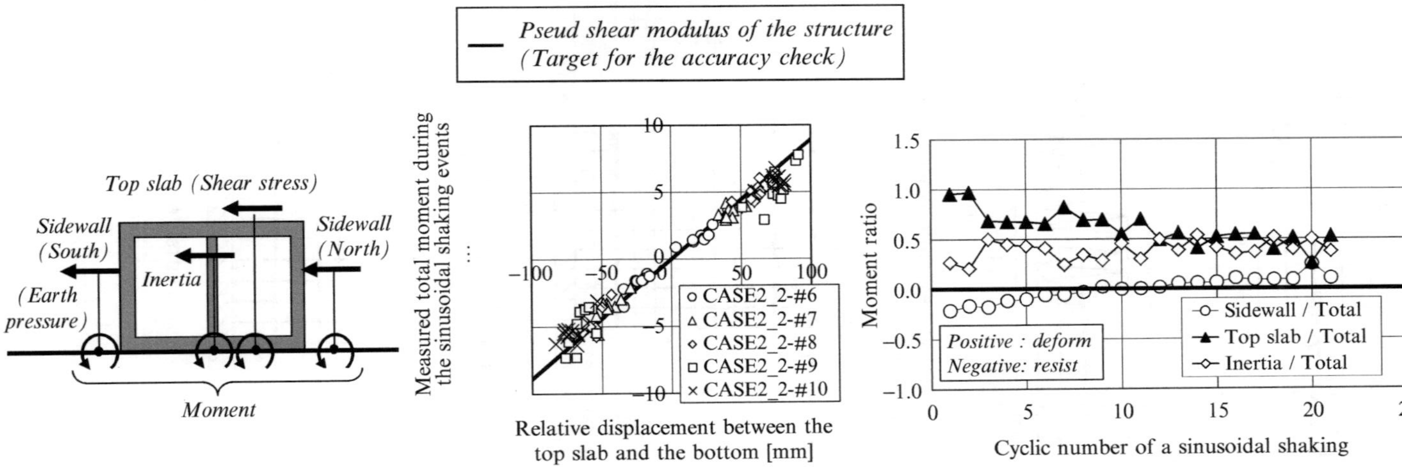

Fig. 24 Confirming the measured data accuracy and the share of the deforming force

The knowledge about underground structures obtained from these results is referred to in the verification manual on the performance-based seismic design of important outdoor civil structures in nuclear power plants [24].

Seismic Stabilities of Model Slopes

For the evaluation of slope stability, a safety factor, in which the sum of the strength along the supposed sliding surface is divided by the force acting on the sliding surface, tends to be used. Since the concept, based on the equilibrium between the strength and the force at the potential sliding surface, is easy to understand, the method has been used in the evaluation of the seismic stability of slopes without any detailed examination. The method which uses the safety factor is valid for the static stability analysis in most cases, where only the equilibrium between the strength and the static force, such as the self-weight, is dominant. In the static equilibrium, if the stress from its self-weight is larger than the strength, sliding blocks are moved by the unbalanced force. The amount of movement can be obtained by subtracting the strength from the force, to cause perfect failure both in the estimation and for the real phenomenon.

However, in order to evaluate the seismic performance of slopes, the deterioration of the safety factor to less than unity for an instant does not mean that the slope will always fail. Consider an object on a slightly inclined plain with friction when shaken. If the object had been standing still before the shaking, it might be moved only during the shaking and then stand still again after the shaking, since the extra force caused by the shaking was transient and the self-weight of the structure was insufficient to move the object against the friction. For the sake of convenience, the dynamic friction was considered to be same as the static friction in this case. The validity of the idea was experimentally confirmed by Ishimaru and Kawai [25]. They conducted a centrifuge test of a man-made rock slope model and concluded that the calculated safety factor during the shaking event, in which the sliding block had fallen, was much smaller than unity. Kawai et al. suggested that using the safety factor in order to evaluate the seismic stability might mislead not only about the timing of the failure but also the location of the failure [26].

Thus, in order to evaluate the seismic stability of a slope properly, it is necessary to examine the mechanism of failure of a slope during an earthquake in detail. For that purpose, a series of slope shaking tests at 1G have been conducted [27, 28] using an imaging device, namely, a high-speed and high-resolution camera, to observe the failure progress during a shaking event. Figure 25 shows several examples of the model slopes and their failure.

An example of data processing is shown in Fig. 26 [29]. In this case, a tension crack at the top of the slope emerged at first, then the strain at the lower part became remarkably large, and the strain in the middle part increased to form the sliding surface. This kind of data processing was adapted to all the tests.

Fig. 25 Examples of a series of slope failure experiments. (**a**) Initial conditions of the slope models. (**b**) Failure after the largest shaking events

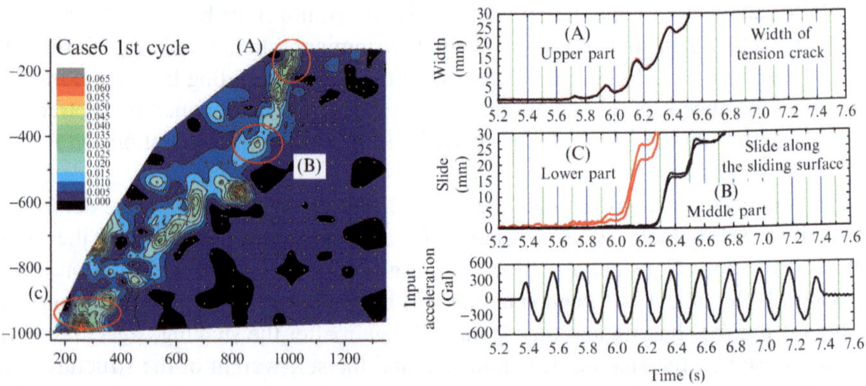

Fig. 26 The measured strain distribution and the time histories

In order to establish the performance-based evaluation method of seismic stability of a slope, both the method to estimate the response and the method to determine the limit are needed. The experiments explained above in detail are very useful for both these purposes.

Summary

Scale model experiments are very important in geotechnical engineering field as well as in other fields. Since soil is a granular material with complex characteristics, scale modeling is a powerful tool which allows us to grasp, experimentally, the essential mechanism of a phenomenon of interest. Further, because those complex characteristics depend on the confining pressure, downscaled models are incapable of reproducing the huge prototype. Therefore a technique referred to as "the centrifuge test" can be adopted to overcome this problem. In this test, centrifugal acceleration is given to increase pseudo-gravity. It enables the model to be regarded as a prototype with regard to stress level. This technique has been successfully used from 1936 [6] when no satisfactory theory of similitude existed. It was not until the 1980s that the scaling laws for centrifuge tests were established.

That situation in the scale modeling of geotechnical engineering may well represent the relationship between engineering and science described by Williams [30]. Giving a centrifugal acceleration to a geotechnical model may be "kufu" that is appreciated as a fourth research tool by Emori [31] as explained by Saito [32]. If it is not considered "kufu," the technique involved in centrifuge testing, is definitely a powerful tool for use in geotechnical engineering.

In geotechnical engineering, the development of the constitutive equations and the corresponding numerical analysis technique have also been extensively attained. However, since the behavior of the soil is too complicated, those numerical simulations can only be used to predict the behavior of the prototype quantitatively well under certain conditions, and the possibility that the model fails to reproduce the essential mechanism of the phenomenon of interest remains. Further, the numerical methods are in some cases too sensitive to enable stable prediction; that is, small differences in just one parameter may lead to a completely different conclusion. This is not the case for the prototype. Therefore, model test methods, including the centrifuge test, are essential tools in the field of geotechnical engineering not only for reproducing a prototype itself but also to cross-check the prediction results acquired in numerical simulations.

References

1. Terzaghi, K.: Theoretical Soil Mechanics, pp. 11–15. Wiley, New York (1943)
2. Rocha, M.: Similarity conditions in model studies of soil mechanics problems. Laboratoria Nacional de Engenharia Civil, Lisbon Publ. 35 (1953)
3. Rocha, M.: The possibility of solving soil mechanics problems by the use of models. In: Proceedings of the 4th International Conference on Soil Mechanics, London, 1 (1957)
4. Kagawa, T.: On the similitude in model vibration tests of earth-structures. In: Proceedings of the Japan Society of Civil Engineers, vol. 275, pp. 69–77 (1978, in Japanese)
5. Iai, S.: Similitude for shaking table tests on soil-structure-fluid model in 1g gravitational field. Report of the Port and Harbour Research Institute, vol. 27(3), pp. 3–24 (1988)
6. Pokrovsky, G.I., Fedorov, I.S.: Studies of soil pressures and soil deformations by means of a centrifuge. In: Proceedings of the 1st ICSMFE, vol. 1, pp. 70–71 (1936)
7. Kimura, T., Kusakabe, O.: Koza Centrifuge. Tuchi-to-kiso, 35(11), pp. 68–74 (1987, in Japanese)
8. Shibata, T., Kita, K., Kobayashi, S., Yashima, A.: Performance of shaking table tests and measurement of shear wave velocities in a centrifuge. Centrifuge **91**, 391–398 (1991)
9. Sato, M., Minowa, C., Saito, Y., Kagawa, T., Abe, A.: Reproduction of large-scale 1g test on dry sand deposits and pile foundations using centrifuge modeling. Physical Modelling in Geotechnics, 723–728 (2002)

10. Allard, M.A., Schenkeveld, F.M.: The Delft geotechnics model pore fluid for centrifuge tests. Centrifuge **94**, 133–138 (1994)
11. Hayashi, K., Fujii, N., Muramatsu, T., Houjyou, K.: Direct comparison of gravity model and centrifuge model for the seismic problem. In: Proceedings of the Japan Society of Civil Engineers 582/III-41, pp. 207–216 (1997, in Japanese)
12. Kutter, B.L., Chang, J., Davis, B.C.: Collapse of cavities in sand and particle size effects. Centrifuge **94**, 809–815 (1994)
13. White, R.J., Stone, K.J.L., Jewell, R.J.: Effect of particle size on localisation development in model tests on sand. Centrifuge **94**, 817–822 (1994)
14. Corte, J.-F., Garnier, J., Cottineau, L.M., Rault, G.: Determination of model soil properties in the centrifuge. Centrifuge **91**, 607–614 (1991)
15. Renzi, R., Corte, J.F., Rault, G., Bagge, G., Gui, M., Laue, J.: Cone penetration tests in the centrifuge: experience of five laboratories. Centrifuge **94**, 77–82 (1994)
16. Goodings, D.J., Gillette, D.R.: Grain and model size effects in centrifuge models of granular slope instability. Centrifuge **91**, 583–590 (1991)
17. Kawai, T., Kanatani, M., Tanaka, Y., Stewart, D.P., Kutter, B.L., Settgast, R.R., Ishikawa, H., Takeda, T., Higuchi, S., Goto, Y.: Seismic performance of a caisson type seawall with an armored embankment. In: Proceedings of the International Conference Centrifuge, vol. 98, pp. 351–358 (1998)
18. Stewart, D.P., Settgast, R.R., Kutter, B.L., Kawai, T., Higuchi, S., Ishikawa, H., Takeda, T.: Experimental performance of a seawall model under seismic conditions. Soils Found. **40–6**, 77–91 (2000)
19. Kanatani, M., Kawai, T., Tochigi, H.: Prediction method on deformation behavior of caisson-type seawalls covered with armored embankment on man-made islands during earthquakes. Soils Found. **41**(6), 79–96 (2001)
20. Kawai, T., Kanatani, M., Tochigi, H., Tanaka, M.: Study on seismic stability of seawall in man-made island (part2)—experimental study of seismic performance of seawall using a centrifuge—, CRIEPI report, U98023 (1999, in Japanese)
21. Kanatani, M., Tochigi, H., Kawai, T., Sakakiyama, T., Kudo, K.: Study on seismic stability of seawall in man-made island (part5)—deformation of actual seawall during earthquake and estimation of seismic performance—, CRIEPI report, U99011 (1999, in Japanese)
22. Kawai, T., Asaoka, A., Noda, T.: Evaluation of seismic performance of a box culvert buried in saturated sand with centrifuge and its numerical simulation, IS-Tokyo, pp. 1087–1093 (2009)
23. Kawai, T., Horie, M.: A study of centrifuge tests of box culvert structures buried in the saturated sand. JSCE J. Earthq. Eng. **28**, 1–6 (2005)
24. Committee on Nuclear Civil Engineering JSCE, The verification manual on performance based seismic design of important outdoor civil structures in the nuclear power plants, JSCE, 2005 (in Japanese)
25. Ishimaru, M., Kawai, T.: Basic study on the evaluation of seismic stability of rock slope using centrifuge model test. In: Proceedings of the Japan Society of Civil Engineers 67-1C, pp. 36–49 (2011, in Japanese)
26. Kawai, T., Ishimaru, M.: A development of an evaluation flow chart for seismic stability of rock slopes based on relations between safety factor and sliding failure, CRIEPI report, N09030 (2010, in Japanese)
27. Nakamura, H., Murata, M., Shinoda, M., Watanabe, K., Sanagawa, T., Kawai, T., Nakamura, S.: Study on characteristics of damages on rock slope models in accordance with different weak layers in a shaking table test. In: Proceedings of the 66th JSCE annual meeting, Matsuyama, pp. 573–574 (in Japanese)
28. Abe, K., Shinoda, M., Watanabe, T., Sanagawa, T., Nakajima, S., Nakamura, S., Kawai, T., Murata, M., Nakamura, H.: Numerical simulation of landslides after slope failure using MPM with SYS Cam-clay model in shaking table test. In: Proceedings of the 15 WCEE, LISBOA (2012)

29. Murata, M., Nakamura, H., Watanabe, K., Shinoda, M., Sanagawa, T., Kawai, T., Nakamura, S.: Investigation of strain progress of rock slope model under seismic force. In: Proceedings of the 46th JGS annual meeting, Kobe, pp. 1607–1608 (in Japanese)
30. Williams, A.F.: Foreword: significance of scale modeling in engineering science. Progress in Scale Modeling, Springer, pp. vii–xvii (2010)
31. Saito, K.: Preface: the art of scale modeling. Progress in Scale Modeling, Springer, pp. xix–xxv (2010)
32. Emori, R.I.: Toys and scale models. Progress in Scale Modeling, Springer, pp. 419–427 (2010)

Section A
Earthquake - Supercomputing and Scale Modeling the Effect of Flotsam Mixed Tsunami: Implications for Tsunami Generated by the 2011 Great East Coast Earthquake

Jun Ishimoto and Kozo Saito

Abstract The behavior of flotsam mixed tsunami is investigated by a new type of integrated super computation using arbitrary Lagrangian–Eulerian method (ALE method) and smoothed particle hydrodynamics method (SPH method). The fully hydrodynamic governing equations without shallow-water theory were used to calculate tsunami characteristics of water flow with flotsam and debris. Our ALE model predicted the effect of fluid–solid coupled interaction in a limited region, and the model predictions were favorably compared with the scale modeling analysis. This study, our first attempt to simulate the degree of damage caused by the flotsam mixed tsunami, can help optimize the strength of seashore buildings and structures against future tsunami threats. This study also can help estimate structural damage that can be caused by large-scale natural disasters, like hurricanes, storms, and tornados, and help to develop effective mitigation tools and systems.

Introduction

A recent numerical study of tsunami simulation conducted by oceanic scientists predicted damage in a relatively large, approximately several 100 km, square area [1–3]. Their model used some assumptions including the shallow-water theory [4–7]. This model may be acceptable to estimate the relative degree of damage over a relatively large area [3–5]. In the shallow-water approximation, the vertical

J. Ishimoto (✉)
Institute of Fluid Science, Tohoku University, 2-1-1, Katahira, Aoba-ku, Sendai 980-8577, Japan
e-mail: ishimoto@alba.ifs.tohoku.ac.jp

K. Saito
IR4TD, Department of Mechanical Engineering, University of Kentucky, 179 RGAN Building, Lexington, KY 40506-0503, USA

K. Saito et al. (eds.), *Progress in Scale Modeling, Volume II*,
DOI 10.1007/978-3-319-10308-2_3

velocity profile is assumed to be uniform; however, it has limitations for accurately estimating the impact forces on specific structures and land locations where the vertical velocity profile of water flow is important.

The magnitude 9.0 earthquake (the Great East Coast Earthquake) hit off the Sendai coast area in Japan, March 11, 2011, and created a huge tsunami which claimed more than 20,000 lives. This high casualty number was partly caused by the enhanced destructive forces of the tsunami that contained floating debris and flotsam. This mixed effect of water and floating debris has not been accurately estimated by any currently existing conventional tsunami simulation models because they only estimate the impact force of single-phase water. Our current model clearly showed that the destructive force was significantly enhanced by the leading wave and the outflow of flotsam as compared to a normal tsunami without floating debris and flotsam. To that end, we created a fully hydrodynamic approach model without shallow-water approximation to compute the flow characteristics of the mixed tsunami accompanied with flotsam and debris. In addition, we calculated the impact forces of the mixed tsunami acting on structures by the leading wave and the outflow by backwash of the flotsam against an ideal geometric structure placement.

We are focusing on a relatively limited region to be analyzed, such as power plants and industrial plants, taking into account the interaction of the structure and the tsunami, fracture behavior, and structural deformation. We are also developing a simultaneous simulation method for the flotsam mixed tsunami behavior of its interface causing deformation when in collision with structures. In addition, the scale modeling analysis for flotsam mixed tsunami is conducted and compared with numerical results.

Coupled Computation (ALE-FEM) of Tsunami-Vehicle Drifting Behavior

For the first computation, the arbitrary Lagrangian–Eulerian method (ALE method) [8] in application to the tsunami simulation with the FEM vehicle model for hydraulic collision analysis was used. Figure 1 shows the schematic of the computational model and the FEM vehicle model. In the present computation, the Euler element is applied to analyze the hydrodynamic behavior of the tsunami getting over the breakwater, and the FEM vehicle model is applied to analyze the collision behavior between vehicles and tsunami or between two vehicles. For the numerical condition, the height of the breakwater was 2.0 m and the initial velocity of tsunami was 5.0 m/s.

The FEM model of the vehicles can tolerate collision analysis. A rigid body model is assumed for shortening the computational time. For small and large vehicles, a GEO Metro (15,000 elements) and Chevrolet C1500 (10,000 elements) were assumed to model the FEM objects.

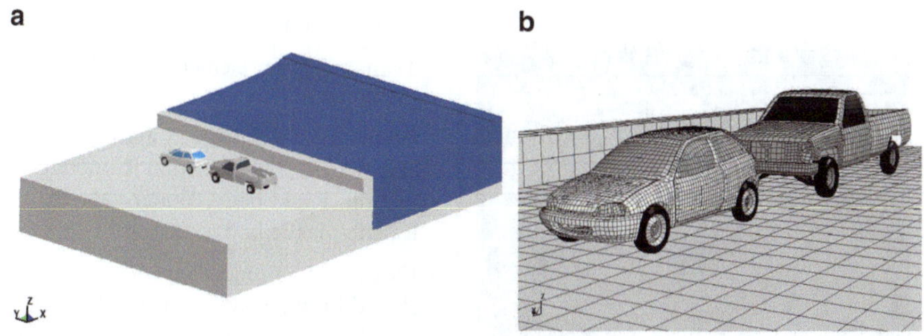

Fig. 1 Schematic of computational model and used FEM vehicle model. (a) Computational domain. (b) FEM two-vehicle model

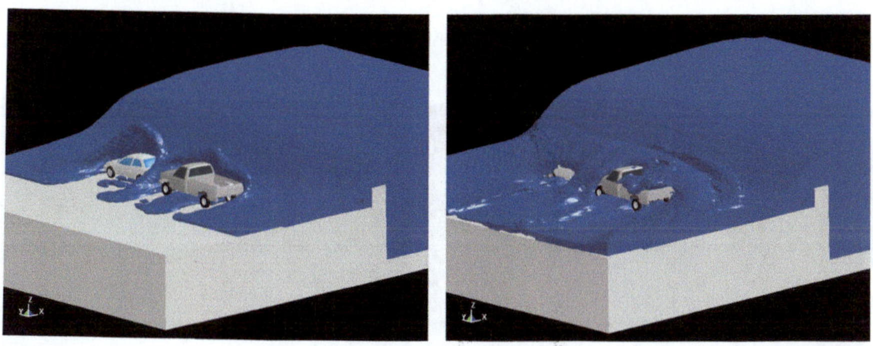

Fig. 2 Computational results of the tsunami behavior getting over the breakwater and its attacking behavior on two vehicles

Figure 2 shows the computational result of the tsunami behavior getting over the breakwater and its attacking behavior on the two vehicles. The hydrodynamic impact behavior of the tsunami on the vehicles is reasonable simulated by the drifting behavior of the vehicles. It was found that the vehicles were covered by the tsunami and were washed away by the tsunami's inertia.

The Lumber Mixed Tsunami's Hydrodynamic Impact Behavior for a Water Gate

Next, the effect of flotsam mixing with the impact force of the tsunami was numerically predicted. In the present computation, the lumber mixed tsunami's hydrodynamic impact behavior for a water gate was performed. Figure 3 shows the computational geometry for lumber mixed tsunami impacting a water gate and the specifications of computational model items including the initial numerical

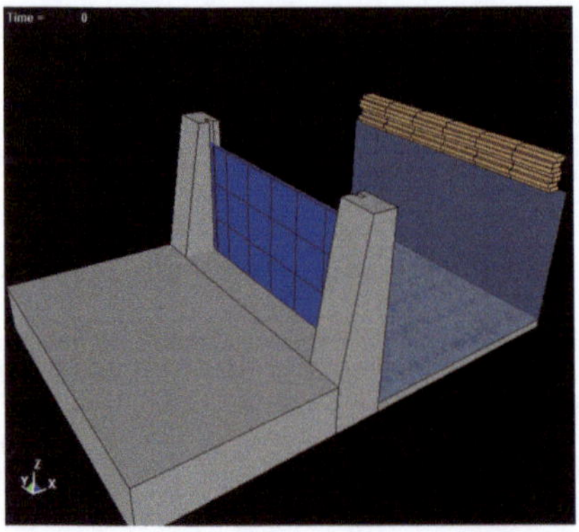

Specifications for numerical model

- Tsunami: Euler element (260,288 elements)
- Water gate and prop : Lagrange element (22,728 elements) [Concrete section: Rigid body, Water gate : Elastic-plastic body]
- Flotsam: Lagrange elements (Rigid body, Specific weight: 0.9, Total mass 1.1 ton) (1,260 el elements)
- Initial velocity of tsunami: 3.0 m/s

Fig. 3 Computational geometry for lumber mixed tsunami impacting a water gate

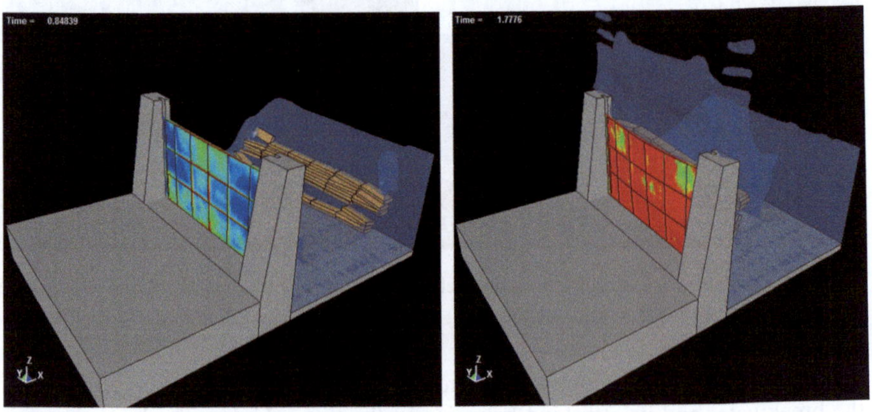

Fig. 4 Lumber mixed tsunami's hydrodynamic impact behavior on the water gate

conditions. For the dynamic material characteristics, it was assumed that the concrete section had a rigid body and the water gate had an elastic–plastic body.

Figure 4 shows the numerical results of the lumber mixed tsunami's hydrodynamic impact behavior on the water gate. The computation was performed with lumber and without lumber to study the differences of tsunami impact behavior. In Fig. 4, the color contours in the water gate represent different magnitudes of stress. Interestingly, the change in flexibility and impact stress of the water gate by the impact of lumber mixed tsunami was numerically reproduced. Figure 5 shows the effect of flotsam (lumber) in the tsunami on the impact force against the water gate. It was found that the impact force of lumber mixed with a tsunami was two times greater than that of single-phase tsunami (without flotsam case).

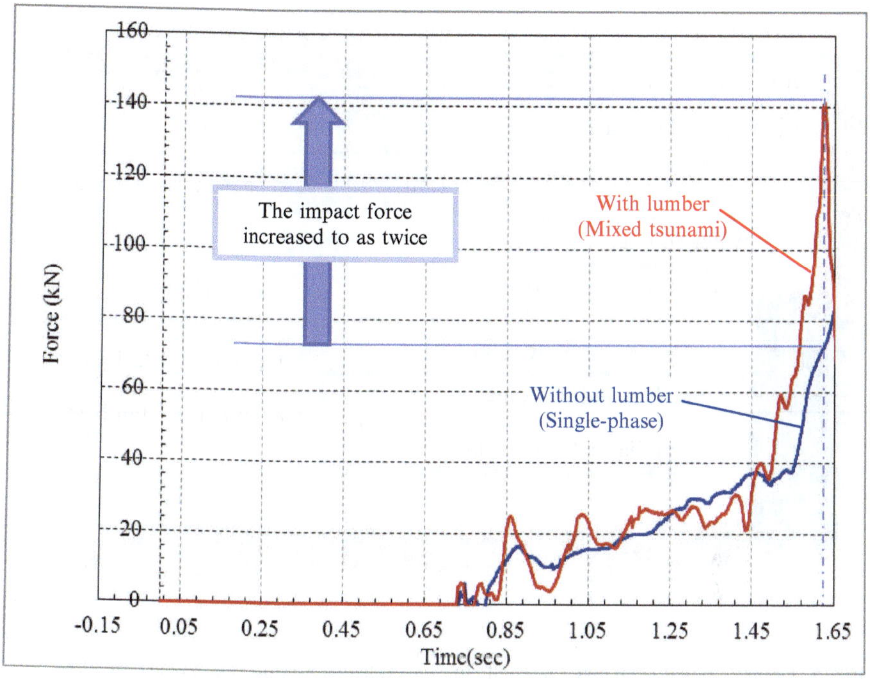

Fig. 5 Effect of flotsam (lumber) mixing with tsunami on impact force for the water gate

Computation of a Flotsam Mixed Tsunami Behavior by Smoothed Particle Hydrodynamics (SPH) Method

In the final computation, we constructed the SPH method [9] to investigate the interaction between a tsunami and structures and also to investigate the damage of a flotsam mixed tsunami on land structures. For the numerical modeling, we assumed the following mixed tsunami flow conditions:

- A flotsam or obstacle was carried away by the backwash of the first wave of the tsunami and then pushed by a second wave of the tsunami.
- The flotsam went onto the shore and collided with the land structure.

To analyze these phenomena, we used the computational geometry which demonstrates two small tanks impacting a large tank, as shown in Fig. 6. The tsunami wave was assumed to break down within a collapsing rectangular water column as it collided with the small tank and began to move the small tank because of the impact force of the tsunami. These small tanks in the tsunami wave then collide with the large tank which is fixed to the ground. The small cylindrical tanks were assumed to be filled with oil and were regarded as obstacles or flotsam within the tsunami with a slip boundary condition relative to the ground. The large

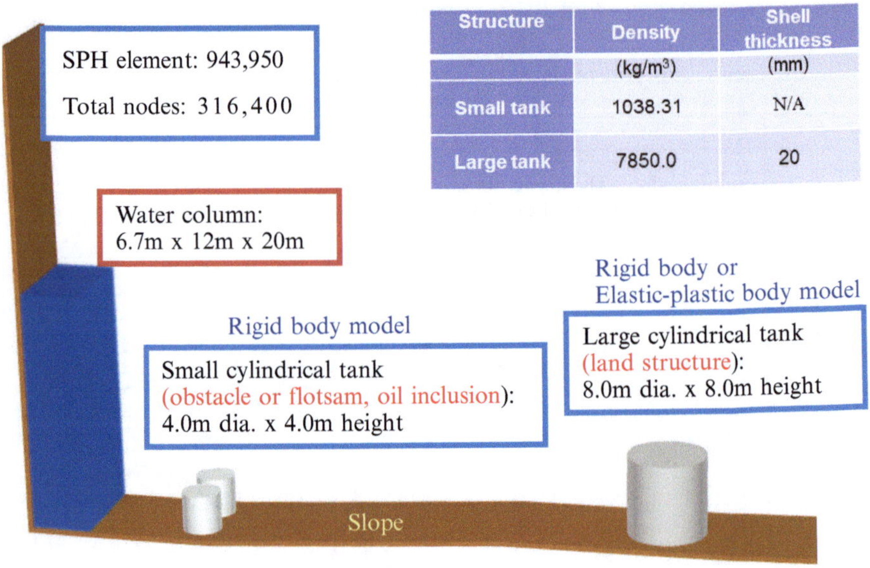

Structure	Density	Shell thickness
	(kg/m³)	(mm)
Small tank	1038.31	N/A
Large tank	7850.0	20

Fig. 6 Computational geometry for small tank mixed tsunami impacting a land structure

cylindrical tank was fixed to the ground and was assumed to be hollow (empty) with a shell thickness of 20 mm.

Figure 7 shows the SPH numerical results of the flotsam mixed tsunami. The flow characteristics and behavior of its interface deformation and collision with the land structures were clearly simulated.

Figure 8 shows the impact stress profile of the land structure with flotsam mixed tsunami impingement; an elastic–plastic body assumption was applied for the land structure. When the two small tanks impinge on the larger-tank land structure, the maximum impact stress was 100 MPa, and a corresponding maximum plastic strain of 0.0015 was numerically obtained.

Figure 9 shows the effect of flotsam mixing with the tsunami on the impact force to the land structure. In the case of a single-phase tsunami (without flotsam, regarded as a leading wave), the maximum impact force was 3.5 MN. In the case of a flotsam mixed tsunami (regarded as secondary wave), the maximum impact force reached 42 MN. In other words, it was found that the impact force magnitude of flotsam (small tank) mixed with the tsunami was over 10 times that of single-phase (without flotsam case) tsunami.

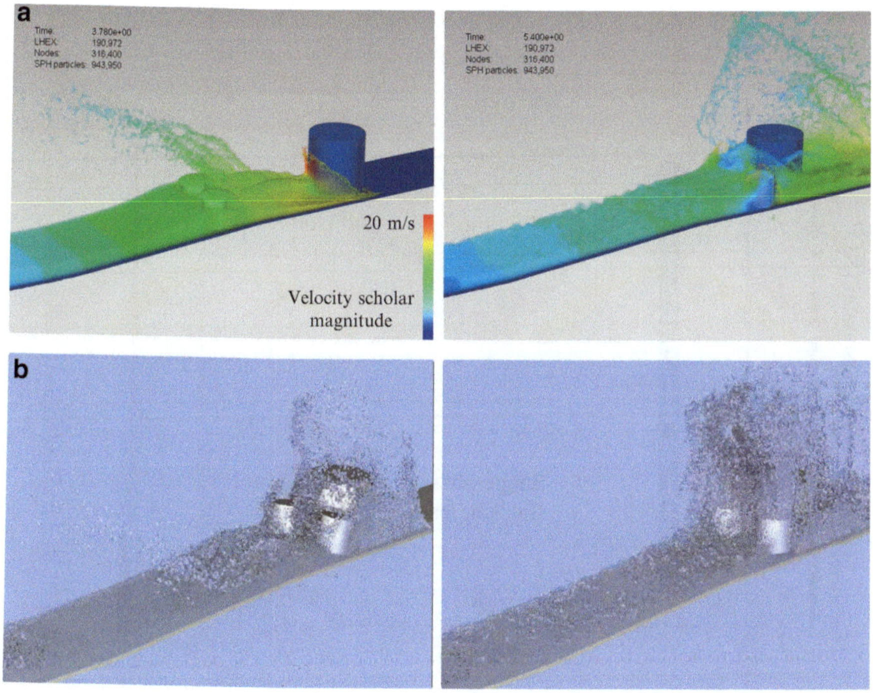

Fig. 7 Flotsam mixed tsunami flow characteristics and behavior of its interface deformation and collision with a land structure by SPH method. (**a**) Small tank mixed tsunami flow characteristics and its impact behavior on a land structure. (**b**) POV-Ray rendering result

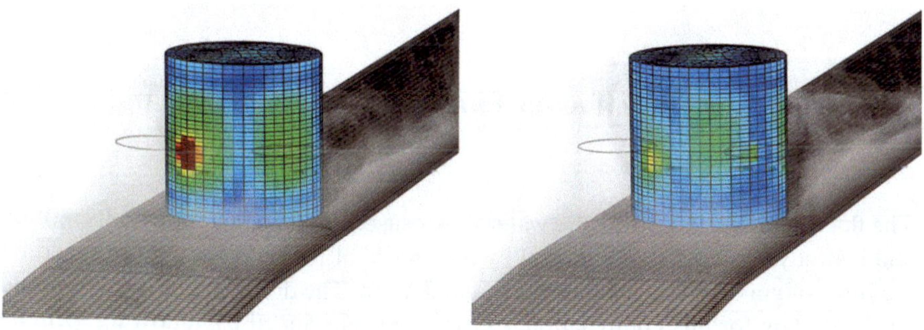

Fig. 8 Impact stress profile of land structure while flotsam mixed tsunami impingement by SPH method (color contour denotes the scalar magnitude of stress in land structure)

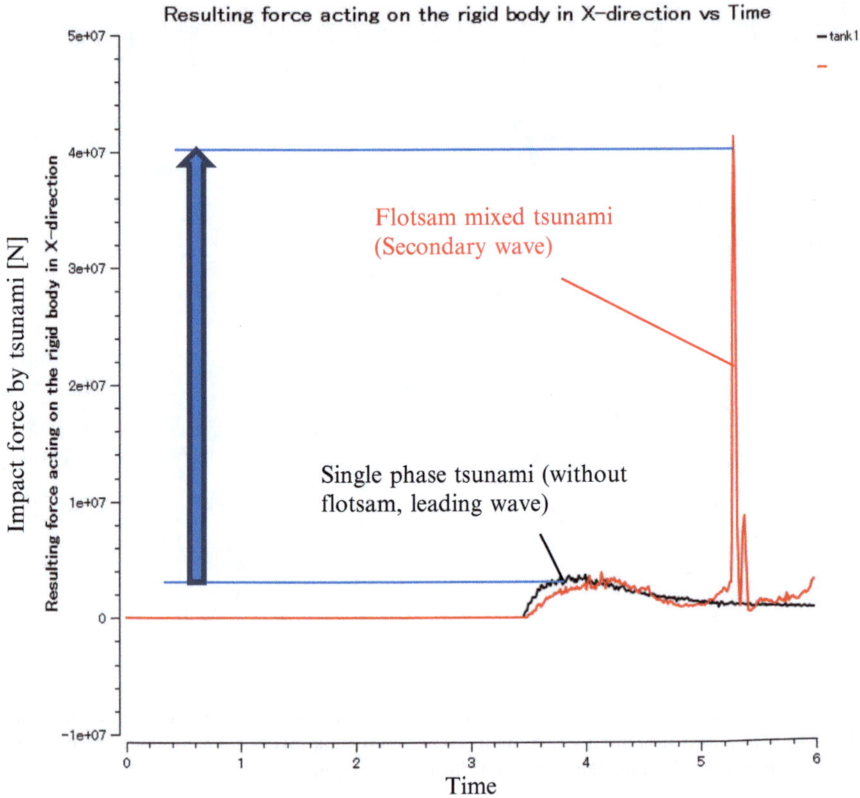

Fig. 9 Effect of flotsam mixing with tsunami on the impact force with a land structure

Scale Modeling the Effect of Flotsam Mixing on Tsunami Damage

The flotsam is accelerated by the water flow caused by the tsunami, with its motion and hydrodynamic force dominated by the inertia of the water and flotsam and by the force of gravity acting on the water and flotsam. The dominant physical law and similarity law for this computational system are introduced by following formula [10, 11].

Dominant physical law:

Buoyancy: $F_b = \Delta\rho g l^3$

Inertia: $F_i = \rho_m l^3 \dfrac{l}{t^2}$ $\xrightarrow{\; v = \frac{l}{t} \;}$ $\rho_m l^2 v^2$

Similarity law:

$\pi_1 = \dfrac{F_i}{F_b} \rightarrow \dfrac{\rho_m v^2}{\Delta\rho g l}$ (Froude number of flotsam mixed tsunami)

$\pi_2 = \dfrac{v^2}{gl} \rightarrow$ (Froude number of single-phase tsunami)

$\dfrac{\pi_1}{\pi_2} = \dfrac{\rho_m}{\Delta\rho} \rightarrow$ (Effect of flotsam mixing on tsunami damage)

where $\Delta\rho$ is the density difference between flotsam and seawater, ρ_m is the mixture density of tsunami, g is the gravitational acceleration, v is the velocity, l is the characteristic length, and t is the time. According to the above analysis, it was found that Froude number was suitable to evaluate the effect of mixing of flotsam on the tsunami impact damage. As the mixture density increased, namely, the amount of flotsam became larger and the density difference between flotsam and seawater became smaller, the damage of the mixed tsunami to the land structure increased. The scale effect of the mixture density and the density difference between flotsam and seawater became the dominant factors for tsunami impact force and damage prediction.

Therefore, quite reasonable results have been obtained by the present computations in which the land structure received much greater damage by the impingement of a small tank mixed tsunami as compared to that of the lumber mixing case.

Experimental Approach for Tsunami Scale Modeling

To validate the numerical results on the height and traveling velocity of a tsunami wave, the University of Kentucky students [12] conducted the 1/1,000th and 1/2,000th scale model during the ME 565 course for the Great East Cost Earthquake generated tsunami in March 2011. A commercially available detergent was added to water to reduce the surface tension force for the scale model to approximately satisfy the Weber number [11], a ratio of the inertia and the surface tension forces. A series of color photographs including Fig. 10 were taken during the experiments, which showed the scale model tsunami shape and behavior surprisingly similar to the full-scale tsunami reported by a Japanese TV news station [13]. The moving speed of the tsunami was roughly correlated by the Fr number scaling although the source term was not scaled, since the full-scale data was not available at that time. This is another surprise for the validity of this simple scale model experiment.

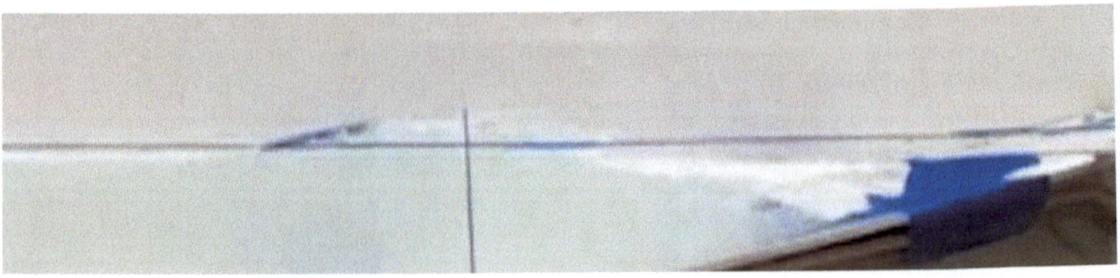

Fig. 10 A color photo taken from the side for the 1/1,000th scale model generated tsunami [12]

Conclusions

1. Two different supercomputing approaches, including the ALE method and SPH method, were used to investigate the flow characteristics of a flotsam mixed tsunami.
2. The flotsam mixed tsunami behavior associated with interface deformation and collision with land structures was reasonably reproduced by the supercomputing methods.
3. The impact stress profile of land structures with flotsam mixed tsunami impingement was computationally predicted. As a result, it was shown that the impact force of a flotsam mixed tsunami would be over 10 times greater than that of single-phase tsunami (without flotsam case). It was also found that the scale effects of mixture density and density differences between flotsam and seawater are the dominant factors for tsunami impact force and damage prediction.

Acknowledgments We thank LANCEMORE Co. and Explosion Research Institute Inc. for their cooperation and assistance on the tsunami code programming. This study was partially sponsored by the Institute of Fluid Science, Tohoku University, under the Multiple Collaborative Research Project 2013 (Budget code: 55059970). We thank John Stencel for his invaluable comments.

References

1. Goto, K., Takahashi, J., Oie, T., Imamura, F.: Remarkable bathymetric change in the near shore zone by the 2004 Indian Ocean tsunami: Kirinda Harbor, Sri Lanka. Geomorphology **127** (1–2), 107–116 (2011)
2. Goto, K., Okada, K., Imamura, F.: Numerical analysis of boulder transport by the 2004 Indian Ocean tsunami at Pakarang Cape, Thailand. Mar. Geol. **268**(1–4), 97–105 (2010)
3. Fujita, K., Goto, C.: Truncation error of numerical simulation of linear dispersive wave theory. Proceedings of 2nd International Conference on Asian and Pacific Coasts [CD-ROM] (2003)
4. Goto, C., Shuto, N.: Numerical simulation of tsunami propagations and run-up. In: Iida, K., Iwasaki, T. (eds.) Tsunamis- Their Science and Engineering, pp. 439–451. Terra Scientific, Tokyo (1983)

5. Goto, C., Shuto, N.: Effects of large obstacles on tsunami inundations. In: Iida, K., Iwasaki, T. (eds.) Tsunamis- Their Science and Engineering, pp. 511–525. Terra Scientific Publishing, Tokyo (1983)
6. Masden, P.A., Sørensen, O.R.: A new form of the Boussinesq equations with improved linear dispersion characteristics, part 2, a slowly-varying bathymetry. Coast. Eng. **18**, 183–204 (1992)
7. Nwogu, O.: Alternative form of Boussinesq equations for near shore wave propagation. J. Waterway Port Coast. Ocean Eng. **119**(6), 618–638 (1993)
8. Bazilevs, Y.: Computational Fluid–Structure Interaction: Methods and Applications. Wiley Series in Computational Mechanics. Wiley, Chichester, UK (2013)
9. Violeau, D.: Fluid Mechanics and the SPH Method: Theory and Applications. Oxford University Press, Oxford, UK (2012)
10. Saito, K. (ed.): Progress in scale modeling: Summary of the First International Symposium on Scale Modeling (ISSM in 1988) and Selected Papers from Subsequent Symposia (ISSM II in 1997 through ISSM V in 2006), Springer (2008)
11. Emori, R.I., Saito, K., Sekimoto, K.: Scale Models in Engineering (Mokei Jikken no Riron to Ohyou), 3rd edn. Gihodo, Tokyo (2000). ISBN:4-7655-3252-6C3053 (Second Print in 2008 in Japanese)
12. Greenwell, C., Dieball, L.: Tsunami Scale Model (Wave Amplitude and Velocity). ME 565 Final Report in the Department of Mechanical Engineering, University of Kentucky (2011)
13. The Great East Coast Earthquake. Asahi Shimbun Press, Japan (2011)

6. [illegible]
7. [illegible]
8. [illegible]

Section B
Fire and Explosion - Scale Modeling of Biomass Fire Associated with Hydrogen-Producing Bacteria

Tadashi Konishi, Yuta Kawamura, Akira Narumi, and Tomek Ziemba

Abstract This paper details a fundamental study for developing a safe handling system for waste biomass fuels, especially for the prevention of fires and explosions associated with refuse-derived fuels (RDF). Unlike conventional solid fuel, RDF is a living fuel that contains various kinds of bacteria among which flammable gases such as hydrogen and methane may be produced during biological fermentation in the proper circumstances. The RDF storage silo at the Mie prefectural RDF power plant unexpectedly exploded in 2003, and firefighter deaths and injuries were reported. Though flammable gas production was the suspected cause of the explosion, there is disagreement about whether the flammable gas was produced by the biological fermentation or by thermal decomposition of the RDF materials. Due to the difficulties in conducting both the experiments and numerical simulations in a full-scale mock-up silo, the detailed explosion mechanism has not been explained since 2003. On the basis of scale modeling principles, this study begins with determining the physical or chemical laws which govern the phenomena of the accident. For this purpose, flammable gas production tests were conducted using two kinds of RDFs, which were provided from two manufacturers: Kuwana and Sakura. The RDF hydrogen production ability was affected by pH, moisture, and ambient temperature. The microbial colony count method revealed that aerobes occupied a dominant position in RDF on the second or third day from the beginning, whereas anaerobes occupied a dominant position hereafter. *Bacillus* sp. was superior to *Clostridium* sp. in Kuwana, whereas *Clostridium* sp. was superior to *Bacillus* sp. in Sakura after 3 days of cultivation. Scale effects on the temperature rise of RDF due to biological activities were examined using three different amounts of RDFs. It was found that the higher the amount of RDF, the higher the maximum temperature obtained after 50–85 h cultivation.

T. Konishi (✉) • T. Ziemba
Department of Mechanical Engineering, Institute of National Colleges of Technology, Oita College, Oita 870-0152, Japan
e-mail: konishi@oita-ct.ac.jp

Y. Kawamura • A. Narumi
Department of Mechanical Engineering, Kanagawa Institute of Technology, Kanagawa 243-0292, Japan

K. Saito et al. (eds.), *Progress in Scale Modeling, Volume II*,
DOI 10.1007/978-3-319-10308-2_4

Introduction

Current climate changes are dominated by the influence of human beings, which are now large enough to exceed the limits of natural variability. The main source of these changes is human induced in atmospheric composition. The environmental consequences of extensive use of fossil fuels have already begun to surface worldwide. The excessive uses of these fuels are one of the primary causes of global warming and acid rain [1, 2]. In addition, the Fukushima Daiichi Nuclear Power Plant Accident after the Great East Japan Earthquake in March 2011 has promoted increased utilization of renewable energy. Bio-refining is one of the promised solutions for global warming. Instead of fossil fuels, biomass fuels are expected to be a long-term and fundamental solution for large-scale production systems of chemical products and energy from renewable sourced biomass. Biomass conversion to ethanol as a liquid fuel by the thermophilic and anaerobic clostridia offers a potential partial solution to the problem of the world's dependence on petroleum for energy. Coculture of a cellulolytic strain and a saccharolytic strain of *Clostridium* on agricultural resources, as well as on urban and industrial cellulosic wastes, is a promising approach to an alternate energy source from an economic viewpoint [3]. However, in order to put bio-refining to practical use, fire and explosion accidents must be avoided at each stage of manufacturing, ensiling process, storage, distribution, and operation. Biomass-associated fire and explosions have frequently occurred in the United States [4–6] and Japan [7–10]. In 2003, the Mie RDF Electric Power Plant in Japan, which burns 200 t/day of RDF, the town of Tado about 80 miles east of Osaka, had an accident which resulted in the deaths of two firefighters. Five others were also injured in the blaze. The RDF storage facility at the power plant showed generation of heat and an outbreak of fire on July 27, 2003. Water was sprayed inside and outside the storage facility, and the RDF was removed from the facility. During this process, a small explosion occurred on August 14 in the lower part of the facility. On August 19, another explosion occurred in the upper part of the facility, resulting in seven causalities, including two fatalities. The fire was extinguished on September 27 and the RDF facility was removed in mid-November. Readers may refer to the literature [7, 11–15] to obtain more information about the accident. As will be later discussed in further detail, the primary cause of this incident was the generation of heat and hydrogen originating from the fermentation of microbes. From safety and economic perspectives, further investigation of the actual machinery involved in the accident complicates this case. Furthermore, after a breakdown of the CFD model, as well as considering the movement of gaseous substances and the heat transfer in the kiln, creation of a credible model is hard to do. This is due to the complications that arise during decision on parameters for the rate of chemical reactions, such as the growth rate of microbes and other organic matters. In the cases of complex phenomena like this, it is essential to be familiar with the essence of these phenomena and use scale modeling experiments through fundamental research. These uses of scale modeling yield potential applications for new fields and domains of research and reinforce scale modeling as the most suitable method for testing.

Plausible Mechanism of Microorganism-Derived Explosions in Storage Facilities

Safety characteristics of biomass storage facilities vary depending on the origin and ingredients of biomass used, the structure and materials of the facilities, and the aspects of the surrounding environment such as temperature and humidity, as is also the case for causes of fire or explosion related to biomass storage facilities. In the case of the explosion of the Mie RDF Electric Power Plant, the cause of explosion remains in dispute; some claim it originates in cellulose gasification [16]; others insist on heat generation by the water adsorption to an organic substance [17, 18] or the fermentation of a microorganism [19–21]. There have been many attempts to develop mathematical models for predicting temperature and distribution in storage bins or silos [22–24]. Most of these mathematical models attempted to simulate the distribution of moisture and temperature with heat generated by latent heat of moisture content. Recently, a mathematical model to predict the heating up in open-air wood chip piles was reported, in which the heat production from chemical, physical, and microbial exothermal processes was included [25, 26]. In this model, the microbial heat production is demonstrated by multiplying the oxygen consumption rate by the oxy-caloric coefficient, i.e., the amount of heat produced per mole oxygen during aerobic respiration.

In this paper, on the basis of scale modeling principles, we will first determine the physical or chemical laws which govern the phenomena of the accident. For this purpose, flammable gas production tests were conducted using two kinds of RDFs, which were provided from two manufacturers: Kuwana and Sakura. The RDF hydrogen production ability was investigated by changing the conditions of moisture and ambient temperature. The pH transition, which is one of the important factors to characterize the hydrogen production possibility, was also measured. The microbial colony count method revealed the dominant microbial flora during the cultivation of Kuwana and Sakura RDFs. To discover and characterize microbial diversity in RDFs, new sequencing technologies and novel isolation techniques have been developed; here, metagenomic analysis was used. Metagenomics [27] is a genomic analysis of microorganisms by direct extraction and cloning of DNA from assembled microorganisms. These approaches have contributed to classify the hydrogen-producing bacteria and non-hydrogen-producing bacteria in RDFs. Finally, we considered how scale affects the silo explosion. Due to low thermal conductivity of RDFs decreasing the heat release from RDF storage silos, scale effects indicate the larger the silo size, the less heat released. We also examined the scale effects on the temperature rise of RDFs due to biological activity by changing the amounts of RDFs in the insulated containers.

Experimental Apparatus and Procedures

Hydrogen Gas Production Apparatuses and Procedures [28]

The Mie RDF power plant uses seven kinds of RDFs produced in seven different plants: Kuwana, Sakura, Kahada, Miyama, Hamazima, Minamimuro, and Kii, as shown in Fig. 1. As the abilities of hydrogen gas production are different among these RDFs, we selected the two RDFs on the basis of our previous experiments: Kuwana for the maximum hydrogen production and Sakura for the minimum hydrogen production. Figure 2 describes briefly the experimental procedures for determining the abilities of the hydrogen production. This system consists of a 185-L incubator (inner width, 45 cm; depth, 45 cm; height, 90 cm), a gas chromatograph (Shimadzu Ltd. GC-2014), and a pH meter (Dkk-Toa Corp.). The measurements of hydrogen production were obtained on triplicate samples. To secure the productivity and reproducibility of the experiments, the ingredient heterogeneity of RDFs was avoided. The RDF pellets (300 g in a plastic bag) were crushed with a hammer to produce coarse aggregates. The metal and plastic piece was removed from these RDFs, and a 10 g aliquot of these RDFs was moved to a sterile vial. Then, 50 mL of pure water was added and sealed tightly with an aluminum screw cap and a silicone/PTFE septum. Then the vial was kept in an incubator at 37 °C for several days. A sampling syringe was used to collect sampling gases, and these gases were introduced to a gas chromatograph for the analysis of disseminating the chemical composition and concentration. The GC system is capable of measuring hydrogen, oxygen, nitrogen, methane, and carbon dioxide. Gas analysis was made from 0 to 168 h at 24-h intervals. The pH measurement was also made after the GC analysis by taking away the aluminum screw cap and a silicone/PTFE septum. It is known that pH significantly influences bacterial activity [2]. The slaked lime is used in order to control the activity of bacteria in manufacturing process of RDF.

Microbial Colony Count and Metagenomic Analysis Procedures [29, 30]

The microbial colony count—the popular method for counting the number of living bacteria in aerobic and anaerobic atmospheres, respectively—determined the dominant microbial flora during the RDF cultivation. After the gas analysis and pH were finished, the samples filtered through an 80-μm filter were centrifuged at 3,000 rpm for 10 min and then decanted the supernatant. Then 10 ml of a sodium chloride solution (0.85 wt%) was added and mixed thoroughly. Serial tenfold dilutions of a working culture in sodium chloride solution (0.85 wt%) were made to obtain suspensions containing 10^{-5}–10^{-6} CFU/mL. The plate count agar (10 g/L peptone, 5 g/L beef extract, 5 g/L sodium chloride, 15 g/L agar), three per serial tenfold

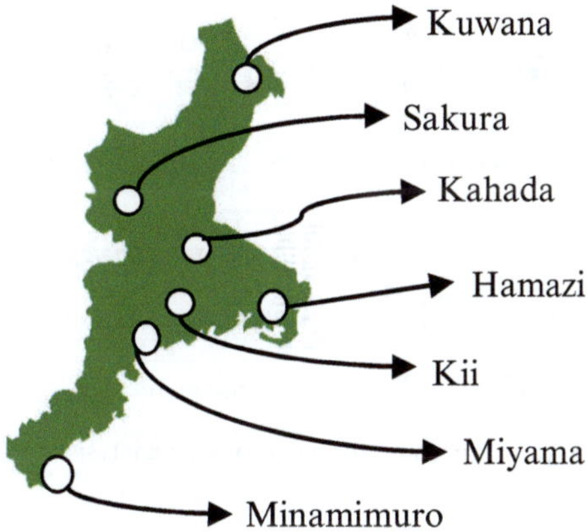

Fig. 1 Locations of RDFs collected and manufactured in Mie Prefecture

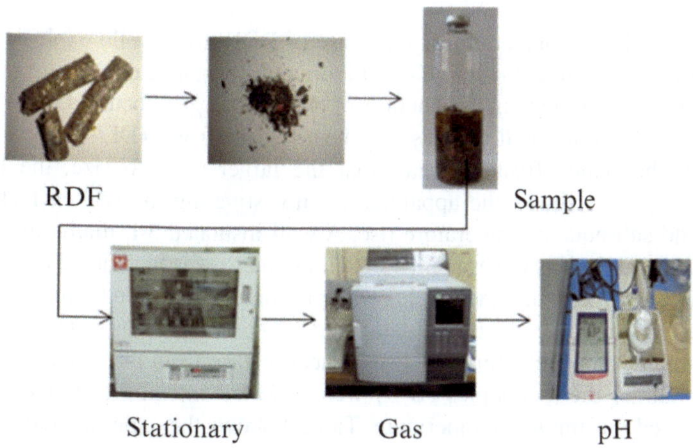

Fig. 2 Measuring procedure of gas analysis and pH

dilution for aerobic and anaerobic atmospheres, respectively, inoculated with the bacteria were subjected to 30 °C for 48–72 h. After incubation, colonies on the agar were enumerated and described. Metagenomics is the genomic analysis of microorganisms by direct extraction and cloning of DNA from assembled microorganisms. It provides a community-wide assessment of metabolic function and bypasses the need for the isolation and the cultivation of individual species. It should be noted that metagenomics could not evaluate the distinction between living bacteria and dead bacteria. Figure 3 shows the procedure of metagenomic analysis: extraction of DNA mixture from RDF, multiplication of ligation of 16SrDNA, transformation of component, bacteria growth on agar, blue/white color screening for recombinants, multiplication of 16SrDNA of each bacteria, sequence of 16SrDNA of each bacteria, and finding query sequence and data match [31].

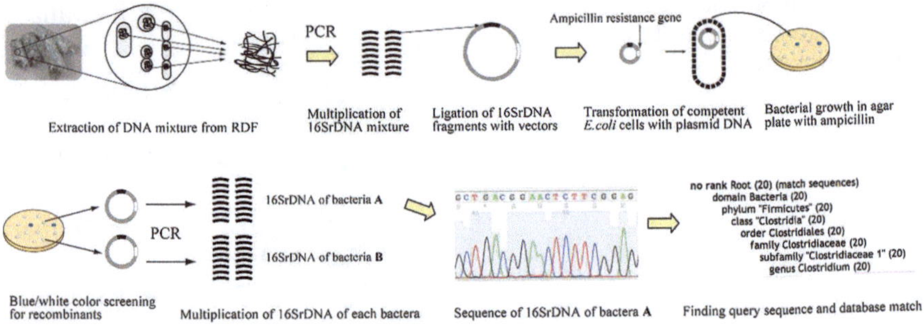

Fig. 3 Procedure for metagenomic analysis

Scale Modeling Experimental Apparatuses and Procedures

Another plausible explanation for the cause of RDF storage silo explosion at Mie may be that the chemical reaction of an inorganic substance due to biological heat release and the consequent production of flammable gas occurred. Low thermal conductivity of RDFs will decrease the heat release from RDF storage silos, and therefore, the scale effect indicates that the larger the silo size, the less heat released. Figure 4 shows the apparatus for investigating the scale effect on heat release and subsequent temperature rise. A well-insulated 3-L diathermic thermos bottle was used for this experiment. A stainless mesh was installed 2 cm above the bottom of the flask so that the surplus water accumulated in the bottom. The top of the container was covered with insulator, and 2 kg of weight was installed on the top of the container for preventing the loss of heat release from the top. Four sheathed chromel-alumel thermocouples (B-E) were inserted into the container, and one (A) measured the ambient temperature. Table 1 shows the experimental conditions for three independent experiments. The ambient temperature was set at 30 °C, the mean temperature around the Mie RDF power plant in August of 2003.

Results and Discussion

The Effect of Environmental Conditions on Hydrogen Gas Production

The effects of temperature and water content on the hydrogen gas production were investigated.

Temperatures varied from 20 to 50 °C, whereas water contents varied from 50 to 75 %. Figure 5 shows the results of 20, 30, 40, and 50 °C at constant 75 % water content and 50, 67, and 75 % at a constant temperature of 40 °C. It was found that the higher the water contents, the higher the hydrogen gas production. The

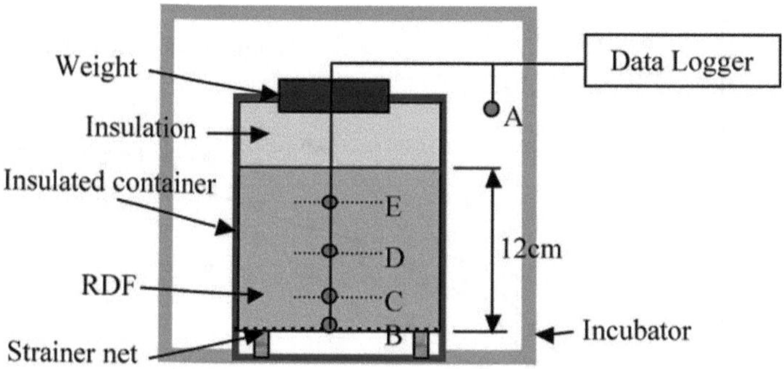

Fig. 4 Apparatus for scale modeling experiment

Table 1 Experimental conditions for scale modeling experiments

Experimental conditions			
RDF [g]	100	750	1,500
Water [mL]	37	280	550
Water content [%]	27.0	27.1	26.8
Ambient temperature [°C]	30		

hydrogen gas production was at a maximum at 40 °C, followed by 30 °C and then 50 °C, with minimum hydrogen production at 20 °C. Figure 6 shows the changes of hydrogen, oxygen, and pH with cultivation time. The oxygen and pH decreased with time, whereas hydrogen increased with time. The oxygen of Kuwana approached steeply to zero within 48 h, whereas Sakura gradually decreased and reached zero at 121 h. The initiation of hydrogen production is different between Kuwana and Sakura. Although Kuwana started to generate hydrogen at 24 h, Sakura generated it at 72 h.

Microbial Colony Count and Metagenomic Analysis

In order to clarify these differences between Kuwana and Sakura, the microbial colony count method was used to disclose the dominant microbial flora during the cultivation of Kuwana and Sakura. The changes of hydrogen and oxygen and the numbers of aerobic, anaerobic, and total bacteria were shown in Fig. 7a for Kuwana and in Fig. 7b for Sakura, respectively. The numbers of aerobic bacteria decreased rapidly as the oxygen concentration reached zero after 48 h, and the numbers of anaerobic bacteria increased and reached maximum at 72 h. This indicated that the dominant bacteria shifted from aerobic to anaerobic after 48 h for Kuwana. On the contrary, Sakura showed a different tendency as shown in Fig. 7b. Though the aerobic and anaerobic bacteria reached the maximum value at 72 h, no hydrogen was produced.

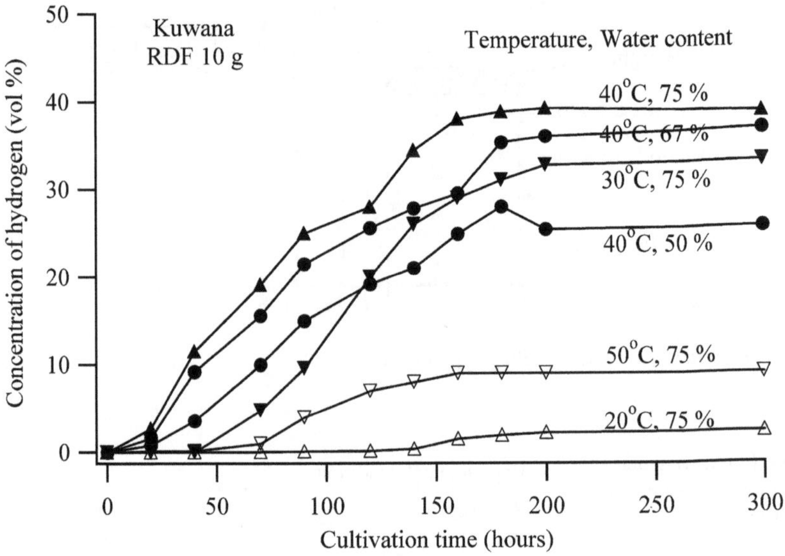

Fig. 5 The effects of temperature and water content on the hydrogen gas production

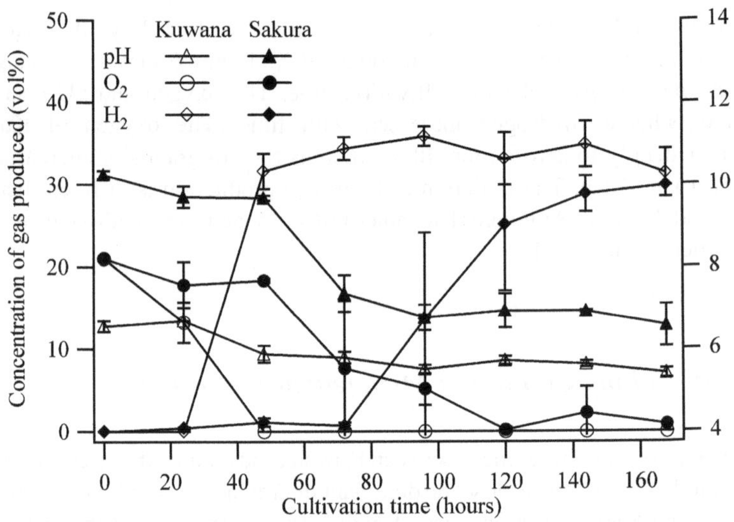

Fig. 6 Time-dependent changes of hydrogen, oxygen, and pH

Oxygen decreased and hydrogen increased slowly with time after 72 h. The distinct shift of aerobic to anaerobic was not seen for Sakura. Figure 8a shows the ratio of hydrogen and non-hydrogen bacteria for the non-culture sample and 3-day culture samples of Kuwana and Sakura. The hydrogen-producing bacteria increased from 34 to 61 %, and the non-hydrogen-producing bacteria decreased from 66 to

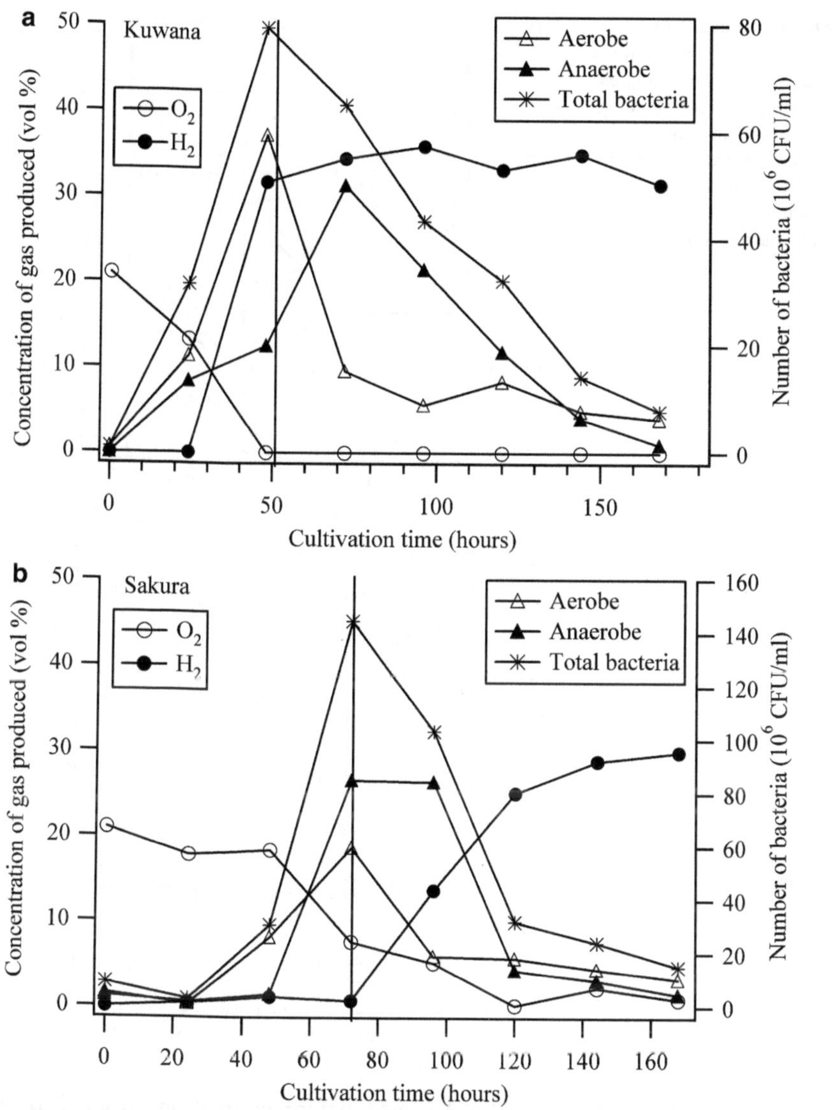

Fig. 7 The changes of hydrogen and oxygen and the numbers of aerobic/anaerobic bacteria and total bacteria for (**a**) Kuwana and (**b**) Sakura, respectively

39 % after 3 days of cultivation for Kuwana. The same tendency was observed for Sakura with the hydrogen-producing bacteria increasing from 29 to 62 % and the non-hydrogen-producing bacteria decreasing from 71 to 38 % after 3 days of cultivation. Figure 8b shows the species of hydrogen-producing bacteria. The hydrogen-producing bacteria were classified into three categories: *Bacillus* sp., *Clostridium* sp., and others. *Bacillus* sp. occupied 50 %, *Clostridium* sp. occupied

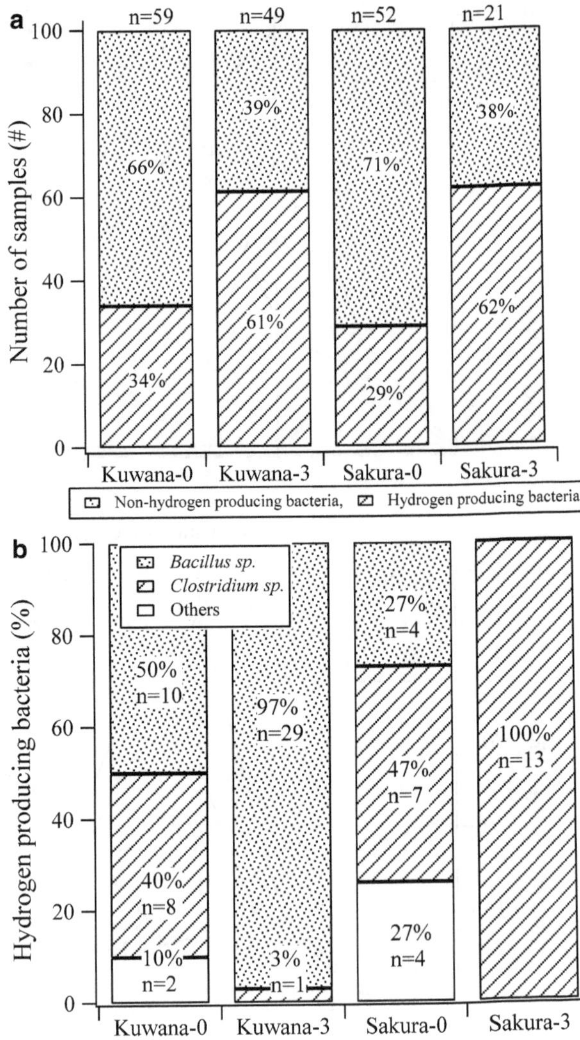

Fig. 8 Metagenomic analysis for the non-culture and 3-day culture samples of (**a**) hydrogen-producing and non-hydrogen-producing bacteria and (**b**) species of hydrogen-producing bacteria for Kuwana and Sakura

40 %, and others are 10 % for Kuwana before cultivation. After 3 days of cultivation, *Bacillus* sp. occupied 97 % and *Clostridium* sp. occupied only 1 %. Although *Bacillus* sp. occupied 27 % and *Clostridium* sp. occupied 47 % and others are 27 % for Sakura before cultivation, *Clostridium* sp. completely occupied after 3 days of cultivation. The *Bacillus* sp. is the facultative anaerobe that can survive under both aerobic and anaerobic atmospheres. Therefore, the facultative anaerobe is expected to be counted as both aerobic and anaerobic by the microbial colony count shown as Fig. 7. *Paenibacillus* sp., which relates closely to *Bacillus* sp., was found by metagenomic analysis, and it was confirmed by the experiments that *Paenibacillus* sp. generated hydrogen gas under both aerobic and anaerobic atmospheres, whereas *Clostridium* sp. is strictly anaerobic and intolerant to oxygen, replicating at low oxidation-reduction at potential sites.

The Scale Effects on the Temperature Rise Due to Biological Activities

Figure 9a shows the temperature changes examined by three different amounts of RDFs. The maximum temperatures were 38.1 °C for 100 g, 51.6 °C for 750 g, and 57.7 °C for 1,500 g. The higher the amount of RDF, the higher the maximum temperature obtained after 50 h to 85 h cultivation. Figure 9a shows the maximum temperature, total heat released, and heat released per mass against the weight of RDFs. The maximum temperature and total heat released increased with the weight of RDFs. The coarse extrapolation shows that the maximum temperature expected to be reached was 75 °C for 10 kg and 95 °C for 100 kg, respectively. In this experiment, RDFs were put into a vacuum bottle, and this controlled the heat release going outside the container. The maximum temperature of RDFs could be determined by the calorific values brought about by microbial fermentation. The pH, temperature, oxygen concentration, and water volumes influence the amount of heat generation. The growth rates, in proportion to microbe multiplication, are also affected by these environmental conditions. It is also thought that differences in container size will affect internal distribution and that the contents will not be distributed uniformly inside the container. Further research into the influences of material distribution inside a container via scale modeling is necessary.

Conclusions

This research detailed the characteristics of hydrogen production and temperature rise by biological activities in relation to the development and safe handling systems for waste biomass fuels, especially for the prevention of fires and explosions associated with refuse-derived fuels (RDF). Three kinds of experiments were conducted to determine the physical or chemical laws by which a scale modeling of this phenomenon would be constructed. Three main points were examined: (1) the effect of environmental conditions on hydrogen gas production, (2) the determination of the microbial diversity and species in RDFs, and (3) the scale effects on the temperature rise due to biological activities. The results are summarized as follows:

(1) The hydrogen gas production was at a maximum at 40 °C, followed by 30 °C and then 50 °C, and had minimum hydrogen production at 20 °C. The water content of 50, 67, and 75 % at constant temperature of 40 °C was also investigated. The results showed that the higher the water contents, the higher the hydrogen gas production.
(2) The numbers of aerobic bacteria decreased rapidly as the oxygen concentration reached zero after 48 h; the numbers of anaerobic bacteria increased and reached a maximum after 72 h. This indicated that the dominant bacteria shifted from aerobic to anaerobic after 48 h for

(continued)

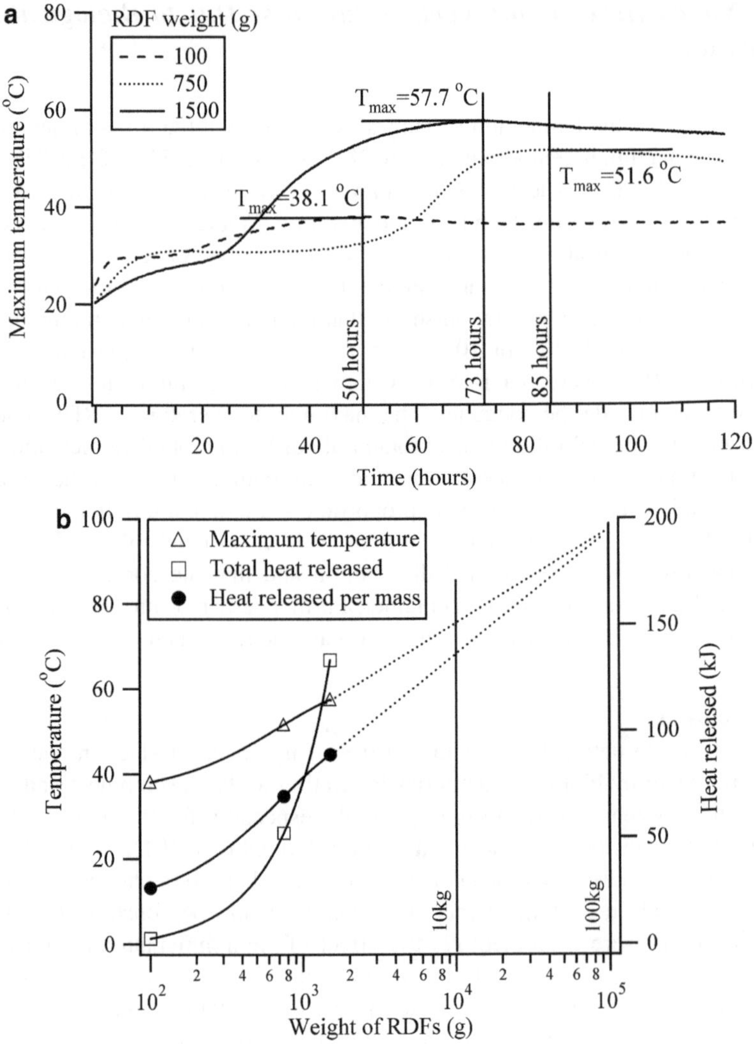

Fig. 9 (**a**) Temperature change examined by three different amounts of RDFs and (**b**) the effects of size on temperature rise and heat release

(continued)

Kuwana. On the contrary, Sakura showed that the aerobic and anaerobic bacteria reached a maximum value after 72 h without hydrogen production. Oxygen decreases and hydrogen increased slowly with time after 72 h. The distinct shift of aerobic to anaerobic was not seen for Sakura.

(continued)

(continued)

(3) The hydrogen-producing bacteria increased from 34 to 61 %, and the non-hydrogen-producing bacteria decreased from 66 to 39 % after 3 days of cultivation for Kuwana. The same tendency was observed for Sakura. *Bacillus* sp. occupied 50 %, *Clostridium* sp. occupied 40 %, and others made up 10 % for Kuwana before cultivation. After 3 days cultivation, *Bacillus* sp. occupied 97 % and *Clostridium* sp. occupied only 1 %. Although *Bacillus* sp. occupied 27 % and *Clostridium* sp. occupied 47 % and 27 % of others for Sakura before cultivation, *Clostridium* sp. completely occupied the space after 3 days of cultivation.

(4) The maximum temperatures were 38.1 °C for 100 g, 51.6 °C for 750 g, and 57.7 °C for 1,500 g. The higher the amount of RDF, the higher the maximum temperature obtained after 50 h to 85 h cultivation.

References

1. Karl, T.R., Trenberth, K.E.: Modern global climate change. Science **302**(5651), 1719–1723 (2003)
2. Khanal, S.K., Chen, W.H., Li, L., Sung, S.: Biological hydrogen production: effects of pH and intermediate products. Int. J. Hydrogen Energy **29**(11), 1123–1131 (2004)
3. Demain, A.L., Newcomb, M., Wu, J.H.D.: Cellulase, clostridia, and ethanol. Microbiol. Mol. Biol. Rev. **69**, 124–154 (2005)
4. Kimball, J.: Industrial silo fire and explosion. Technical report series (United States Fire Administration) p. 122 (1999)
5. Kimball, J.: The hazards associated with agricultural silo fires. Technical report series (United States Fire Administration) USFA-TR-096 (1998)
6. Panther, T.G.: Silo fires-prevention and control. The University of Tennessee Extension PB 1307 (1988)
7. Tsuruda, T.: Explosion in an RDF-5 fired power plant in Mie. J. Jpn. Assoc. Fire Sci. Eng. **54** (5), 15–18 (2004) (in Japanese)
8. Tokyo Fire Department: Fire Report: fire caused by an imported garbage disposer. J. Jpn. Assoc. Fire Sci. Eng. **54**(6), 57–60 (2004) (in Japanese)
9. Osaka City Fire Department: A fire started in a waste storage of RPF production facility. J. Jpn. Assoc. Fire Sci. Eng. **57**(2), 5–9 (2007). in Japanese
10. Hara, K., Nakamura, T.: A fire started in a waste storage of RPF production facility series of fires in a waste disposal plant. J. Jpn. Assoc. Fire Sci. Eng. **57**(2), 10–14 (2007) (in Japanese)
11. Tsuruda, T.: The fire and explosion accident on refuse derived fuel facilities. In: Information of the 7th Lecture on Fire Prevention Research, National Research Institute of Fire and Disaster, Mitaka, pp. 21–38 (2004) (in Japanese)
12. Gao, L.-J., Tsurada, T., Suzuki, T., Ogawa, Y., Liao, Ch.-H., Saso, Y.: Possibility of refused derived fuel fire inception by spontaneous ignition. In: Kim, E.-S., Kim, J.-D., Park, Y.-H., Dlugogorski, B.Z., Kennedy, E.M., Hasemi, Y. (eds.) Proceedings of the 6th Asia-Oceania Symposium on Fire Science and Technology, Daegu, pp. 102–107 (2004)
13. Special Committee for Investigation on the Accident of Refuse Derived Fuel Power Station in Mie Prefecture: The final report on investigation for the accident of Refuse Derived Fuel power station in Mie Prefecture, Mie Prefecture (2003) (in Japanese)

14. Fuji Electric Holdings co. Ltd.: The report on cause investigation of a refuse derived fuel fire accident (2003) (in Japanese)
15. Suzuki, T., Tsuruda, T., Ogawa, Y., Liao, C.: A study on extinction of RDF (refuse derived fuel) pile, Fire Safety Science. In: Proceedings 8th International Symposium. International Association for Fire Safety Science (IAFSS), Beijing, 18–23 Sept 2005
16. Gao, L., Hirano, T.: Process of accidental explosions at a refuse derived fuel storage. J. Loss Prev. Process Ind. **19**(2–3), 288–291 (2006)
17. Fu, Z., Li, X., Koseki, H.: Heat generation of refuse derived fuel with water. J. Loss Prev. Process Ind. **18**(1), 27–33 (2005)
18. Murata, S.: A considerations of the RDF fire accident in storage from the viewpoint of heat generation by water vapor adsorption to over dried organic matter. Refrigeration **79**(922), 36–42 (2004) (in Japanese)
19. Sakka, M., Kimura, T., Sakka, K., Ohmiya, K.: Hydrogen gas generation from refuse-derived fuel (RDF) under wet conditions. Biosci. Biotechnol. Biochem. **68**(2), 466–467 (2004)
20. Sakka, M., Kimura, T., Ohmiya, K., Sakka, K.: Detection of hydrogen gas-producing anaerobes in refuse-derived fuel (RDF) pellets. Biosci. Biotechnol. Biochem. **69**(11), 2081–2085 (2005)
21. Sakka, M., Kimura, T., Sakka, K.: Comparison of microbial consortia in refuse-derived fuel (RDF) preparations between Japan and Germany. Biosci. Biotechnol. Biochem. **70**(12), 2868–2873 (2007)
22. Jian, F., Jayas, D.S., White, N.D.G.: Temperature fluctuations and moisture migration in wheat stored for 15 months in a metal silo in Canada. J. Stored Prod. Res. **45**(2), 82–90 (2009)
23. Rastikian, K., Capart, R.: Mathematical model of sugar dehydration during storage in a laboratory silo. J. Food Eng. **35**(4), 419–431 (1998)
24. Khatchatourian, O.A., de Oliveira, F.A.: Mathematical modeling of airflow and thermal state in large aerated grain storage. Biosyst. Eng. **95**(2), 159–169 (2006)
25. Ferrero, F., Lohrer, C., Schmidt, B.M., Noll, M., Malow, M.: A mathematical model to predict the heating-up of large-scale wood piles. J. Loss Prev. Process Ind. **22**(4), 439–448 (2009)
26. Tremier, A., de Guardia, A., Massiani, C., Paul, E., Martel, J.L.: A respirometric method for characterising the organic composition and bio degradation kinetics and the temperature influence on the biodegradation kinetics, for a mixture of sludge and bulking agent to be co-composted. Bioresour. Technol. **96**, 169–180 (2005)
27. Medini, D., Serruto, D., Parkhill, J., Relman, D.A., Donati, C., et al.: Microbiology in the post-genomic era. Nat. Rev. Microbiol. **6**, 419–430 (2008)
28. Nagatsuka, N., Narumi, A., Iida, Y.: The influence of temperature, wet, and production locality on hydrogen gas generation from refuse derived fuel and the mechanism of generation. In: CD-ROM Proceedings of the International Conference of Thermal Treatment Technologies and Hazardous Waste Combustors, #29, pp. 1–7 (2009)
29. Jacquelyn, G.B.: Microbiology -principles and explorations, 6th edn. Wiley, pp. 148–153 (2005)
30. Konishi, T., Kohara, Y., Narumi, A., Saitou, H.: Bull. Jpn. Assoc. Fire Sci. Eng. **61**(2), 25–35 (2011) (in Japanese)
31. Cole, J.R., Chai, B., Farris, R.J., Wang, Q., Kulam, S.A., McGarrell, G.M., Garrity, G.M., Tiedje, J.M.: The ribosomal database project (RDP-II): sequences and tools for high-throughput RNA analysis. Nucleic Acids Res. **33**(1), 294–296 (2005)

Section B
Fire and Explosion - A Study of Flame Spread in Engineered Cardboard Fuel Beds Part I: Correlations and Observations of Flame Spread

Mark A. Finney, Jason Forthofer, Isaac C. Grenfell, Brittany A. Adam, Nelson K. Akafuah, and Kozo Saito

Abstract Wind-aided laboratory fires spreading through laser-cut cardboard fuel beds were instrumented and analyzed for physical processes associated with spread. Flames in the spanwise direction appeared as a regular series of peaks and troughs that scaled directly with flame length. Flame structure in the stream-wise direction fluctuated with the forward advection of coherent parcels that originated near the rear edge of the flame zone. Thermocouples arranged longitudinally in the fuel beds revealed the frequency of temperature fluctuations decreased with flame length but increased with wind speed. The downstream extent of these fluctuations from the leading flame edge scaled with Froude number and flame zone depth. The behaviors are remarkably similar to those of boundary layers, suggesting a dominant role for buoyancy in determining wildland fire spread.

Nomenclature

- D Horizontal flame zone depth (m)
- f Frequency (Hz)
- g Acceleration of gravity (9.81 ms^{-2})
- L Flame length (m)
- R Fire spread rate (ms^{-1})
- t Time (s)
- U Horizontal wind speed (m/s)
- w Fuel loading ($\mathrm{kg\ m}^{-2}$)
- X Horizontal stream-wise distance (m)

M.A. Finney (✉) • J. Forthofer • I.C. Grenfell
US Forest Service, Missoula Fire Sciences Lab, 5775 Hwy 10W, Missoula, MT 59808, USA
e-mail: mfinney@fs.fed.us

B.A. Adam • N.K. Akafuah • K. Saito
Department of Mechanical Engineering, University of Kentucky, Lexington, KY 40506, USA

K. Saito et al. (eds.), *Progress in Scale Modeling, Volume II*,
DOI 10.1007/978-3-319-10308-2_5

Y Transverse width of fuel bed (m)
Z Vertical fuel bed depth (m)
λ Transverse wavelength of flames (m)

Subscripts

r Pearson correlation coefficient
f Flame residence time

Introduction

In the study of wildland fire spread, the heating and ignition of fuel particles by flame contact has been largely neglected in favor of radiation. This is unfortunate because research has recently suggested that radiant heating is insufficient alone to ignite fine fuel elements at fluxes common to wildland fires [1]. Fine particles (<1 mm diameter), such as grasses and pine needles that comprise most wildland fuel beds, cool efficiently by free (and forced) convection within the thin boundary layers of small particles [2, 3]. Experiments imply that fine fuels remain well below nominal ignition temperature until impinged by flames, which takes place within the final centimeters before ignition [4–6]. Flame structure and variability in and near the fuel bed are, therefore, critical to the preheating and ignition process.

Flame structure in spreading fires, particularly near the leading edge, has received scant attention. Stationary fires of liquid or solid fuel (pool fires, crib fires) in the absence of wind are perhaps the most commonly researched and exhibit regular pulsing [7, 8]. Similar pulsatile behavior has been observed in fires on sloping surfaces [9–11]. Here we report observations of coherent flame structures in spreading wind-driven laboratory fires, finding that the flame dynamics were responsible for spread.

Methods

Fire spread experiments were conducted in uniform fuel beds made of laser-cut cardboard. This new technique is more practical at larger scales than laboratory fuel beds made of matchsticks [12–14] or toothpicks [15], and with more regular particle spacing than excelsior [16] or pine needles [4]. Cardboard and paper strips have been used previously [17] and offer advantages of known homogenous properties such as density and customizable physical dimensions of discrete particles (length, surface area, etc.). Fuel beds were burned in the wind tunnel at the Missoula Fire Sciences Laboratory. The wind profiles of the 3 m cross section have been described previously [4, 16] and are laminar except along the bottom surface where an upstream trip fence produces a turbulent boundary layer. Wind speeds were varied from 0.22 ms^{-1} to 1.5 ms^{-1} with relative humidity of approximately 25 %.

Fig. 1 Picture of laser-cut cardboard laboratory fuel bed. Note that the scale is graduated in inches

We employed a commercial CO_2 laser system for cutting cardboard fuel elements that are connected at regular spacing along a common spine. The cards or "combs" could then be arranged in rows at various spacing to form a fuel bed with vertically standing particles (Fig. 1). The cardboard used was brown "chip board" which was 1.27 mm (0.05 in.) thick with approximately 60 % recycled content. Fuel particles were created at different lengths and widths and arranged at different row spacing to achieve specific fuel bed properties (Table 1). The laser cutter/engraver system was a Universal Laser Systems, Inc. ILS12.150D model equipped with two 60 W laser cartridges. The beams from both lasers were collimated for cutting. The table accommodates sheets of cardboard 0.61 m × 1.22 m (2 ft × 4 ft) so that multiple combs can be cut from the same sheet in one operation.

Fuel beds constructed of these cardboard combs were 1.22–2.45 m in width and 3.05–6.1 m in length (Table 1). The combs were supported and arranged on a foundation of cement-board strips (hardy board) 0.635 cm × 5.08 cm (1/4 in. × 2 in.) each separated by a steel spacer 0.158 cm × 2.54 cm (1/16 in. × 1.0 in.). The steel spacers rested on the floor to preserve a slot at the upper surface which pinched the spine of the fuel combs such that only the vertical tines were exposed (Fig. 1). Tine lengths of 2.54 cm (1 in.), 10.1 cm (4 in.), 20.3 cm (8 in.), and 35.6 cm (14 in.) were used in the burns. The longitudinal spacing of the combs could be adjusted every 1.43 cm (5/16 in.). To limit inflow to the combustion zone along the lateral edges during burning, the sides of the beds were lined with paper that was treated with the flame retardant diammonium phosphate ($(NH_4)_2HPO_4$). This technique was described by Byram et al. in a USDA report, where fire retardant limits independent flaming combustion but allows the paper to burn in conjunction with the advancing fire front [18]. The consumption of the paper sideliners at the trailing edge of the burning zone avoids channeling of air inflow to the rear of the fire which has been shown to affect fire spread on slopes [19]. Cutouts of the sideliner permitted filming of the ignition process within the fuel bed.

A series of preliminary burns of the cardboard fuel beds in the wind tunnel were used to refine the instrumentation. The fires were filmed with digital video cameras from the top, sides, front, back, and various oblique angles. This footage revealed two principal dynamic features of the flame zone. First the flame zone became

Table 1 Data obtained from experimental burns in the wind tunnel using cardboard fuel

Experiment	D [m]	Z [m]	L [m]	U [m/s]	Y	Fr	Str	λ [m]	W [kg/m^2]	R [m/min]	t [s]
1	0.250	0.102	0.300	0.500	0.914	0.085	1.272	0.457	0.708	1.201	12
2	0.250	0.102	0.300	0.500	0.914	0.085	1.079	0.457	0.708	1.252	12
3	0.600	0.102	0.600	1.000	0.914	0.170	2.323	0.610	0.708	3.103	7
4	0.800	0.102	0.800	1.500	0.914	0.287	1.178	0.610	0.708	2.938	12
5	0.250	0.102	0.300	0.500	0.914	0.085	1.465	0.457	0.708	1.317	12
6	0.400	0.102	0.600	1.000	0.914	0.170	1.737	0.457	0.708	2.258	12
9	0.100	0.102	0.300	0.220	1.219	0.016	2.431	0.305	0.389	0.734	10
10	0.200	0.203	0.600	0.220	1.219	0.008	3.462	0.406	0.653	0.719	12
11	0.100	0.025	0.100	0.220	1.219	0.049	1.440	0.203	0.144	1.118	4
12	0.200	0.025	0.200	0.440	1.219	0.099	1.575	0.152	0.144	1.606	4
13	0.300	0.025	0.300	0.670	1.219	0.153	1.903	0.406	0.144	3.445	4
14	0.200	0.102	0.500	0.440	1.219	0.039	1.994	0.406	0.546	1.374	7
15	0.500	0.102	0.800	0.670	1.829	0.057	2.331	0.610	0.409	3.595	7
16	0.788	0.102	1.000	0.890	2.438	0.081	3.482	0.813	0.409	5.077	8
17	0.915	0.102	1.000	0.890	2.438	0.081	2.528	0.813	0.409	7.333	7
18	0.200	0.203	1.000	0.440	1.829	0.020	2.717	0.610	1.092	1.798	6
19	0.400	0.203	1.500	0.670	1.829	0.031	3.678	0.914	1.092	3.254	7
20	0.800	0.102	1.200	1.341	2.438	0.153	1.854	1.219	0.546	6.867	8
21	0.400	0.025	0.300	1.341	1.829	0.611	0.770	0.366	0.158	4.058	5
22	0.400	0.025	0.300	1.341	1.829	0.611	0.847	0.366	0.158	3.786	5
23	1.200	0.356	2.500	0.670	2.438	0.018	4.396	1.219	1.430	5.240	12
24	1.200	0.203	2.000	0.890	2.438	0.040	3.054	1.219	1.092	6.464	15
25	0.500	0.203	1.000	0.440	2.438	0.020	5.566	0.813	0.653	5.330	7
26	0.300	0.203	0.600	0.440	2.438	0.033	1.585	0.610	0.466	3.969	8
27	0.900	0.356	1.500	0.220	2.438	0.003	9.258	0.813	1.430	4.322	12
28	0.300	0.102	0.400	1.341	2.438	0.458	0.851	0.610	0.570	4.336	7

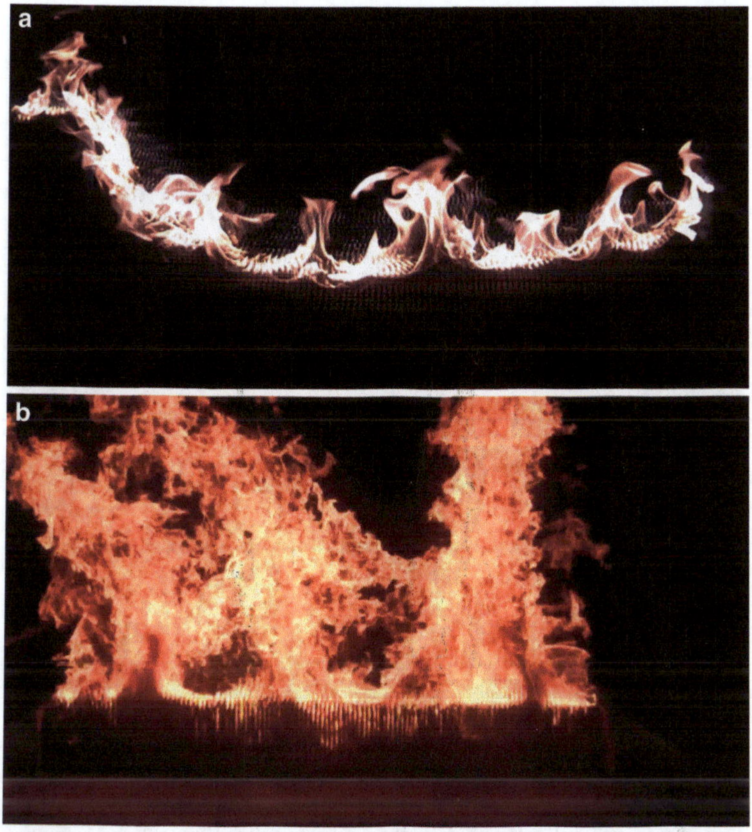

Fig. 2 Peak and valley structure of flames looking upwind. In the *top image* (**a**), flames are approximately 0.15 m tall. In the *bottom image* (**b**), flames approximately 2 m tall

divided in the transverse or spanwise direction into convective peaks and troughs at fairly regular spacing. The ignition interface (at the leading edge of the combustion zone) was convoluted in association with this flame structure, with a concave segment located directly beneath these peaks (Fig. 2) and convex segments in the troughs. Second, the flame zone exhibited clear instabilities, which, when viewed at an angle from behind and above the bed, appeared as patches originating near the rear of the burning zone. The patches produced dish-shaped depressions on the upper flame surface which expanded in horizontal area as they advected coherently downwind through the troughs (Fig. 3). After reaching the ignition interface at the leading edge of the burning zone, they impinged new fuels ahead. When the flame zone was viewed normal to the stream-wise direction, eddies appeared on the upper and lower flame edges which rotated in the opposite direction (Fig. 4). Regardless of the local geometry of the fire edge, it progressed at the same rate across the fire front (the convolutions did not noticeably increase or decrease the local spread rate).

To record temperature signals within the fuel bed produced by the observed flame fluctuations, a series of 64 thermocouples (0.012 mm, type K, bead welded)

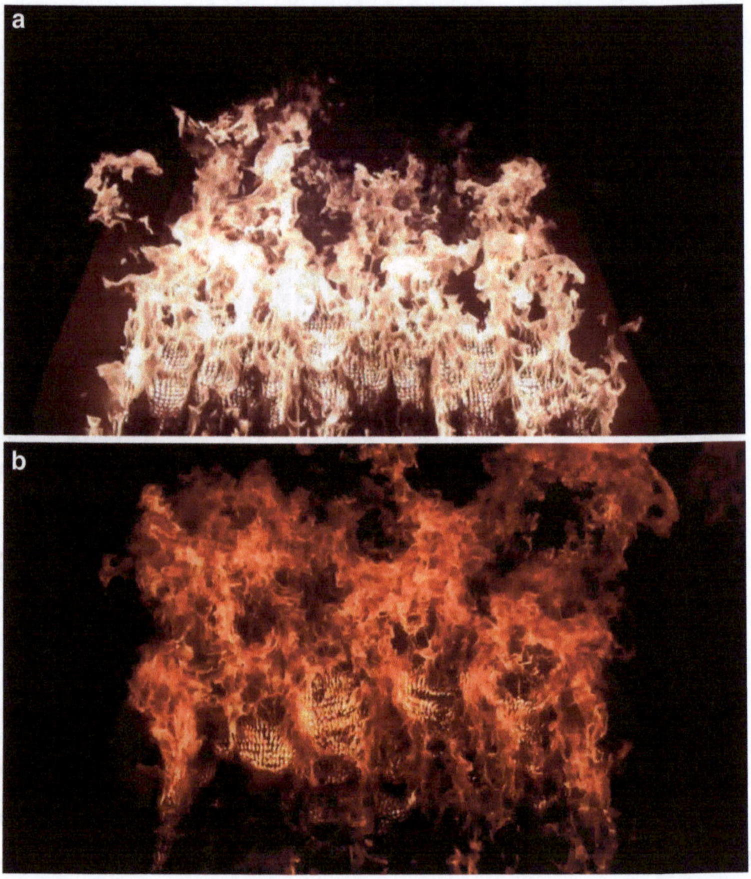

Fig. 3 Flame zone structure viewed looking downwind (fire spreading away from camera) shows patches of instabilities (**a**) soon after the fire starts when the Görtler vortices have shorter wavelength, and (**b**) after larger flames the steady fire front develops. Note the unstable portions expand in horizontal dimension as they advect downwind through the flame troughs

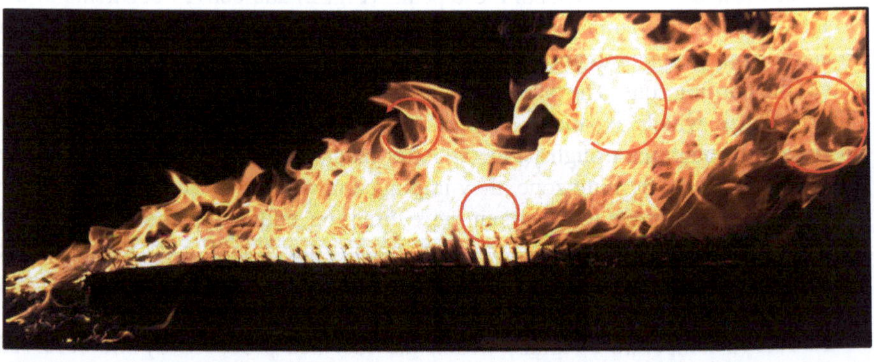

Fig. 4 Flame structure normal to spread direction showing rotation of flame eddies

were arranged in a single line in the direction of fire spread. The thermocouple junction was located at the same height as the upper surface of the fuel bed. The first 32 thermocouples were spaced 1.5 cm, and the second 32 thermocouples spaced 3.0 cm apart. Data logging occurred at 500 Hz using a National Instruments, Inc. data acquisition system. The range of burning conditions reported here include wind speeds from 0.22 to 1.5 m/s with relative humidity about 25 %.

Temperature time series recorded by the 64 thermocouples were analyzed for frequency and correlation of the temperature signals among thermocouples. Each time series was divided into three periods, preignition, burning, and glowing (Fig. 5). Preignition was defined from the first crossing of the 350 °C temperature to where a spline smoothing of the temperature data reached its peak. The burning phase then continued until the spline dropped below the 350 °C level. The begin-

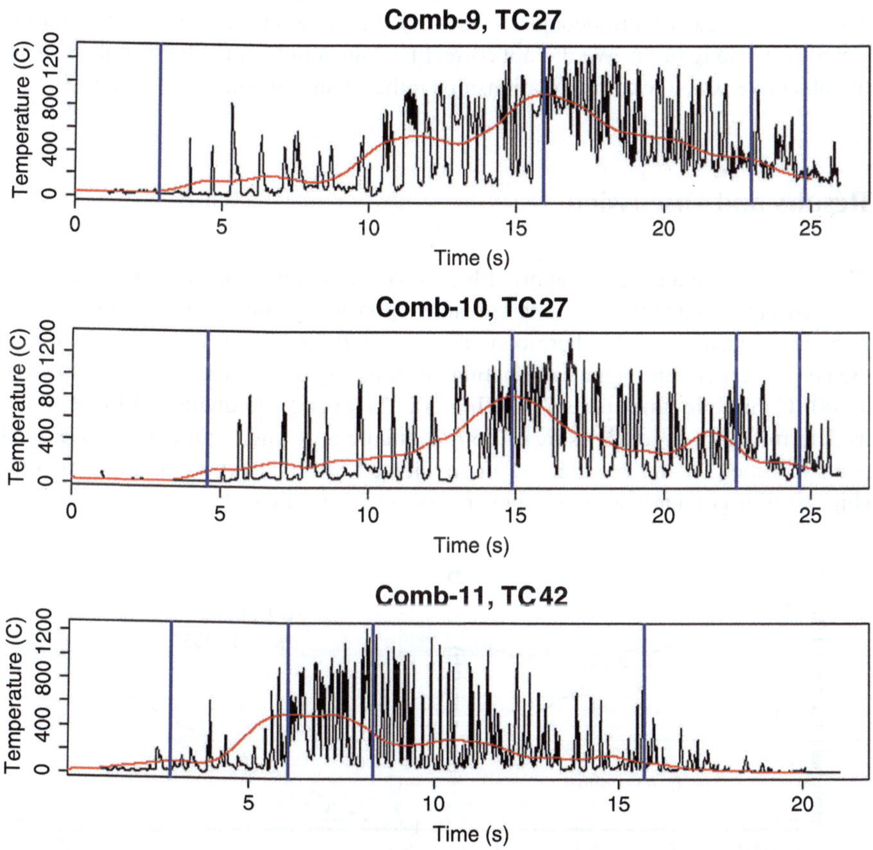

Fig. 5 Temperatures recorded by individual thermocouples show nonsteady and quasiperiodic flame contacts prior to ignition. *Blue vertical lines* delimit preignition, burning, and glowing phases

ning time of the burning phase was used to calculate fire spread rate based on the fixed distances among thermocouples.

Frequency analysis of raw temperature signals for each period was conducted using a level crossing of 350 °C and summarized as the average frequency (number of up crosses divided by the time period) for all 64 thermocouples. Despite temperature pulsing apparent in thermocouple traces (Fig. 5), level crossing was used to estimate average frequency of each period because Fourier spectral analysis could not identify peak frequencies consistently among thermocouples. This may be because the peaks were obviously not sinusoidal and apparently not regular enough to produce consistent spectral peaks. The horizontal downwind distance that the temperature signals extend from the ignition interface prior to ignition was estimated by correlating the temperature time series of the preignition period with the temperatures on each downwind thermocouple at that time interval. To find this *flame correlation distance*, the correlation coefficient was plotted by downwind distance from each thermocouple to identify the horizontal extent of temperature signals prior to ignition. We did not correct for transit time of the flame pulses since the objective was to explore relationships rather than estimate exact distances.

Results and Discussion

The 26 experimental burns reported here revealed features of the flame zone that were strongly indicative of flame spread by nonsteady convective heating of fuel particles. Specifically, the thermocouples recorded fluctuating flame presence that creates a temperature signal alternating from nearly ambient temperature to over 1,200 °C multiple times per second (Fig. 5). The average frequency of fluctuations as determined by level crossing of 350 °C (number of high temperature crossings per unit time) for the preignition phase showed Strouhal-Froude number scaling (Fig. 6). For comparison, the exponent −0.39 is less than the −0.5 for diameter

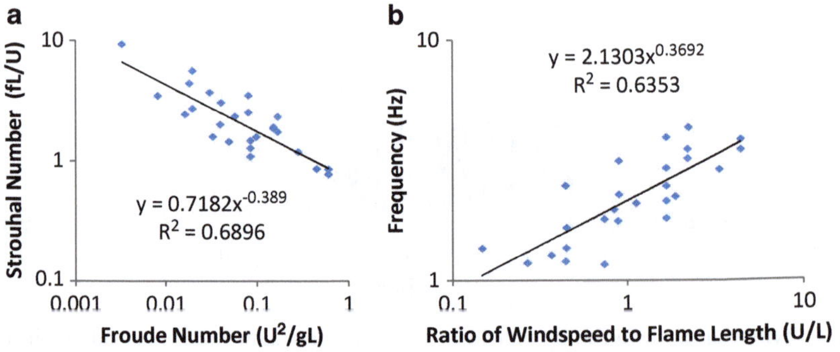

Fig. 6 Average temperature fluctuation frequency from level crossing analysis suggests (**a**) Strouhal-Froude number relationship with negative power and (**b**) positive relationship with ratio of wind speed to flame length

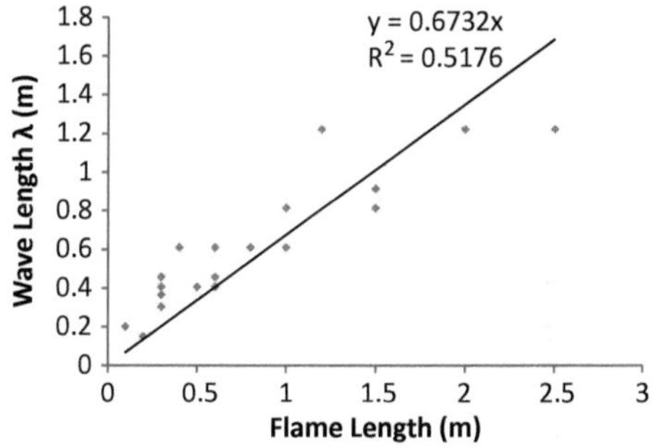

Fig. 7 Linear relationship between average flame length and spanwise wavelength from Görtler vortices

scaling of pulse frequency in axisymmetric pool fires [8, 20]. A single flame zone dimension did not correlate well to frequency in our spreading fires, but frequency did show a strong relation to the ratio of horizontal wind speed and flame length (Fig. 6). Our difficulty in detecting a dominant frequency with Fourier spectral analysis suggests more complex buoyant instabilities similar to those of boundary layers during transition to turbulence [21–23].

As with boundary layer instabilities [22], we observed a regular transverse (spanwise) spacing of flame "towers" or alternating peaks and troughs (Fig. 2) with a wavelength that was proportional to visually estimated flame length (Fig. 7). In the flame zones of spreading fires, these structures may be explained by the lift imparted to the ambient wind stream by the buoyant flame zone (1,000–1,200 C) which creates instabilities driven by centrifugal forces. Similarly, centrifugal forces in flows over concave surfaces result in counterrotating longitudinal (stream-wise) vortices at regular spacing, known as Görtler vortices [24, 25]. The vortex spacing in our fires increases with flame zone dimensions (similar to the radius of curvature of concave surface flows). The longitudinal vortices tilt somewhat to the vertical in the flame zone apparently due to buoyancy and partition the flame front into alternating zones of downwash and upwash where the vortex pairs converge. The downwash zones channel the ambient air flow down and forward through the flame zone and downwind of the burning region as described by Beer [26]. The downwash also forces flame contact with fuel particles as the dish-shaped parcels related to secondary instabilities journey forward from the rear of the flame zone to the ignition interface where they impinge and ignite new fuel particles. Quasiperiodic frequencies of roller vortices in plane mixing layers [27–29] are remarkably similar to the flame fluctuations recorded by thermocouples in our spreading fires (Figs. 3 and 5). Flame towers move sideways erratically, similar to the sinuous instabilities noted for Görtler vortices [30] and possibly in association with the passing of the flame pulses in the troughs (contributing a lateral source of variability to the temperature signals recorded by each thermocouple).

The longitudinal sequence of thermocouples permitted temperature patterns of the preignition phase to be statistically correlated with thermocouples located at increasing distance downwind (Fig. 8). Plots from all burns suggested that the downwind distance over which correlations remain positive was proportional to

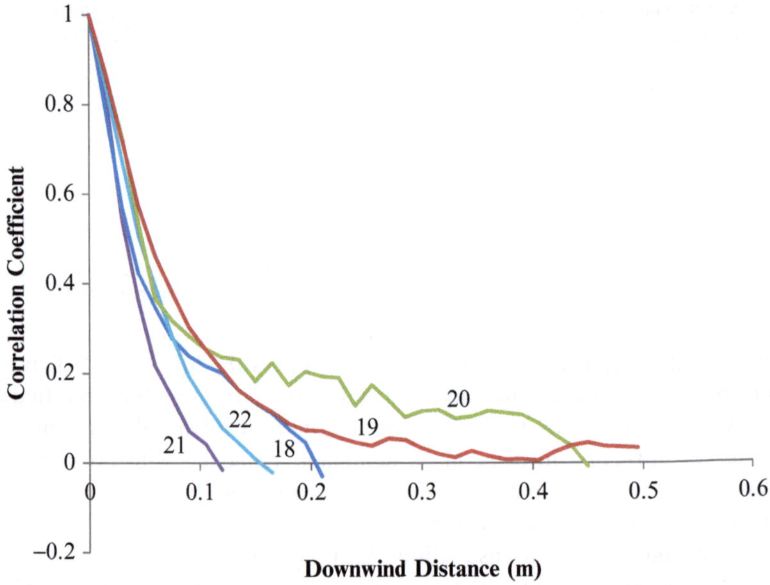

Fig. 8 Temperature time series during preignition were found to be correlated among thermocouples for different distances ahead of the fire. Shown here are average correlations from 32 thermocouples for five fires with very different overall spread rates, wind speeds, and fuel configurations (see Table 1)

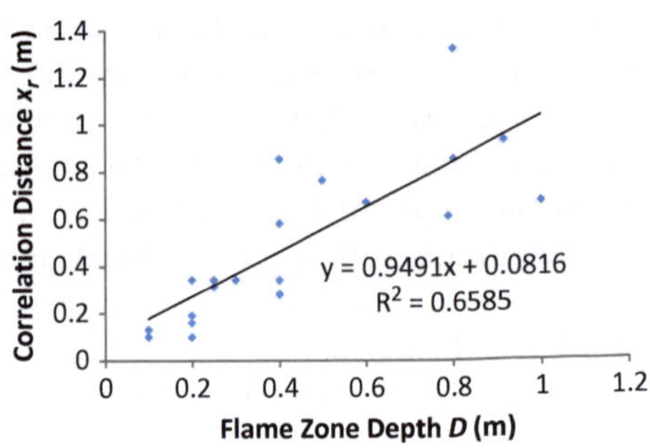

Fig. 9 Relationship between flame correlation distance (distance downwind of ignition interface where flame temperatures are statistically correlated) and horizontal flame zone dimension (depth)

flame zone depth (Fig. 9). This is consistent with analyses of flame drag (flame trailing) downwind of pool fires which typically scale with the diameter of the flame source [31, 32]. The implication here is that the downwind extent to which flames are intermittently heating fresh fuel ahead of the ignition interface is related to factors producing flame drag. Also, the reported relationships of Froude number to the tangent of flame angle [33, 34] were supported by our data where the tangent was estimated from the ratio of correlation distance to fuel bed depth (Fig. 10).

The upshot of this research is the suggestion that wildland fire spread results from the trajectory and frequency of flame contacts which produce fuel particle

Fig. 10 Relationship between the tangent of correlation distance (X_r/Z) and the Froude number

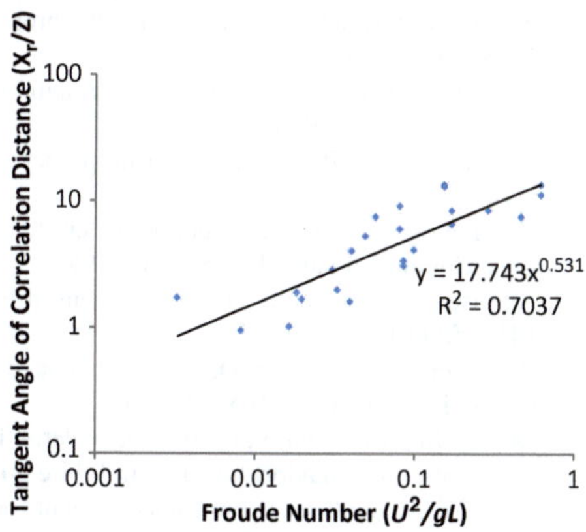

ignition. Quasiperiodic flame behavior in spreading fires has been noticed previously for trench fires for pine needle beds, but the role of such nonsteady flame impingement in igniting fuel particles and flame spread has not been considered [9–11]. The Strouhal-Froude number scaling suggested by our data is consistent with buoyant dynamics of stationary fire phenomena but new to spreading fires. These findings suggest that the difficulty of identifying an integral length scale for convection related to ignition [35] comes from obviating the time dependency of particle heating. It also suggests by virtue of the inverse dependency of average pulse frequency on flame dimensions that slower buoyant dynamics of larger flame zones compensate for the increases in energy release and convective heating distance to avoid runaway spread. If buoyant instabilities are responsible for the flame behaviors and particle ignition, then it strongly suggests that laboratory-scale fire spread processes should extend readily to field proportions because flame temperature in diffusion flames (and thus buoyancy) remains approximately the same regardless of fire size. Much work is yet to be done to understand useful scaling relationships, but the ultimate goal is to someday incorporate these findings into practical tools for wildland fire managers.

References

1. Finney, M., Cohen, J., McAllister, S., Jolly, W.: On the need for a theory of wildland fire spread. Int. J. Wildl. Fire **22**, 25–36 (2013)
2. Alvares, N., Blackshear, J., Perry, L., Murty, K.: The influence of free convection on the ignition of vertical cellulosic panels by thermal radiation. Combust. Sci. Technol. **1**, 407–413 (1970)
3. Martin, S.: Diffusion controlled ignition of cellulosic materials by intense radiant energy. Symp. (Int.) Combust. **10**, 877–890 (1965)
4. Rothermel, R., Anderson, H.: Fire spread characteristics determined in the laboratory. US For. Serv. (1966)

5. Baines, P.: Physical mechanisms for the propagation of surface fires. Math. Comput. Model. **13**, 83–94 (1990)
6. Fang, J., Steward, F.: Flame spread through randomly packed fuel particles. Combust. Flame **13**(4), 392–398 (1969)
7. Cox, G., Chitty, R.: Some stochastic properties of fire plumes. Fire Mater. **6**(3–4), 127–134 (1982)
8. Cetegen, B., Ahmed, T.: Experiments on the periodic instability of buoyant plumes and pool fires. Combust. Flame **93**, 157–184 (1993)
9. Atkinson, G., Drysdale, D., Wu, Y.: Fire driven flow in an inclined trench. Fire Saf. J. **25**, 141–158 (1995)
10. Woodburn, P., Drysdale, D.: Fires in inclined trenches: time-varying features of the attached plume. Fire Saf. J. **31**, 165–172 (1998)
11. Dupuy, J.L., Marecha, J., Portier, D., Valette, J.-C.: The effects of slope and fuel bed width on laboratory fire behavior. Intl. J. Wildl. Fire **20**, 272–288 (2011)
12. Fons, W.L.: Analysis of fire spread in light forest fuels. J. Agric. Res. **72**(3), 93–121 (1946)
13. Vogel, M., Williams, F.A.: Flame propagation along matchstick arrays. Combust. Sci. Technol. **1**(6), 429–436 (1970)
14. Prahl, J.M., Tien, J.S.: Preliminary investigations of forced convection on flame propagation along paper and matchstick arrays. Combust. Sci. Technol. **7**(6), 271–282 (1973)
15. Carrier, G., Fendell, F., Wolff, M.: Wind-aided firespread across arrays of discrete fuel elements. II. Experiments. Combust. Sci. Technol. **75**, 261–289 (1991)
16. Catchpole, W., Catchpole, E., Butler, B., Rothermel, R.C., Morris, G.A., Latham, D.J.: Rate of spread of free-burning fires in woody fuels in a wind tunnel. Combust. Sci. Technol. **131**, 1–37 (1998)
17. Emmons, H., Shen, T.: Fire spread in paper arrays. Proc. Combust. Inst. **13**(1), 917–926 (1971)
18. Byram, G., Clements, H., Elliott, E., George, P.: An experimental study of model fires. USDA Forest Service, Southeastern Forest Experiment (1964)
19. Smith, D.: Measurements of flame length and flame angle in an inclined trench. Fire Saf. J. **18**, 231–244 (1992)
20. Malalasekera, W.M.G., Versteeg, H.K., Gilchrist, K.: A review of research and an experimental study on the pulsation of buoyant diffusion flames and pool fires. Fire Mater. **20**(6), 261–271 (1996)
21. Hall, P.: The linear development of Goertler vortices in growing boundary-layers. J. Fluid Mech. **130**, 41–58 (1983)
22. Herbert, T.: Secondary instability of boundary layers. Annu. Rev. Fluid Mech. **20**, 487–526 (1988)
23. Kachinov, Y.: Physical mechanisms of laminar-boundary-layer transition. Annu. Rev. Fluid Mech. **26**, 411–482 (1994)
24. Sparrow, E., Husar, R.: Longitudinal vortices in natural convection flow on inclined plates. J. Fluid Mech. **37**, 251–255 (1969)
25. Maughan, J., Incropera, F.: Secondary flow in horizontal channels heated from below. Exp. Fluids **5**, 334–343 (1987)
26. Beer, T.: The interaction of wind and fire. Boundary-Layer Meteorol. **54**(3), 287–308 (1991)
27. Hussain, A., Zaman, K.: An experimental study of organized motions in the turbulent boundary layer. J. Fluid Mech. **159**, 85–104 (1985)
28. Bernal, L., Roshko, A.: Streamwise vortex structure in plane mixing layers. J. Fluid Mech. **170**, 499–525 (1986)
29. Robinson, S.: Coherent motions in the turbulent boundary layer. Annu. Rev. Fluid Mech. **23**, 601–639 (1991)
30. Li, F., Malik, M.: Fundamental and subharmonic secondary instabilities of Görtler vortices. J. Fluid Mech. **297**, 77–100 (1995)
31. Oka, Y., Kurioka, H., Satoh, H., Sugawa, O.: Modeling of unconfined flame tilt in cross-winds. In: Proceedings of 6th International Symposium on Fire Safety Science, pp. 1101–1112 (1999)

32. Raj, P.: A physical model and improved experimental data correlation for wind induced flame drag in pool fires. Fire. Technol **46**, 579–609 (2010)
33. Weise, D., Biging, G.: Effects of wind velocity and slope on flame properties. Can. J. Forest Res. **26**(10), 1849–1858 (1996)
34. Nelson, R., Adkins, C.W.: Flame characteristics of wind-driven surface fires. Can. J. Forest Res. **16**, 1293–1300 (1986)
35. Anderson, W., Catchpole, E., Butler, B.: Convective heat transfer in fire spread through fine fuel beds. Intl. J. Wildl. Fire **19**, 284–298 (2010)

Section B
Fire and Explosion - A Study of Flame Spread in Engineered Cardboard Fuel Beds Part II: Scaling Law Approach

Brittany A. Adam, Nelson K. Akafuah, Mark A. Finney, Jason Forthofer, and Kozo Saito

Abstract In this second part of a two-part exploration into the dynamic behavior observed in wildland fires, time scales differentiating convective and radiative heat transfer are further explored. Scaling laws for the two different types of heat transfer were considered: radiation-driven fire spread and convection-driven fire spread, which can both occur during wildland fires. A new interpretation of the inertial forces introduced a downstream, time-dependent frequency ω, which captures the dynamic, vortex shedding behavior of flames due to the unstable nature of the turbulent flow created in the wake of the fire. Excelsior and paper strip experiments suggest many wildland fire scenarios fall into the convection-driven spread regime.

Nomenclature

$\Delta\theta_1$	Temperature rise of air and gas
$\Delta\theta_2$	Temperature rise of fuel
$\Delta\rho_1$	Density change of air and gas
c_2	Specific heat of fuel
c_p	Specific heat of a gas at atmospheric pressure
E	Irradiance received by radiometer
F_b	Buoyancy force of air and gas
F_i	Inertial force
F_i	Inertial force of air and gas
Fr	Froude number
g	Acceleration due to gravity
H	Fuel height
H_f	Flame height

B.A. Adam • N.K. Akafuah (✉) • K. Saito
Institute of Research for Technology Development, University of Kentucky, Lexington, KY 40506, USA
e-mail: nkakaf0@engr.uky.edu

M.A. Finney • J. Forthofer
USDA Fire Science Laboratory, Missoula, MT 59808, USA

K. Saito et al. (eds.), *Progress in Scale Modeling, Volume II*,
DOI 10.1007/978-3-319-10308-2_6

I	Fire intensity
l_2	Fuel bed width
L_a	Height of flame and fire plume
L_e	Effective length over which the majority of heat transfer occurs (also effective length for vortex shedding)
l_H	Characteristic length, height of fuel
L_w	Flame zone thickness
Ø	Ratio of consumed fuel to the total available fuel
Q	Heat generated
Q_{c1}	Heat stored in the air and gas associated with the temperature rise
Q_{c2}	Heat stored in the unburned fuel
q_f	Heat value per unit mass of fuel
Q_r	Radiant heat received by the unburned fuel
Q_λ	Latent heat of fuel (heat value per unit mass of fuel)
R	Spread velocity of flame front
St	Strouhal number
t	Characteristic time
u	Horizontal wind velocity
α	Fuel bed angle
π	pi number
ρ_1	Gas density
ρ_f	Fuel density
ω	Downstream, shedding frequency

Introduction

Although significant progress has been made in wildland fire study over the last decade, it continues to present interesting academic challenges, as pointed out by Pyne et al., in *Introduction to Wildland Fires*, "the fire phenomenon in the wildland setting, however, has not been and may never be explained to the level of first principles" [1]. Spreading and ignition mechanisms are still poorly understood. Finney et al. further detailed our main target for exploration, "In the study of wildfire spread, heating and ignition of fuel particles by flame contact has been largely neglected in favor of quantifying heat transfer by radiation. This focus on only one method of heat transfer is unfortunate as research has recently suggested that radiant heating alone is frequently insufficient to ignite fine fuel elements at fluxes common to wildland fires" [2].

A study of wildland fires leading to a correct understanding of flame spread in wildland fire must be firmly situated on an understanding of the governing mechanisms, processes, and threshold of ignition. It is, therefore, very important for effective firefighting efforts and safety reasons to identify the roles of radiative and convective heating.

Over the years, the United States Department of Agriculture (USDA) Missoula Fire Sciences Laboratory has conducted a series of experiments in their unique wind tunnel fire experimental facility. This rich database provides years of numerical data and video from burns conducted under a wide range of well-specified conditions. After identifying the need to explore the roles of both convective heat transfer and radiative heat transfer in the ignition process, the USDA's well-documented line fire data provided an opportunity to observe ignition and, subsequently, fire spread phenomenon, through a uniform fuel bed of laser-cut cardboard combs under controlled conditions.

The USDA Missoula Fire Sciences Laboratory and University of Kentucky Institute of Research for Technology Development (IR4TD) teams conducted two explorations to further understanding ignition on a wildland fuel bed. The team first began by identifying and exploring flame behaviors in the lab environment that were previously only observed in large-scale wildland fires [3]. The first part of the study demonstrated the importance of the dynamic behavior of diffusion flames; we believe the buoyancy instabilities causing time-dependent flickering behavior increase convective heat transfer and must be considered. The goal of the second paper of the two-part study is to identify features distinguishing radiative heat transfer from convective heat transfer. The team worked to explain scaling laws, determine key parameters to support the development of scaling laws, and begin a comparison of the scaling law predictions with USDA data.

When the above scaling laws are validated, it would be reasonable to design medium-scale prescription fire experiments, which fall between the USDA experiments and the full-scale wildland fires. We believe that this step-by-step approach guided by the scaling analysis in cooperation with different size-scale model experiments will eventually allow us to understand the governing physics that control the mechanism of flame spread through and ignition on the wildland fuel bed.

Scale Modeling

Williams' pioneering study, Scaling Mass Fires [4]; the two studies introducing scaling laws for pool and crib fires, Emori and Saito 1983, and a study on flame spread, Emori et al. 1988, can all provide the guidance for development of scaling laws for flame spreading through wildland fuel bed [5, 6] (Fig. 1). Only a brief summary of this previous work is provided here. Assumptions were made as follows: flow pattern is generally turbulent suggesting the inertial force is dominant over the viscous force. Near and in the flame region, temperature of air and gaseous combustion by-products can easily exceed 1,000 K, creating a strong buoyancy effect that will interact with the inertia force of air. In the upstream location, there are a horizontal wind and a fire induced flow, both of which are governed by the inertial force. The fire can act like a wall to generate vortex shedding in the downstream location. The two separate roles of the inertial force, in the upstream and the downstream, can be written separately. Radiation and convection heat transfer may both play important roles in determining spread rate; as the relative

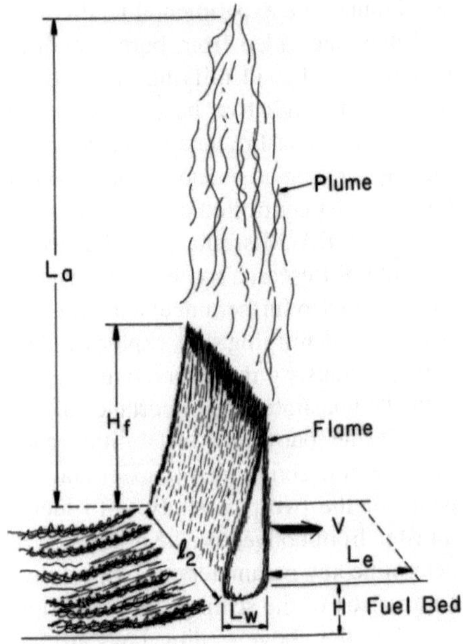

Fig. 1 Schematic of flame spread over fuel bed and dimensions of flame height, plume height, and fuel bed [6]

significance of these two methods of heat transfer is not certain at this point, we assume both are significant. Conduction heat transfer is assumed negligible over radiation and convection. As a result, the following key forces and types of heat are identified: the buoyancy force of air and gas (F_b), inertial force of air and gas (F_i), heat generated (Q), heat stored in the air and gas associated with the temperature rise (Q_{c1}), heat stored in the unburned fuel (Q_{c2}), radiant heat received by the unburned fuel (Q_r), and latent heat of fuel (Q_λ). Those two different forces and five separate types of heat can be written as follows using characteristic parameter [6]. They are the following:

$F_{i,\mathrm{up}} = \rho_1 l_2 (L_a)^2 u\omega$	$F_{i,\mathrm{down}} = \rho_1 l_2 L_a u^2$	$F_b = \Delta\rho_1 l_2 L_w L_a g$	$Q = \varnothing\, q_f \rho_f l_2 H L_w = I L_w t$
$Q_r = E l_2 L_e t$	$Q_{c1} = c_p \rho_1 L_a l_2 L_e \Delta\theta_1$	$Q_{c2} = c_2 \rho_f l_2 H L_e \Delta\theta_2$	$Q_\lambda = \rho_f l_2 H L_e$

These identified two different forces and five different types of heat that constitute the following seven independent pi numbers:

$\pi_1 = \frac{F_{i,\mathrm{down}}}{F_b} = \frac{\rho_1 u^2}{\Delta\rho_1 L_w g}$	$\pi_2 = \frac{F_{i,\mathrm{down}}}{F_{i,\mathrm{up}}} = \frac{L_e \omega}{u}$	$\pi_3 = \frac{Q_r}{Q} = \frac{E l_2 L_e}{I L_w}$
$\pi_4 = \frac{Q_{c1}}{Q} = \frac{c_p \rho_1 L_a R \Delta\theta_1 l_2}{I L_w}$	$\pi_5 = \frac{Q_{c2}}{Q} = \frac{L_e}{L_w}$	$\pi_6 = \frac{Q_\lambda}{Q} = \frac{\lambda L_e}{\phi q_f L_w}$
	$\pi_7 = \frac{F_i u t}{Q} = \frac{\rho_1 L_a u^3 l_2}{I L_w}$	

For the equations

$$V = \frac{L_e}{t} \quad I = \frac{Oq_f \rho_f l_2 H}{t},$$

the scaling criteria demand $\pi_i = \pi_i^{'}$ for similarity, where i 1 to 7 of the left hand π_i represents a full-scale scenario and the right hand $\pi_i^{'}$ represents a corresponding scale model. Note that π_2 is the ratio of the inertial force causing vortex shedding behind a flame, $F_{i,\,\mathrm{down}}$ to the inertial force of the wind, $F_{i,\,\mathrm{up}}$ is unique to the current wildland fire problem, where a flame acts like a vertical solid cylinder to generate a wake in the downstream against an upcoming horizontal flow [7]. Finney et al. [2] suggested the *St–Fr* correlation, which is equivalent to the (π_2–π_1) correlation, based on their careful observations of the engineered cardboard fuel bed experiments. Using the same fuels for both the full scale and the model and assuming the same temperature at the corresponding points, π_5 and π_6 can be easily satisfied, and the above scaling criteria yield the following (Eq. (1)):

$$\phi\left[\frac{u^2}{L_w}, \frac{L_e \omega}{u}, \frac{El_2}{I}, \frac{L_a R}{I}, \frac{L_a u^3}{I}\right] = 0 \tag{1}$$

Based on previous studies [5, 6] there are two different types of fire: radiation driven and convection driven. Scaling laws for each type of fires were developed separately; these scaling laws can be applied to the current flame spread study, since the governing forces and heat balances described in the above references also can be applied to the current problem. The following provides a summary of the relevant scaling studies [5, 6]:

(1) Radiation-driven fire type
There is little effect of the fluid dynamics on the overall heat balance, and radiation is the main source of the solid fuel evaporation and ignition leading to the flame to spread. For this type of fire, Eq. (1) can yield the specific scaling relationships seen in Eq. (2):

$$\frac{u}{u'} = \sqrt{\frac{L_w}{L_w'}} \quad \frac{R}{R'} = \frac{E}{E'} \tag{2}$$

$$t = t' \tag{3}$$

Note that the fire intensity, I, was set to satisfy $I/I' = (L_w/L_w')^2$, and π_7 has no effect on this angle.

(2) Convection-driven fire type
For this type of fire, contrary to the radiation-driven type, fluid dynamics forces the heat transfer mechanism in the form of heat convection, creating a coupling between the force and heat balances and leading to the following Eq. (3):

$$\frac{u}{u'} = \frac{R}{R'} = \frac{E}{E'} = \sqrt{\frac{L_w}{L'_w}} \tag{4}$$

Time scale also can be obtained as:

$$\frac{t}{t'} = \sqrt{\frac{L_w}{L'_w}} \tag{5}$$

(3) Consideration for ignition time scaling
It would be reasonable to assume that the diameter of fuel particles, d, is much smaller when compared to other characteristic lengths, allowing scale separation to occur when $d \ll L$. Under this condition, the inner and surface temperature of the fuel particles can achieve steady-state temperature across the particle diameter within a very short time after they are exposed to an external heat source, either radiation or convection. This assumption can simplify the ignition time scaling, to follow Eq. (3) under a radiation heating and Eq. (5) under a convection heating.

(4) Distinguishing between convective and radiative heat transfer
To test the scaling laws, a condensed-phase characteristic length (the fuel depth, l_H) of several burns was plotted against the recorded spread rate, R, as fire spread along a fuel bed of paper combs at four different inclinations. The inclinations ranged from a horizontal bed ($\alpha = 0°$) to a bed with a positive slope of 45°. The paper strips were coated in wax so that they burned at a moderate speed and released enough heat to create turbulent flames [6]. Figure 2 is a plot of the different spread rates with respect to that characteristic length, the fuel depth, l_H.

Flame spread behavior during exploratory paper comb burns on both a horizontal bed ($\alpha = 0°$) and a bed with a slope of 15° best fits the radiation-driven spread profile identified by the scaling law given in Eq. (3). Flame spread behavior during paper comb burns on a slope of 30° and a slope of 45° best fits the convection-driven spread profile identified by the scaling law given in Eq. (5).

Excelsior trials yielded surprising results. Spread rates recorded for excelsior burns on both a horizontal bed ($\alpha = 0°$) and a bed with a slope of 25° fit identified convection spread profile, with the ½ slope spread pattern as shown in Fig. 3. This suggests that flame spread over excelsior fuel beds is convection driven, irrespective of fuel bed slope up to at least 25°, indicating that full-scale wildland fire spread is likely convection driven. Excelsior better mimics litter and duff of the actual wildland fuel bed than paper comb indicating that the flame spread mechanism in wildland fires could be significantly convection driven. However and unfortunately, past studies [2] have found that excelsior experiments demonstrate poor reproducibility because excelsior's packing density and material properties are not uniform like those of paper combs, creating

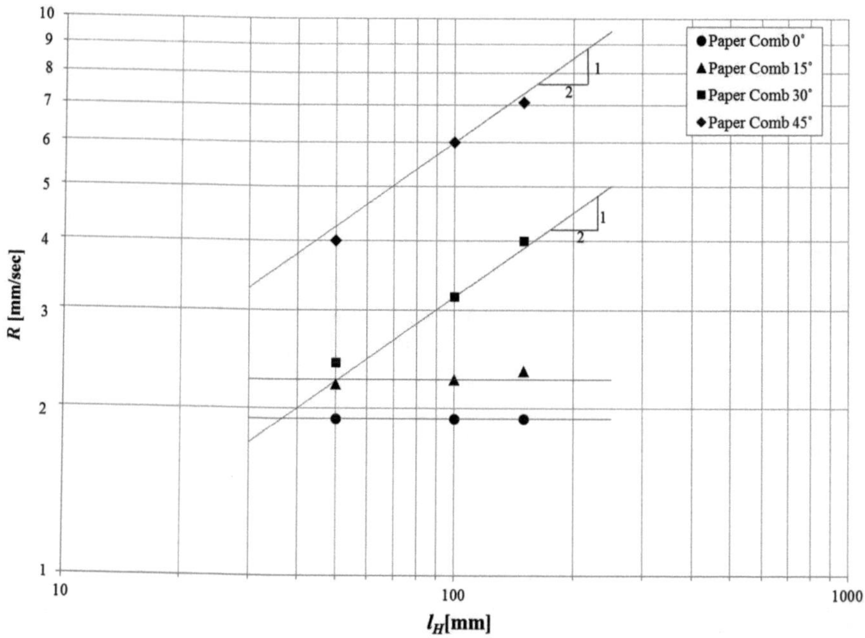

Fig. 2 Convection-dominated fire spread and radiation-dominated fire spread obey different power law relationships creating different lengthy scale dependencies. The relationship between flame height (l_H) and spread rate (R) for the burns on horizontal beds ($\alpha = 0°$) and beds at an inclination of 15° corroborates the proposed radiation-driven scaling law suggesting that the rate of spread is proportional to a condensed-phase characteristic length. The relationship between flame height (l_H) and spread rate (R) for the burns on beds at an inclinations of 30° and 45° corroborates the proposed convection-driven scaling law suggesting that the rate of spread is proportional to the square root of a condensed-phase characteristic length [5]

inconclusive results and preventing researchers from identifying hidden mechanism.

Discussion

Recent studies by Finney et al. [2, 3] found that forest fires are inherently dynamic, while the source and mechanisms of that dynamic nature have not been clearly understood. Time-dependent flame behaviors like flickering, pulsing, and vortex shedding have all been observed when diffusion flames interact with flow or pressure waves, though studies noting these behaviors have been primarily focused on buoyant diffusion flames originating from circular nozzles or jets [8–10]. Wild-land fires, particularly when they interact with wind, exhibit these behaviors as well [11–13]. Time-averaged analyses of these fires, while more mathematically suitable [14], disregard key time-dependent mechanisms, like vortex shedding, a behavior that has been found to significantly increase convective heat transfer in a variety of

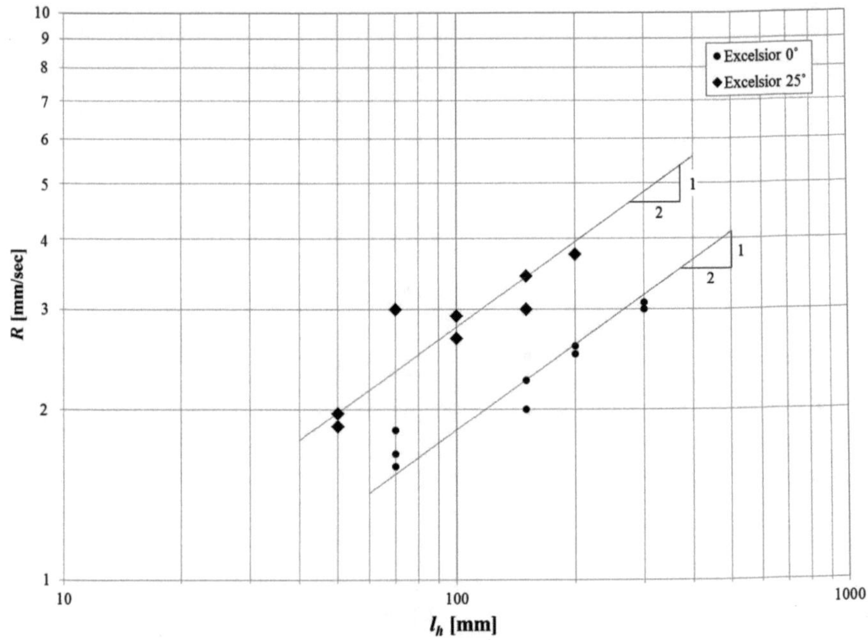

Fig. 3 Excelsior trials at two different slopes fit identified convection spread profile

applications [15, 16]. A time-averaged, static analysis would not capture the entirety of convective heating and thus would not accurately predict ignition.

To compensate for this deficiency, a preliminary set of assumptions for a typical flame-spreading pattern (i.e., flame spreading through pine forest under relatively moderate wind speed condition over rather moderate terrain geometry) is considered. At this point, no consideration is given to whether the fuel is alive or dead; rather, the fuel is treated as an average wildland fuel with typical thermophysical properties. It is understood that later this analysis would be modified when applied to specific scenarios and fuels.

Using those assumptions, the following new idea is proposed to describe this dynamic system. The wind-driven wildland fire spread is a "Strouhal number"--controlled dynamic system where instabilities break the spreading fire front flame structure into several turbulent fire columns. The instability which generates smaller stable fire columns is controlled by the buoyancy force acting on the flame and the pressure difference created inside the flame [17].

Each fire column generates a strong upward buoyant flow and acts like a solid wall against the horizontal wind [18, 19]. The columns are then similar to a pool fire. When the wind flows around the column, it picks up heat from the flame. When wind passes around the column, it sheds a von Karman vortex street in the downstream direction [7]. This combination of the heated air/combustion gas and von Karman vortices can effectively transfer the convective heat to the unburned fuel.

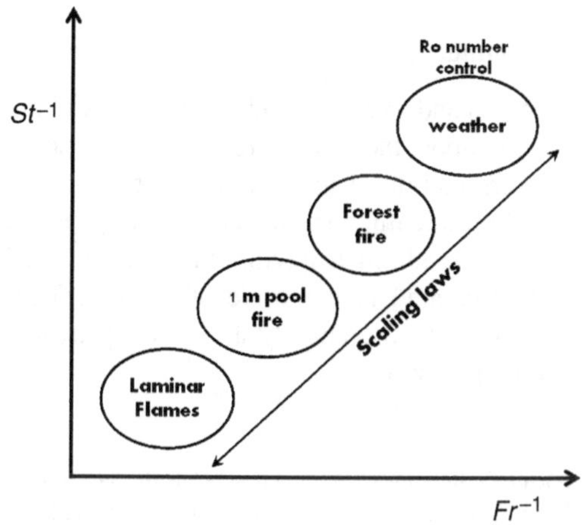

Fig. 4 Proposed St^{-1}–Fr^{-1} number relation map, including relations of small-scale laminar diffusion flames, large-scale wildland fires, and global-scale weather phenomena

To simplify this concept, a flame column could be treated as a pool fire governed by Froude number [5] defined as the ratio of the inertia force of air to the buoyancy force of hot combustion by-products in the flame. When this oscillating flame interacts with the inertia force of wind, it forms a St^{-1}–Fr^{-1} number-controlled dynamic system Fig. 4.

St^{-1}

The Strouhal number defined in the scaling law section can be repeated here: *St* number is the ratio of the inertial force causing vortex shedding behind a flame to the inertia force of wind. It is unique to the current wildland fire problem, where a flame acts like a vertical solid cylinder to generate a wake in the downstream flow against an upstream horizontal flow [7].

The idea of the *St*–*Fr* number dynamic system also can help connect seemingly disparate phenomena for the benefit of wildland fire research. Figure 4 shows a possible scaling correlation that requires validation. If this correlation works, then we should be able to correlate various size-scale phenomena, which include flame sizes as small as a 10 mm diameter candle-like laminar flame and a laminar spreading flame over a few centimeter-wide paper strip, a medium-scale 1 m diameter turbulent liquid pool fire, a typical wildland fire with a fire front length on the order of 10–100 m, and up to as large as atmospheric weather, which is governed by the Rossby number.

Conclusion

In the pioneering work by Emori and Saito [5], scaling laws for both radiation-driven pool fire and convection-driven crib fires were developed. Later, Emori and Iguchi applied those scaling laws to flame spread over paper

(continued)

(continued)

combs and excelsior and identified six pi numbers [6]. In this paper, we have built upon their work to establish a seventh pi number based on a new interpretation of the role of the inertial force. This interpretation introduces a downstream, time-dependent frequency ω, which captures the dynamic, vortex shedding behavior of flames due to the unstable nature of the turbulent flow created in the wake of the fire. This downstream inertial term is in addition to the already accepted upstream inertial force due to the wind's initial flow.

Acknowledgements We acknowledge the late Professor Ichiro Emori and his former student Yasuo Iguchi for their pioneering work on scaling laws on flame spread. We also thank Professor Forman Williams for his invaluable discussions on the *St–Fr* correlation on wildland fires. This study was supported by USDA Forest Service under Collaboration Forest Service agreement: 12-CS-11221637-133.

References

1. Pyne, S., Andrews, P., Laven, R.: Introduction to Wildland Fires, 2nd edn. Wiley, New York (2000)
2. Finney, M., Cohen, J., McAllister, S., Jolly, W.: On the need for a theory of wildland fire spread. Int. J. Wildl. Fire **22**, 25–36 (2013)
3. Finney, M., Forthofer, J., Adam, B., Akafuah, N., Saito, K.: A study of flame spread in engineered cardboard fuelbeds, part I: correlations and observations. In: The Seventh ISSM, Submitted, Hirosaki, Japan (2013)
4. Williams, F.: Scaling mass fires. Fire Res. Abst. Rev. **11**, 1 (1969)
5. Emori, R., Saito, K.: A study of scaling laws in pool and crib fires. Combust. Sci. Technol. **31** (5–6), 217–231 (1983)
6. Emori, R., Iguchi, Y, Saito, K., Wichman, I.: Simplified scale modeling of turbulent flame spread with implication to wildland fires. Fire Safety Science–Proceedings of the Second International Symposium, pp. 263–273 (1988)
7. Schlichting, H.: Boundary Layer Theory, 7th edn. McGraw-Hill, New York, NY (1979)
8. Cetegen, B.M., Dong, Y.: Experiments on the instability modes of buoyant diffusion flames and effects of ambient atmosphere on the instabilities. Exp. Fluids **28**(6), 546–558 (2000)
9. Jiang, X., Luo, K.H.: Dynamics and structure of transitional buoyant jet diffusion flames with side-wall effects. Combust. Flame **133**(1), 29–45 (2003)
10. Gotoda, H., Asano, Y., Chuah, K.H., Kushida, G.: Nonlinear analysis on dynamic behavior of buoyancy-induced flame oscillation under swirling flow. Int. J. Heat Mass Transf. **52**(23), 5423–5432 (2009)
11. Banta, R., Olivier, L., Holloway, E., Kropfli, R., Bartram, B., Cupp, R., Post, M.: Smoke-column observations from two forest fires using Doppler lidar and Doppler radar. J. Appl. Meteorol. **31**(11), 1328–1349 (1992)
12. Clark, T.L., Radke, L., Coen, J., Middleton, D.: Analysis of small-scale convective dynamics in a crown fire using infrared video camera imagery. J. Appl. Meteorol. **38**(10), 1401–1420 (1999)

13. Coen, J., Mahalingam, S., Daily, J.: Infrared imagery of crown-fire dynamics during FROSTFIRE. J. Appl. Meteorol. **43**(9), 1241–1259 (2004)
14. Bejan, A.: Plumes. In: Convection Heat Transfer, 3rd edn, pp. 430–439. Wiley, Hoboken, NJ (2004)
15. Kottke, V.: Taylor–Görtler vortices and their effect on heat and mass transfer. In: Proceedings of the Eighth International Heat Transfer Conference, San Francisco, CA (1986)
16. McCormack, P.D., Welker, H., Kelleher, M.: Taylor-Görtler vortices and their effect on heat transfer. J. Heat Transf. **92**, 101–112 (1970)
17. Emori, R., Saito, K., Sekimoto, K.: Scale models in engineering (Mokei Jikken no Riron to Ohyou), 3rd edn. Gihodo, Tokyo (2000)
18. Kuwana, K., Sekimoto, K., Saito, K., Williams, F.A.: Scaling fire whirls. Fire Saf. J. **43**(4), 252–257 (2008)
19. Kuwana, K., Morishita, S., Dobashi, R., Chuah, K.H., Saito, K.: The burning rate's effect on the flame length of weak fire whirls. Proc. Combust. Inst. **33**(2), 2425–2432 (2011)

Section B
Fire and Explosion - Application of Pressure Fire Modeling Under Low Pressure

Zhihui Zhou, Jian Wang, and Richard Kwok Kit Yuen

Abstract Ambient pressure has a significant effect on combustion behavior. Especially for high ambient pressure, numerous studies have been developed. De Ris proposed pressure fire modeling as a suitable method in scaling fire controlled by convection term. However, due to limited availability of data, its application for fires under low ambient pressure remains in dispute. Thus, some experiments on small-scale pool fires were conducted in an "altitude chamber" (i.e., controllable pressure conditions), which was located in the State Key Laboratory of Fire Science, University of Science and Technology, China at Hefei, PRC. Experimental and theoretical modeling results are discussed. The results indicate that pressure fire modeling could be adopted for fires under low pressure. However, underestimation of the value of burning intensity was observed.

Nomenclature

D	Pool diameter
P	Ambient pressure
$\dot{m}''$	Burning intensity
μ	Viscosity

Z. Zhou
State Key Laboratory of Fire Science, University of Science and Technology of China, Hefei 230027, People's Republic of China

Department of Architecture and Civil Engineering, City University of Hong Kong, Tat Chee Avenue, Kowloon, Hong Kong SAR

J. Wang
State Key Laboratory of Fire Science, University of Science and Technology of China, Hefei 230027, People's Republic of China

R.K.K. Yuen (✉)
Department of Architecture and Civil Engineering, City University of Hong Kong, Tat Chee Avenue, Kowloon, Hong Kong SARbckkyuen@cityu.edu.hk

K. Saito et al. (eds.), *Progress in Scale Modeling, Volume II*,
DOI 10.1007/978-3-319-10308-2_7

Subscripts

g Gas phase

Introduction

Most pool fire-type diffusion flames are dominated by buoyancy, especially for small pool fires. The Froude number, which represents the relative importance between momentum and buoyancy in the flame, has always been used to classify fire types and correlate aspects of fire plumes, such as burning rate.

The ambient pressure has significant impact on fire behavior because of its effect on gas density. Many studies have been conducted to understand the influence of ambient pressure on combustion. One effective method to scale the fire is Froude modeling, and the way to preserve the non-dimension group (π-group) of Froude modeling as well as the Reynolds number is by changing the ambient pressure in pressure modeling [1, 2]. Although it was demonstrated to be suitable for scaling convectively controlled fires [3, 4], combustion characteristics such as radiation heat fluxes and chemical kinetics cannot be properly analyzed in this way.

Recently, there has been growing concern about fire safety of heritage constructions (such as the historical ancient temples) in Tibet, China. This has led to a series of systematic and scientific fire researches at high altitude (i.e., under low pressure). This presents new challenges for the application of pressure fire modeling. Tests of n-Heptane pool fires carried out at various altitudes by Li et al. [5] indicated that the mass burning flux $\dot{m}''$ can be correlated with pressure P by $\dot{m}'' \propto P^{\alpha}$ with $\alpha = 1.3$. Previous studies indicated that the power component α varies with the pool sizes [6–8]. Hu et al. [9] suggested that a generalized function of the mass burning rate can be written as $\dot{m}'' = f(D, P)$.

To expand the range of the tested ambient pressures (i.e., by mimicking various altitudes), a reduced-scale "altitude chamber" of medium size was built in the State Key Laboratory of Fire Science (SKLFS), University of Science and Technology, China (USTC) at Hefei, PRC. In this paper, the experiments on fires under various ambient pressures performed in this reduced-scale altitude chamber are reported. The burning mass loss rate of fuel was adopted as the typical parameter in these experiments for the verification of the applicability of pressure fire modeling. During the experiments, video of flame was also recorded to capture the phenomena and appearance of the fires. The objective of this work is to gain a better understanding of the effects of low ambient pressures on pool fires.

Experimental Design

The pool fire experiments were conducted in a pressure chamber located at SKLFS of USTC at Hefei, PRC, as shown in Fig. 1a. The size of altitude chamber is 2 m × 2 m × 3 m with adjustable pressure ranging from 0.3 to 1 atm. To ensure adequate supply of oxygen inside the chamber to ascertain fuel-controlled burning of fuel throughout the experiment, small circular liquid fuel pans of 6, 8, and 10 cm were selected. The ambient pressure might increase at the initial stage of the combustion process due to the rapid rise of internal temperature, and the suction pump was used to control the pressure at the set value when it increased more than 5 kPa. After the short initial stage, the pressure remained stable, and the internal environment was considered as equivalent to an open space under different low ambient pressures.

A liquid fuel, n-Heptane, was used in the experiments. To keep the consistency of thermal parameters of fuel pans, which has a significant effect on fire behaviors [10], the circular liquid pools were designed to be 0.32 cm thick and 2 cm high to maintain the thermal parameters of pool rim. Error was less than ±0.3 mm. Meanwhile, the initial fuel level was maintained with an identical height of 1.5 cm, and the related fuel weight of n-Heptane was 28.8, 51.2, and 80.1 g, respectively. The relative errors in initial fuel weight were ensured to be less than ±5 %.

The fuel pans were placed on top of an electronic scale with the resolution of 0.01 g at the center of the chamber, as shown in Fig. 1b. A high-resolution digital camera was also placed in the chamber to obtain the flame video, at a frame rate of 25 fps. The experimental configuration was presented in Table 1 below.

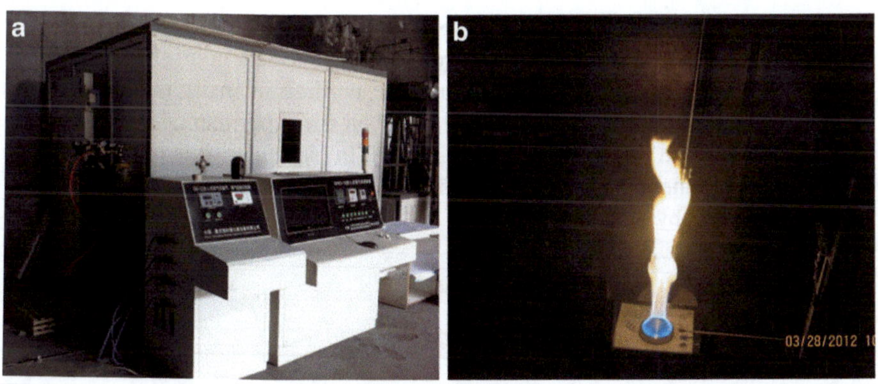

Fig. 1 Configuration of experimental setup: (**a**) altitude chamber; (**b**) pool fire

Table 1 Summary of the experimental configurations

Fuel	Pan size (cm)	Internal pressure (kpa)	Measurement
n-Heptane Industrial purity $\geq 99\%$ Density:683~685 kg/m^3 Boiling range:96.5–98.5 °C	6	40, 60, 80, 100	*Mass burning rate* *Video record*
	8	40, 60, 80, 100	
	10	40, 60, 80, 100	

Results and Discussions

Burning Intensity

An entire burning process contains four stages [10]: (1) pre-steady burning stage, (2) quasi-steady burning stage, (3) boiling stage, and (4) decay stage. The quasi-steady burning stage was selected for detailed observation and analysis owing to its relatively stable burning intensity, as shown in Fig. 2a. Logarithmic curve fitting method was adopted to analyze the correlations between $\dot{m}''$ (denoting the burning intensity or burning rate per unit area) and P. It has been indicated that $\dot{m}''$ is proportional to P^{α} (i.e., $\dot{m}'' \propto P^{\alpha}$) for different values of parameter α of 0.23, 0.40, and 0.50 for 6, 8, and 10 cm fuel pans, respectively. This result, pointed out by Fang et al., may be explained by the transformation of flame heat feedback [6].

In pressure fire modeling, the burning intensity ($\dot{m}''$, burning rate per unit area) can be correlated with the Grashof number (Gr) as shown in the following equation [3, 11]:

$$\dot{m}'' D/\mu = \text{fcn}(Gr)$$

where $Gr \equiv gD^3\beta\Delta T\rho^2/\mu^2 \sim P^2D^3$, the volumetric expansion ratio $\beta\Delta T = 6$, and the viscosity μ are independent of pressure [11]. As plotted in Fig. 2b, from the experimental results (i.e., the dashed line), the burning intensity is dependent on the 0.27 power of Gr. This current model based on our experimental data is slightly different from those previously reported [3, 11]. In particular, it is in contrast with the reported exponential factor of 1/3 for circular pools of polymeric solids under elevated pressure by Kanury [11].

Assessment of Pressure Fire Modeling

Wiser et al. [12] and Niu et al. [13] had found that the burning intensity was proportional to 2/3 power of pressure at high altitude, which is supported by pressure fire modeling [1]. These studies pointed out that pressure fire modeling should be available for cases of fires controlled by convective term. However, due

Fig. 2 Burning intensity: (**a**) logarithmic curve fitting, (**b**) pressure modeling

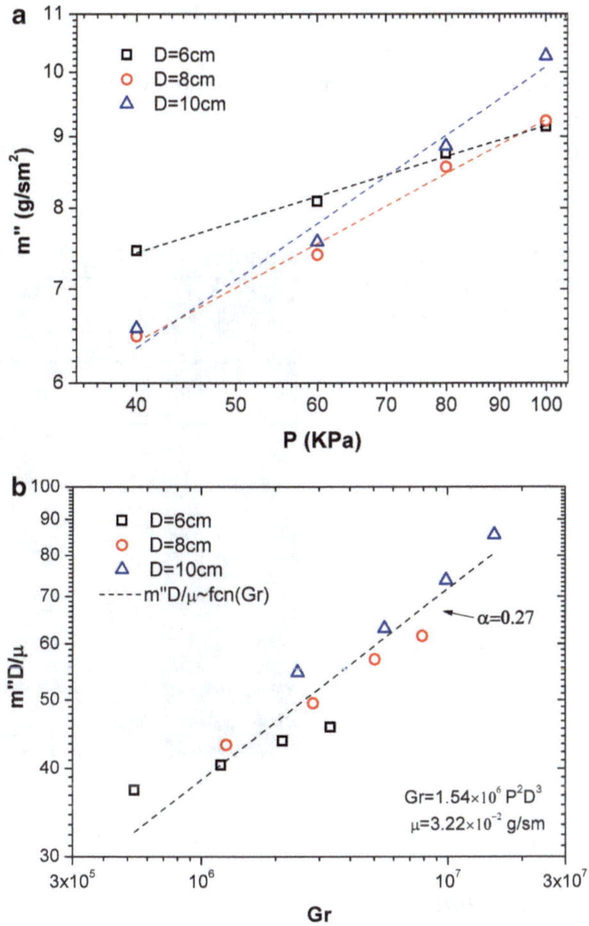

to the lack of experimental data under different pressures, its applicability is still questionable and subject to further investigation.

Based on burning intensity measured in our altitude chamber experiments, pressure fire modeling would be assessed further. If the parameter of *Gr* (i.e., P^2D^3) was set to be equal in the experiments, $D \cdot \dot{m}''_{P,D}$ would be invariable [1]. For instance, the burning rate of 8 cm pool fire under 0.4 atm could be predicted by that of 6 cm pool fire under 0.62 atm, as $\dot{m}''_{P=0.4\ \text{atm},\ D=8\ \text{cm}} = 6 \cdot \dot{m}''_{P=0.62\ \text{atm},\ D=6\ \text{cm}}/8$. The value of $\dot{m}''_{P=0.62\ \text{atm},\ D=6\ \text{cm}}$ is derived from the linear interpolation of experimental data ($D = 6$ cm with $P = 0.6$ atm and $P = 0.7$ atm). The results of the measured and predicted burning intensity are listed in Table 2. It is apparent that pressure fire modeling would underestimate the value of burning intensity, although to within 15 %. This may be due to the transformation of dominated flame feedback term.

Table 2 Assessment of pressure fire modeling

	8 cm (g/m^2)		10 cm (g/m^2)	
Pressure (atm)	Measured	Predicted by 6 cm	Measured	Predicted by 8 cm
0.6	7.41	6.75	7.57	6.95
0.4 atm	6.48	6.11	6.56	5.78

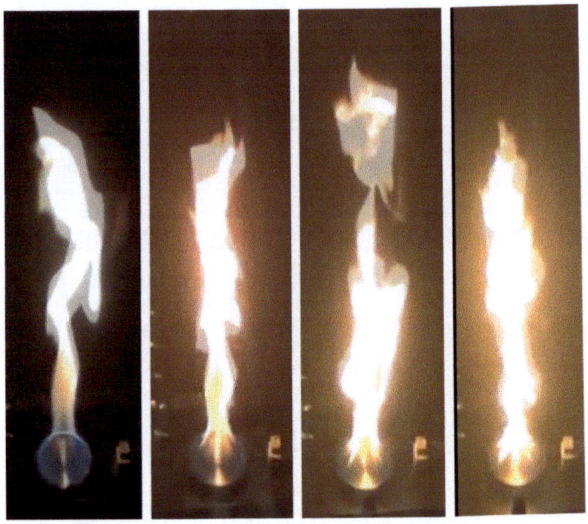

Fig. 3 Flame appearance of 10 cm pool fires: 0.4, 0.6, 0.8, and 1.0 atm (from *left* to *right*)

Flame Appearance

Reynolds number is a critical parameter to characterize the flame turbulence, which can be expressed as $\mathrm{Re} = \rho_g \nu_g D/\mu \propto \dot{m}'' D$ (ρ_g, ν_g-fuel gas density and velocity just above the pans, and $\rho_g \nu_g = \dot{m}''$). Thus, the decreased pressure would lead to the shape of flame being more smooth and tending laminar [14], as shown in Fig. 3. Another obvious variation on flame appearance happens at its bottom. Careful inspection reveals that, just above the burner rim, the flame is blue, similar in appearance to a premixed flame [15]. As ambient pressure decreases, soot particle concentration reduces rapidly [16, 17], and the yellowish luminosity fades away in the flame root region where the carbon zone starts [18]. This observed result also agrees with the study of Amell [19] on premixed flame that the height of blue cone increased with the increasing altitude.

Conclusion

This study has verified the validity of pressure fire modeling applied under low ambient pressure and thus expanded its range of application. However, its underestimation of burning intensity has also been reported. Similar to the premixed flame, the emergence and height elevation of blue zone at flame bottom can be observed.

Acknowledgment The work described in this paper was fully supported by a grant from the Research Grant Council of the Hong Kong Special Administrative Region, China [Project No. CityU 122612].

References

1. De Ris, J., Kanury, A.M., Yuen, M.C.: Pressure modeling of fires. Proc. Combust. Inst. **14**, 1033–1044 (1973)
2. Quintiere, J.G.: Scaling applications in fire research. Fire Saf. J. **15**(1), 3–29 (1989)
3. Alpert, R.L.: Pressure modeling of fires controlled by radiation. Proc. Combust. Inst. **16**, 1489–1500 (1977)
4. Alpert, R.L.: Pressure modeling of transient crib fires. Combust. Sci. Technol. **15**(1–2), 11–20 (1977)
5. Li, Z.-h., He, Y., Zhang, H., Wang, J.: Combustion characteristics of n-heptane and wood crib fires at different altitudes. Proc. Combust. Inst. **32**, 2481–2488 (2009)
6. Fang, J., Tu, R., Guan, J.-f., Wang, J.-j., Zhang, Y.-m.: Influence of low air pressure on combustion characteristics and flame pulsation frequency of pool fires. Fuel **90**(8), 2760–2766 (2011)
7. Jun, F., Yu, C.-Y., Ran, T., Qiao, L.-F., Zhang, Y.-M., Wang, J.-J.: The influence of low atmospheric pressure on carbon monoxide of n-heptane pool fires. J. Hazard. Mater. **154**(1–3), 476–483 (2008)
8. Tu, R., Fang, J., Zhang, Y.-m., Zhang, J., Zeng, Y.: Effects of low air pressure on radiation-controlled rectangular ethanol and n-heptane pool fires. Proc. Combust. Inst. **34**, 2591–2598 (2012)
9. Hu, X.K., He, Y.P., Li, Z.H., Wang, J.: Combustion characteristics of n-heptane at high altitudes. Proc. Combust. Inst. **33**, 2607–2615 (2011)
10. Kang, Q., Lu, S.X., Chen, B.: Experimental study on burning rate of small scale heptane pool fires. Chin.Sci. Bull. **55**(10), 973–979 (2010)
11. Kanury, A.M.: Modeling of pool fires with a variety of polymers. Proc. Combust. Inst. **15**, 193–202 (1975)
12. Wieser, D., Jauch, P., Willi, U.: The influence of high altitude on fire detector test fires. Fire Saf. J. **29**, 195–204 (1997)
13. Niu, Y., He, Y., Hu, X., Zhou, D., Lin, C.-H., Yin, J., Yao, W., Wang, J.: Experimental study of burning rates of cardboard box fires near sea level and at high altitude. Proc. Combust. Inst. **34**, 2565–2573 (2012)
14. Yin, J., Yao, W., Liu, Q., Zhou, Z.H., Wu, N., Zhang, H., Lin, C.-H., Wu, T., Meier, O.C.: Experimental study of n-Heptane pool fire behavior in an altitude chamber. Int. J. Heat Mass Trans. **62**, 543–552 (2013)
15. Drysdale, D.: An Introduction To Fire Dynamics. Wiley, New York (1998)

16. Bento, D., Thomson, K., Gulder, O.: Soot formation and temperature field structure in laminar propane–air diffusion flames at elevated pressures. Combust. Flame **145**(4), 765–778 (2006)
17. Yao, W., Zhang, J., Nadjai, A., Beji, T., Delichatsios, M.: Development and validation of a global soot model in turbulent jet flames. Combust. Sci. Technol. **184**(5), 717–733 (2012)
18. Hirst, R., Sutton, D.: The effect of reduced pressure and airflow on liquid surface diffusion flames. Combust. Flame **5**, 319–330 (1961)
19. Amell, A.A.: Influence of altitude on the height of blue cone in a premixed flame. Appl. Therm. Eng. **27**(2–3), 408–412 (2007)

Section B
Fire and Explosion - Observation of Confined Deflagration Phenomena of Flammable Gas Mixtures Under Elevated Gravity

Hiroyasu Saitoh, Takaaki Mizutani, Sayuri Kurihara, and Tomohito Hori

Abstract In order to investigate the effects of gravity on ignition and deflagration of flammable mixtures, we conducted flammable gas explosion experiments using constant volume combustion vessels under several gravity levels from normal gravity (1 G) to elevated gravity up to 50 G. The elevated gravity field was formed by a large-size centrifuge system. We examined flame kernel growth, flame spreading, and explosion pressure in hydrogen-air, methane-air, and propane-air mixtures in an elevated gravity environment. The flammability ranges were determined by the records of the explosion pressures, and it was confirmed that the upper and lower explosion limits were approximately the same regardless of the gravity level. However, the shapes of the propagating flames changed dramatically with increase of gravity level due to strong buoyancy effect especially in lean and rich mixture conditions with slow burning velocity. Moreover, it was found that similar flame spreading phenomena were observed in large-size deflagration of rich mixture under normal gravity and small-size deflagration of stoichiometric mixture under elevated gravity.

Introduction

Gravity is one of the most important factors for combustion phenomena. In recent years, many researchers have tried to clarify the gravity effects on combustion characteristics of various kinds of flames from microgravity to elevated gravity,

H. Saitoh (✉) • S. Kurihara
Department of Engineering Science and Mechanics, Shibaura Institute of Technology, 3-7-5 Toyosu, Koto-ku, Tokyo 135-8548, Japan
e-mail: hsaito@sic.shibaura-it.ac.jp

T. Mizutani
Chemical Safety Research Group, National Institute of Occupational Safety and Health Japan, 1-4-6 Umezono, Kiyose-shi, Tokyo 204-0024, Japan

T. Hori
Construction Safety Research Group, National Institute of Occupational Safety and Health Japan, 1-4-6 Umezono, Kiyose-shi, Tokyo 204-0024, Japan

K. Saito et al. (eds.), *Progress in Scale Modeling, Volume II*,
DOI 10.1007/978-3-319-10308-2_8

such as flame length, shapes and flickering motion of diffusion flames [1–4], premixed flames [5–7], pool fires [8–10], flames surrounding droplets [11, 12], flame spread on solid surfaces [13, 14], and so on. Under microgravity conditions, we can observe combustion phenomena without buoyancy effects, while we can observe the effects of enhanced natural convection on flames in an elevated gravity environment.

Moreover, buoyancy influences explosion and fire phenomena. For example, in vapor cloud explosions of flammable gases, boiling liquid expanding vapor explosions (BLEVE), forest fires, and so on, formation of updraft due to buoyancy causes upward motion of fire balls and high-temperature gas clouds, which may lead to expansion of damage scale. Scales of the flammable gas deflagration accidents vary in size from small to large. Small-scale gas explosion accidents are relatively easy to reproduce by laboratory scale experiments, whereas it is not easy to verify damage of large-scale deflagration accidents by large-scale experiments (e.g., deflagration volume of several thousand cubic meters) due to issue of cost and safety. Therefore, if we can reproduce actual large-scale deflagrations by small-scale experiments, we believe that it becomes easier to predict explosion damage.

In most cases, small-scale deflagration phenomena have ceased before the influence of buoyancy becomes remarkable. So, in this research, we tried to utilize high gravity of several tens of times earth gravity in order to enhance buoyancy effects on small-scale deflagrations. In the present work, we first confirmed the effects of elevated gravity formed by a centrifuge on the basic characteristics of small-scale deflagrations including flammability limits, flame kernel growth, flame spread, and maximum explosion pressure. Then we conducted deflagration experiments using a larger-size combustion vessel under normal gravity in an attempt to clarify dynamic similarity between large-scale deflagration under normal gravity and small-scale deflagration under elevated gravity. As far as the authors know, this is the first experimental investigation to demonstrate the possibility of simulating different scale-confined deflagrations by changing gravity level.

Experimental Setup and Procedure

Figure 1 shows a schematic diagram of the experimental apparatus. The small- and large-size constant volume combustion vessels used in the present study have cylindrical shapes with inner diameter of 100 and 800 mm; the cavity volumes are 0.00079 and 0.40 m^3, respectively.

A glass window is installed to the side of the small combustion vessel to observe ignition, flame kernel growth, and flame spread phenomena inside the vessel. However, it was difficult to mount a glass window to the large-size vessel due to safety problems; therefore, ion probes installed on the top and bottom walls of the vessel were used to record arrival time of propagating flames in order to examine buoyancy effects on flame spread.

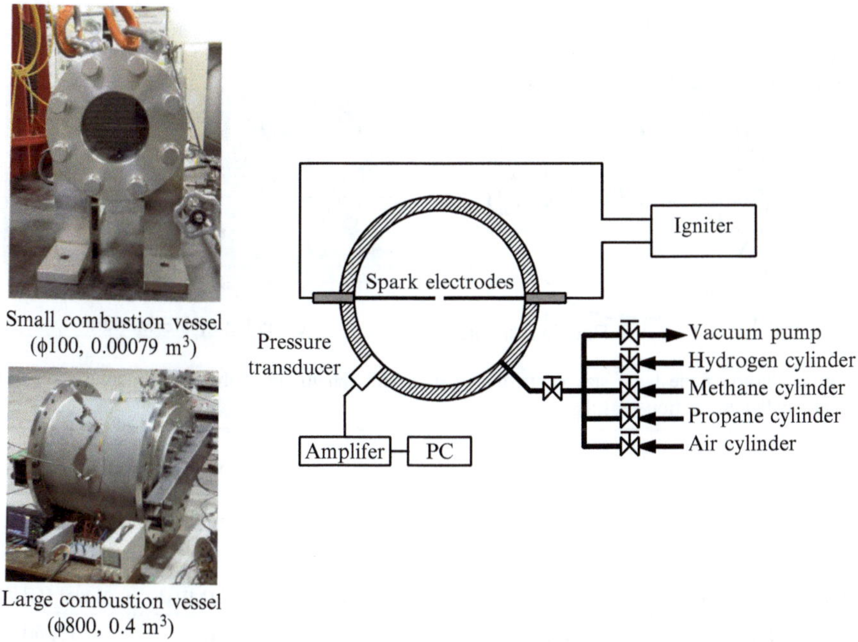

Fig. 1 Schematic diagram of the experimental apparatus and photos of the small- and large-size combustion vessels

The flammable mixtures were ignited by electric spark from the spark electrodes installed at the center of the vessels. The spark energy supplied from the high-voltage igniter system was about 33 J per shot (15 kV, 22 mA, duration time of the spark, 0.1 s), which is much higher energy than the minimum ignition energy of the test gases used here. The explosion pressure was detected with a pressure transducer (Kyowa Electronic Instruments Co., Ltd., PE-30kF), and after amplification the pressure signals were recorded by a data logger or an oscilloscope at the sampling frequency of 10 or 100 kHz, depending on the conditions.

In order to conduct explosion tests under elevated gravity conditions, the small-size combustion vessel and the igniter system were set on the test stand of the centrifuge as shown in Fig. 2. The rotation speed of the arm of the centrifuge and ignition timing were remotely controlled from the operating room. The gravity level formed here was up to 50 times the normal gravity.

The experimental procedure was as follows. First, the insides of the combustion vessels and gas supply lines were vacuumed by a vacuum pump; then gaseous fuels and dry air were supplied from gas cylinders. The flammable gases used here were hydrogen, methane, and propane. The initial pressure of the mixtures was 0.1 MPa. In the deflagration tests under elevated gravity, the mixtures were ignited after the gravity level became a target value by the centrifuge. Table 1 shows the experimental conditions. The spreading flames in the vessel were observed with a digital high-speed video camera (NAC Image Technology, Memrecam fx RX-6, model

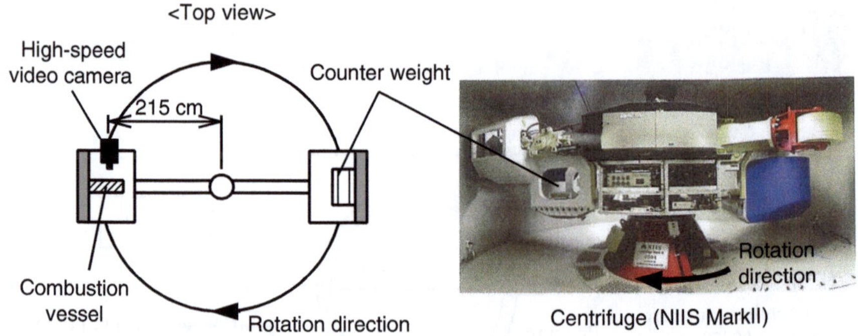

Fig. 2 Photo of the large-size centrifuge system and schematic of the experimental setup for small-scale combustion vessel

Table 1 Experimental condition

Vessel diameter [mm] (volume [m^3])	Flammable gas	Oxidizer	Concentration [vol.%]	Initial pressure [MPa]	Gravity level [G]
100 (0.00079)	Hydrogen	Dry air	4–75	0.1	1, 15, 50
	Methane		5–15		
	Propane		2.5–9		
800 (0.4)	Methane	Dry air	13	0.1	1

ST-662). The frame rate of the high-speed video camera was 1,000 fps, and the frame size was 512 × 512 pixels with bit depth of 10.

Results and Discussion

Flammability Limits

Figure 3 shows the experimental results of flammable ranges obtained in the ignition test under several gravity conditions. We determined whether ignition was successful or not from the pressure records. We used high-energy electric spark in the ignition tests; therefore, the explosion risk from the fuel gas concentration conditions in which we could not confirm ignition proved to be low. As shown in Fig. 3, the flammable ranges and the lower and upper explosion limits are almost the same as in the case of normal gravity. However, the flame motions under elevated gravity become quite different from normal gravity environment, as described later.

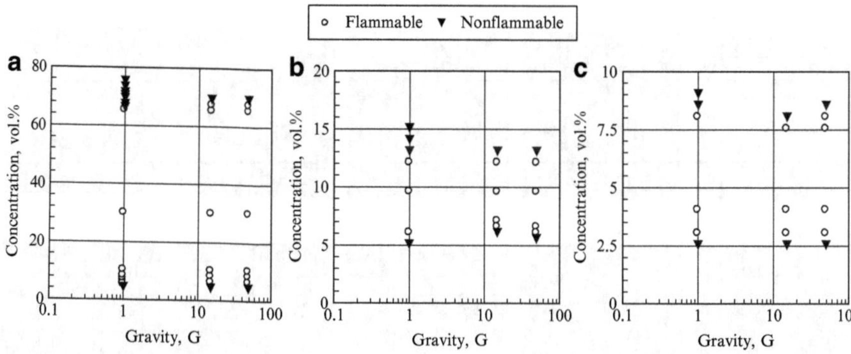

Fig. 3 Flammable ranges under normal and elevated gravity environments. (**a**) Hydrogen-air mixture, (**b**) methane-air mixture, and (**c**) propane-air mixture

Flame Kernel Growth, Flame Spread, and Maximum Explosion Pressure

Figure 4 shows the time series enlarged images of flame kernel growth and initial flame spread under normal gravity and 50 G in the case of stoichiometric mixture and fuel-rich mixture near the upper explosion limit. It is observed that the flame spread is influenced by a strong buoyancy effect, and upward flame spread speed is higher than downward speed, even in the case of stoichiometric mixture with fast burning velocity. This tendency becomes more noticeable in the case of fuel-rich mixture near upper explosion limit under 50 G. As shown in Fig. 4d, the flame cannot propagate downward.

Time series images of the flame behaviors after ignition and flame kernel growth are shown in Figs. 5 and 6. As shown in Figs. 5b and 6b, it is observed that stoichiometric methane-air and propane-air flames propagate spherically under earth gravity and 15 G conditions. However, the shapes of the propagating flames change significantly under elevated gravity field of 50 G. In the case of 50 G, downward flame propagation speed decreases, and consequently the flame front cannot reach the bottom wall of the vessel. The flame motion changes dramatically in the cases of lean and rich flame conditions under 50 G, as shown in Figs. 5a, c, and 6a. These flames cannot propagate downward from the location of spark gap due to strong buoyancy. Moreover, convection generated in the vessel caused by nonspherical flame propagation affects the shapes of downward-propagating flames. Generally, burning velocity is known to be affected by flame stretch [15]. So, in the case of nonspherical propagating flames, the flame stretch effect is different depending on location, which might cause variation of local burning velocity. Moreover, cellular instability is known to appear on flame surface depending on mixture condition [16]. To clarify the effects of elevated gravity on local flame structures and flame instabilities, further experiments are required.

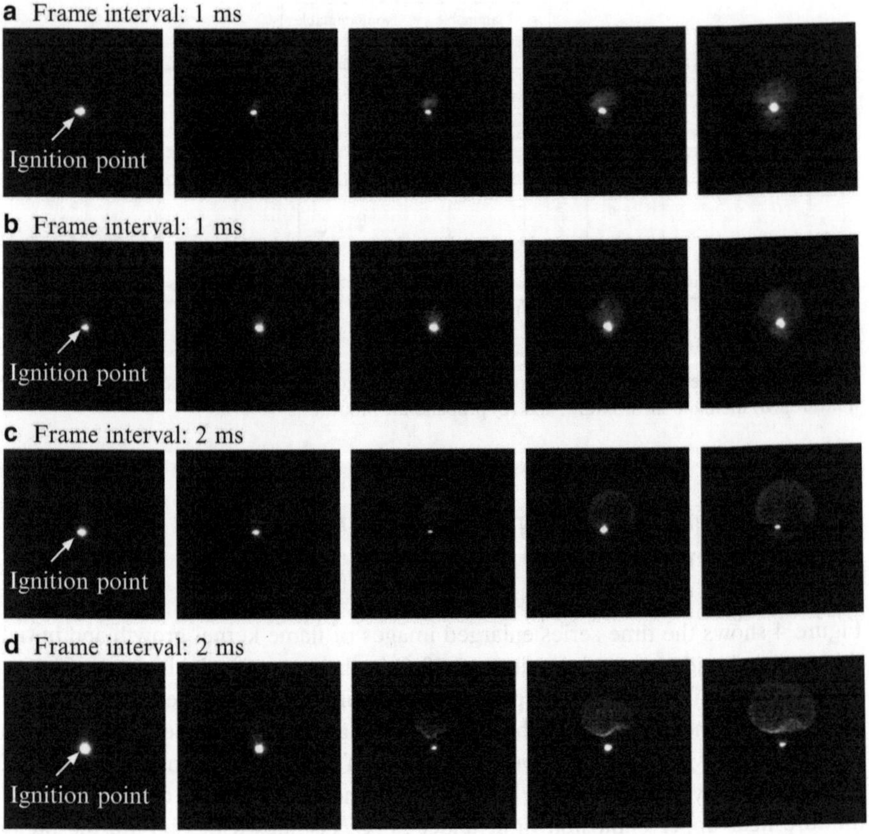

Fig. 4 Time series enlarged images of flame kernel growth under normal and elevated gravities. (**a**) 9.5 vol.%, methane-air mixture, 1 G; (**b**) 9.5 vol.%, methane-air mixture, 50 G; (**c**) 12 vol.%, methane-air mixture, 1 G; and (**d**) 12 vol.%, methane-air mixture, 50 G

Figure 7 shows the effect of gravity level and mixture concentration on maximum explosion pressure P_{max}. In the cases of methane and propane as shown in Fig. 7b, c, it is confirmed that P_{max} decreases with increase of gravity. This is due to variation of flame spread depending on gravity level as mentioned above.

Similarity of Flame Spreading Influenced by Buoyancy Between Large- and Small-Scale-Confined Deflagrations

In order to examine dynamic similarity of flame spreading phenomena between large-scale deflagration under normal gravity and small-scale deflagration under elevated gravity, we compared explosion pressure records and flame motions. Figure 8 shows a typical result of comparison of the explosion pressure records

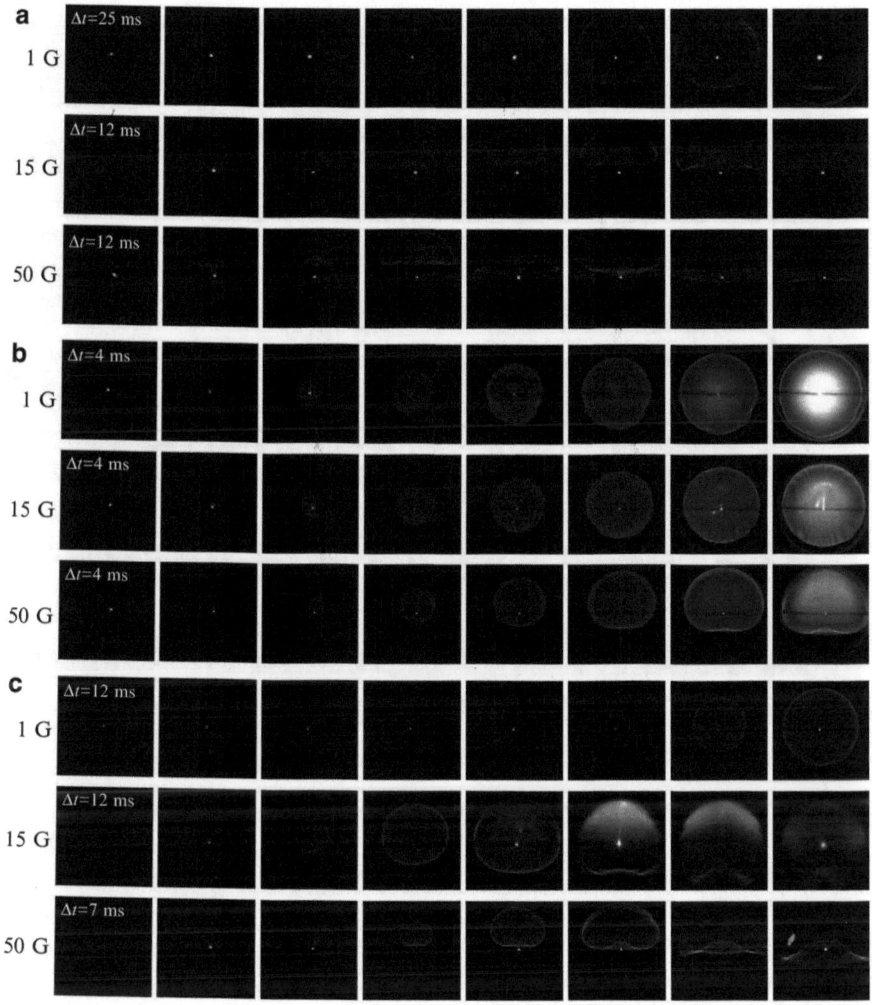

Fig. 5 Time series images of the propagating flames in methane-air mixtures. Δt means time interval of frames. (**a**) Lean mixture (upper, 6 vol.%; middle, 6.5 vol.%; lower, 6.5 vol.%), (**b**) stoichiometric mixture (9.5 vol.%), (**c**) rich mixture (12 vol.%)

obtained by the small- and large-scale deflagration experiments of methane-air mixtures. The horizontal and the vertical axes indicate dimensionless time and explosion pressure, respectively. P_{max} is maximum explosion pressure, and t_{max} is the time when maximum explosion pressure was recorded. The pressure records with flexion point as indicated by an arrow in Fig. 8 were typically obtained in the case where the flame spreading did not occur spherically due to buoyancy effect. It is confirmed that these two dimensionless pressure records are very similar. In addition, we checked the dimensionless arrival time of the flame front at the upper

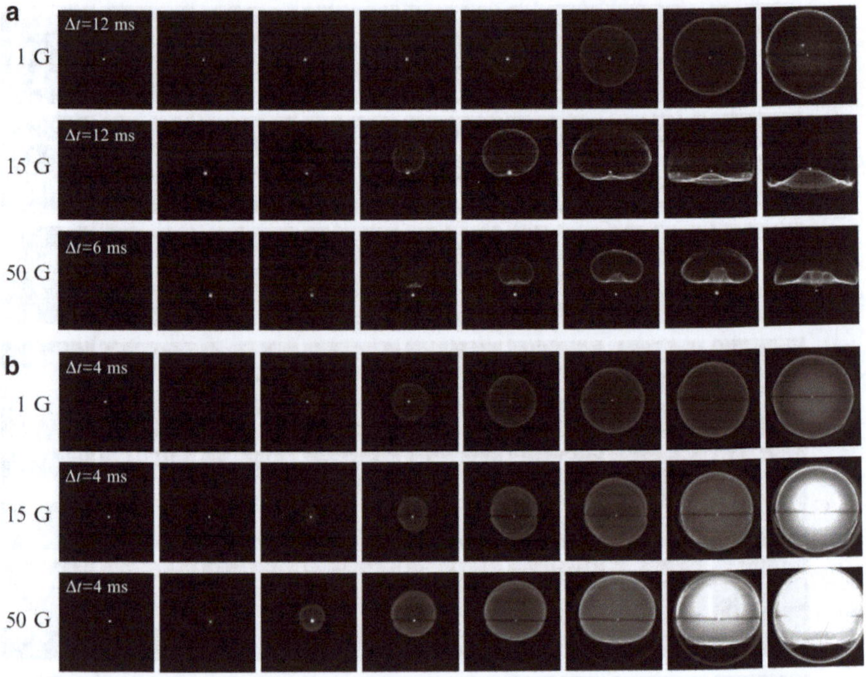

Fig. 6 Time series images of the propagating flames in propane-air mixtures. Δt means the time interval of frames. (**a**) Lean mixture (3 vol.%) and (**b**) stoichiometric mixture (4 vol.%)

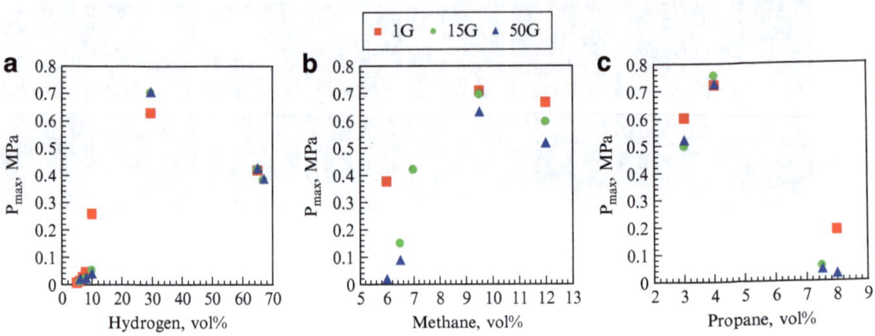

Fig. 7 Effect of gravity on the maximum explosion pressure. (**a**) Hydrogen-air mixture, (**b**) methane-air mixture, and (**c**) propane-air mixture

wall in the two deflagration experiments to compare the flame motion and found that both dimensionless times were almost the same, as shown in Fig. 8. Therefore, from the above comparison results, these two deflagration phenomena are in all likelihood similar.

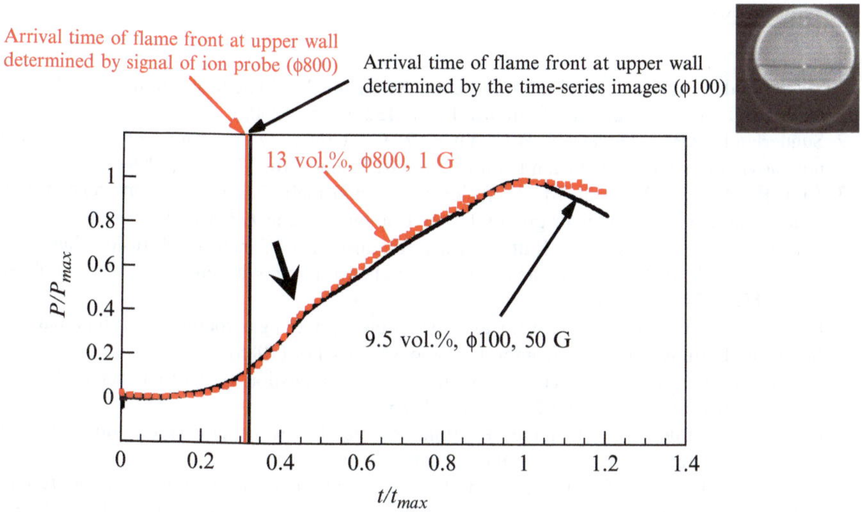

Fig. 8 Comparison of the explosion pressure records obtained by the small- and large-scale deflagration experiments of methane-air mixtures

Conclusions

We conducted flammable gas explosion experiments under gravity level of 1–50 G environments formed by a large-size centrifuge system in order to examine the effects of buoyancy on confined deflagration phenomena. The lower and upper explosion limits of hydrogen-air, methane-air, and propane-air mixtures were approximately the same regardless of the gravity level tested here. However, the shapes of the propagating flames dramatically changed with increase of gravity level due to the strong buoyancy effect especially in lean and rich mixture conditions with slow burning velocity. Moreover, we tried to evaluate dynamic similarity of flame spreading in large- and small-scale-confined deflagrations from explosion pressure records and flame motions. As a result, in the case where the flame spreading did not occur spherically due to buoyancy, it was found that similar flame spreading phenomena were observed in large-size deflagration of rich mixture under normal gravity and small-size deflagration of stoichiometric mixture under elevated gravity.

Acknowledgment The authors would like to thank Dr. Satoshi Tamate of the *National Institute of Occupational Safety and Health, Japan*, for his support of operating the centrifuge. Also, the authors would like to thank Professor Akihiko Ito and Associate Professor Hiroyuki Torikai of *Hirosaki University* for their valuable discussions and comments about the present work.

A part of this work was supported by a grant from *Shibaura Institute of Technology*.

References

1. Sunderland, P.B., Haylett, J.E., Urban, D.L., Nayagam, V.: Length of laminar jet diffusion flames under elevated gravity. Combust. Flame **152**, 60–68 (2008)
2. Sunderland, P.B., Mendelson, B.J., Yuan, Z.-G., Urban, D.L.: Shapes of buoyant and nonbuoyant laminar jet diffusion flames. Combust. Flame **116**, 376–386 (1999)
3. Sato, H., Amagai, K., Arai, M.: Diffusion flames and their flickering motions related with Froude numbers under various gravity levels. Combust. Flame **123**, 107–118 (2000)
4. Takahashi, F., Linteris, G.T., Katta, V.R.: Extinguishment of methane diffusion flames by carbon dioxide in Coflow Air and oxygen-enriched microgravity environments. Combust. Flame **155**, 37–53 (2008)
5. Ronney, P.D., Wachman, H.Y.: Effect of gravity on premixed gas combustion I: flammability limits and burning velocities. Combust. Flame **62**, 107–119 (1985)
6. Ronney, P.D.: Effect of gravity on premixed gas combustion II: ignition and extinction phenomena. Combust. Flame **62**, 121–133 (1985)
7. Durox, D., Baillot, F., Scouflaire, P., Prud'homme, R.: Some effects of gravity on the behaviour of premixed flames. Combust. Flame **82**, 66–74 (1990)
8. Most, J.M., Mandin, P., Chen, J., Joulain, P.: Influence of gravity and pressure on pool fire-type diffusion flames. In: 26th Symposium (International) on Combustion, pp. 1311–1317 (1996)
9. Ross, H.D., Sotos, R.G.: An investigation of flame spread over shallow liquid pools in microgravity and nonair environments. In: 23th Symposium (International) on Combustion, pp. 1649–1655 (1991)
10. Kozo Saito, H., Amagai, K., Arai, M.: Scale modeling of puffing frequencies in pool fires related with Froude number. In: Progress in Scale Modeling, Part I, Springer, pp. 133–147 (2008)
11. Kumagai, S., Isoda, H.: Combustion of fuel droplets in a falling chamber. In: 6th Symposium (International) on Combustion, pp. 726–731 (1957)
12. Mikami, M., Oyagi, H., Kojima, N., Kikuchi, M., Wakashima, Y., Yoda, S.: Microgravity experiments on flame spread along fuel-droplet arrays using a new droplet-generation technique. Combust. Flame **141**, 241–252 (2005)
13. Nakamura, Y., Yamashita, H., Takeno, T., Kushida, G.: Effects of gravity and ambient oxygen on a gas-phase ignition over a heated solid fuel. Combust. Flame **120**, 34–48 (2000)
14. Mell, W.E., Kashiwagi, T.: Dimensional effects on the transition from ignition to flame spread in microgravity. In: 27th Symposium (International) on Combustion, pp. 2635–2641 (1998)
15. Law, C.K.: Dynamics of stretched flames. In: 22nd Symposium (International) on Combustion, pp. 1381–1402 (1989)
16. Groff, E.G.: The cellular nature of confined spherical propane-air flames. Combust. Flame **48**, 51–62 (1982)

Section B
Fire and Explosion - Extinguishment Characteristics of a Jet Diffusion Flame with Inert-Gas Vortex Ring

Yuki Chiba, Hiroyuki Torikai, and Akihiko Ito

Abstract For firefighting, we propose the vortex-ring transport method using a gaseous fire-extinguishing agent. The vortex ring formed with a gaseous extinguishing agent has a possibility to transport the extinguishing gas more effectively and over longer distance than jet flows which issue from conventional extinguishers. To clarify the extinguishing characteristics of the inert-gas vortex ring, blowout experiments using a methane-air jet diffusion flame have been performed. Nitrogen, carbon dioxide, argon, and air are used for forming a vortex ring with a round orifice. The extinguishing velocity limit has been measured. The extinguishing velocity limit is defined as the lowest displacement velocity of the vortex ring which can blow the jet diffusion flame off perfectly. From the experimental results, it is found that all inert-gas vortex rings indicate the lower extinguishing velocity limit than the air vortex ring. This means that the inert-gas vortex ring can travel in air while keeping its extinguishing ability. All inert-gas vortex rings retain extinguishing effectiveness over a distance longer than seven times that of the orifice diameter. Significantly, after a distance larger than four times the orifice diameter, the extinguishing velocity limits of all inert gases show constant values independent of the traveling distance of the vortex ring. Furthermore, the order of the extinguishing effectiveness of the inert-gas vortex ring is $CO_2 > Ar > N_2$. This result is different from the extinguishing effectiveness ranking determined by a cup-burner method, and it is considered that the extinguishing effectiveness ranking is influenced by the transport process of the inert gas to the flame. When the extinguishing velocity limit is scaled by the Peclet number which is defined as the ratio between displacement velocity of vortex ring at extinguishment limit and mass diffusion rate of inert gas, the data of the Peclet number can be expressed as a single curve as a function of the normalized traveled distance of the vortex ring.

Y. Chiba • H. Torikai (✉) • A. Ito
Graduate School of Science and Technology, Hirosaki University, 3- Bunkyo-cho, Hirosaki, Aomori 036-8561, Japan
e-mail: torikai@cc.hirosaki-u.ac.jp

K. Saito et al. (eds.), *Progress in Scale Modeling, Volume II*,
DOI 10.1007/978-3-319-10308-2_9

Nomenclature

d	Diameter
D	Mass diffusivity in oxygen gas
P	Blowout probability
Pe	Peclet number of vortex ring at extinguishing velocity limit ($\equiv V_{ex}\, d_o/D$)
U	Cross-sectional mean flow velocity of methane gas
V	Displacement velocity of vortex ring
Z	Distance from orifice plate

Subscripts

b	Burner ring
o	Orifice ring
v	Vortex ring
ex	Extinguishing velocity limit

Introduction

The donut-shaped vortex called a vortex ring is easily generated by propelling a certain volume of a fluid from a round orifice with one application of force, such as with an air cannon. The vortex ring travels in a straight line through the still surrounding fluid over a distance longer than a hundred times the orifice diameter until decay [1]. The vortex ring can trap fine solid particles, droplets, and bubbles inside and entrain them without losing the basic features of the vortex ring [1–4]. When a vortex ring is made of a gaseous extinguishing agent, the gaseous agent can also be carried by the vortex ring. If the vortex ring is used as a means to transport extinguishing agents to a fire source in firefighting, the targeting delivery method of extinguishing agents will be developed like a drug delivery system which delivers precisely the required amount of drugs to the targeted site in the human body for the necessary period of time [5]. If the extinguishing agent delivery method by using a vortex ring is established, the needed amount of extinguishing agents can be delivered to the targeted fire area at the appropriate timing and more effectively than jet flows of extinguishing agents, such as issuing from conventional extinguishers. However, there are few researches on fire extinguishment with a vortex ring [1], and the fundamental blowout characters of the extinguishing-agent vortex ring have not been clear enough. In particular, the extinguishing-agent vortex ring may lose its fire-extinguishing ability as it travels through the surrounding air because the extinguishing gas diffuses into and mixes with the surrounding air. Therefore, the limitation of the effective range of the extinguishing-agent vortex ring has to be clarified for establishing the new extinguishing-agent delivery method.

Generally, the extinguishing effectiveness of an inert gas is evaluated by a cup-burner method [6]. The cup-burner method determines the minimum extinguishing concentration, which means the lowest inert-gas concentration to be able to blow off a laminar diffusion flame formed on the coflowing burner. The extinguishing effectiveness of CO_2, N_2, and Ar depends on the magnitude of their heat capacities, and the extinguishing effectiveness ranking is $CO_2 > N_2 > Ar$ [6]. In the cup-burner test, the minimum extinguishing concentration is measured by slowly increasing the inert-gas concentration in the oxidizer flow [6]. On the contrary, in practical firefighting, the inert gas is continuously ejected at high velocity from extinguisher nozzles and supplied to the targeted fire area with the velocity fluctuation and the entrained air. Therefore, the transport process of the inert gas to the flame zone in the cup-burner test is fundamentally quasi-steady. On the other hand, the transport process of the inert gas in actual firefighting is highly unsteady. Based on this consideration, it is also important to clarify the influence of the transport process of inert gases from the extinguishing apparatus to flames on the extinguishing effectiveness ranking. By changing the traveled distance and the displacement speed of the inert-gas vortex ring, it is possible to investigate the influence of the transporting process of the inert gas on the extinguishing effectiveness ranking.

In the present study, to investigate the extinguishing characteristics of the vortex ring formed with gaseous extinguishing agents, blowout experiments using a methane-air jet diffusion flame have been performed. Carbon dioxide, nitrogen, and argon gases are used as gaseous extinguishing agents. The blowout probability has been measured by varying the displacement velocity of the vortex ring and the distance between the orifice position and the flame base location. From the experiments, we will clarify the extinguishing velocity limit, beyond which the inert-gas vortex ring can blow the jet diffusion flame off perfectly and the influence of the transport process of the inert gas on the extinguishing effectiveness ranking. Finally, we will determine the scaling law for the blowoff limit of the extinguishment method with the inert-gas vortex ring.

Experimental Setup and Method

Figure 1 shows the schematic of the experimental setup. A methane-air jet diffusion flame was formed with an L-shaped stainless tube burner. The outer and inner diameters of the stainless tube were 3 mm and 2.6 mm, respectively. The axis of the tube burner was set parallel to the direction of the gravitational acceleration, and the vertical part of the L-shaped tube burner has the length of 30 mm. The vortex ring was supplied from just under the burner. Therefore, before the vortex ring reached the flame, the vortex ring was always interfered with by the burner tube. The laminar diffusion flame was formed at the methane flow rate of 1.045 l/min, at which the cross-sectional mean methane flow velocity, U, became 3.28 m/s, and the

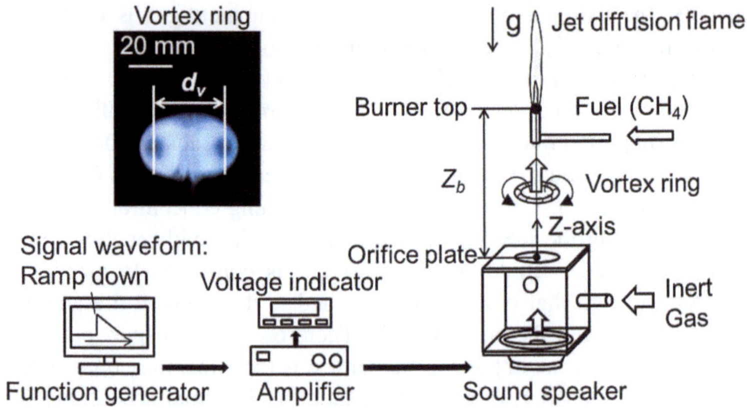

Fig. 1 Experimental setup

average flame height was 190 mm. The Reynolds number of the methane flow in the burner tube was 505.

The toroidal vortex ring was ejected from the acrylic box with a round orifice and a sound speaker (VISATON, WS13E). The visualized image of the vortex ring is shown in Fig. 1. A ramp-down signal generated from the function generator (National Instruments, NI-5401) was amplified by the amplifier (SONY, TA-FE400R) and transmitted to the sound speaker. To vary the displacement velocity of the vortex ring, the input voltage to the sound speaker was controlled with the amplifier and measured with a voltmeter (ADVANTEST, R6552). The orifice diameter, d_o, was 47 mm. The vortex rings were formed with air, CO_2, N_2, and Ar gases. The inert-gas vortex rings were generated as follows. First, the acrylic box, whose volume was 730 mL, was filled with inert gas at a flow rate of 1,000 mL/min. After 1 min, the sound speaker was driven with the ramp-down signal. The center axis of the orifice was adjusted to the center axis of the tube burner. The distance between the orifice position and the burner top, Z_b, was varied as an experimental parameter.

To clarify the moving characteristics of the vortex ring, it was visualized with a metal halide lamp (Photron, HVC-UL) and silicon oil droplets. The lamp had a power of 250 W. The silicon oil droplets were used as light-scattering particles. Silicon oil (Momentive, TSF-451-50) was atomized with a glass nebulizer (Omron, NE-C10-11), and the droplets were seeded into air and inert gases. The silicon oil droplet had the nominal diameter of 2 μm [7]. The visualized images were recorded with a high-speed digital camera (Casio, EX-F1) at the exposure time of 1/320 s and frame rate of 300 fps. From the analysis of the recorded images, the displacement velocity of the vortex ring, V, and the diameter of the vortex ring, d_v, were measured in the range of $Z = 50$–350 mm. The vortex-ring diameter was defined as the length from center to center of the vortex core in the vortex ring as shown in Fig. 1. Figure 2 shows the displacement velocity of the vortex ring and the vortex-ring

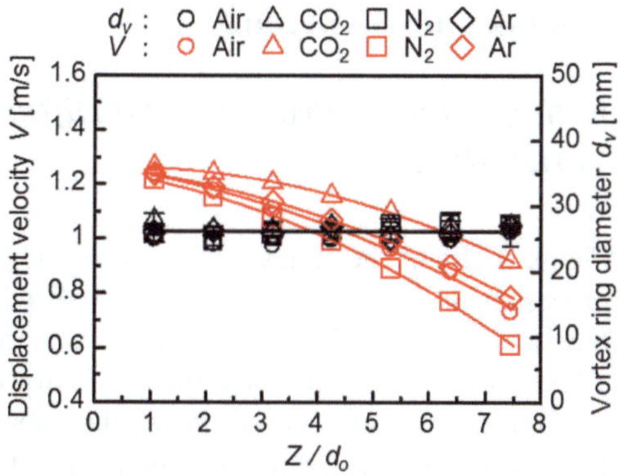

Fig. 2 Relationship between displacement velocity of vortex ring V, vortex-ring diameter d_v, and nondimensional distance Z/d_o at the input voltage of 2 V (*Red color* plots show the displacement velocity of vortex ring and *black color* plots show vortex-ring diameter)

diameter at the speaker input voltage of 2 V. The graph shows typical characters of the vortex ring in this experiment. The horizontal axis indicates the nondimensional distance, Z/d_o. From the graph, all displacement velocities of the vortex rings decrease gradually as the traveled distance of the vortex ring increases. On the other hand, the diameters of the vortex rings show a nearly constant value of about 26 mm regardless of gas species. The d_v increased with the increase in the speaker input voltage.

The extinguishment experiments with the air and inert-gas vortex rings were performed in the following way. First, the location of the burner top was set at a certain distance from the orifice, and the stable jet diffusion flame was formed. Second, the vortex ring was generated and collided with the flame base. Then, we checked whether the flame was extinguished or not visually. When the flame was blown off perfectly, we recorded it as successful extinguishment by the vortex ring. The probability of extinguishment, P, was computed as the ratio of the number of successful extinguishments to the number of total experiments (20). Moreover, the displacement velocity of the vortex ring at the burner top location was calculated from the relationship between the displacement velocity of vortex ring and its traveled distance as shown in Fig. 2.

In the extinguishment process, the vortex rings were visualized with laser-light sheet method, by which the cross-sectional area of the vortex ring could be observed. A diode-pumped Nd:YAG/YVO_4 laser (Kato-koken co., Ltd., PIV Laser G450) was used as a light source. The laser power and wavelength were 450 mW and 532 nm, respectively. The sheet thickness was approximately 1 mm. The images were recorded with a high-speed digital camera (CASIO, EX-FH25) at the exposure time of 1/250 s and frame rate of 240 fps. The light-scattering particles were silicon oil droplets.

Results and Discussion

Extinguishment Process of Jet Diffusion Flame with Inert-Gas Vortex Ring

Figure 3 shows the series of the sequential images of the typical extinguishment process with the vortex ring. The normalized distance from the orifice plate to the burner top was 4.3. The displacement velocity of the vortex ring at $Z_b/d_o = 4.3$ was 1.59 m/s, at which the jet diffusion flame was blown off perfectly, that is, the blowout probability showed unity. The cross-sectional area of the vortex rings in Fig. 3 was visualized with laser-light sheet method. The green color region in the visualized images is considered to show inert gas.

At 0 ms in Fig. 3, the vortex ring coming up from the bottom of the image just arrives at the burner top, that is, the flame base. At the same time, it can be seen that the base flame starts to lift off from the burner rim locally. Although the vortex ring already interacts with the burner tube at 30 mm below the burner top, the vortex ring keeps the donut-shape to the flame base. At 17 ms, the vortex ring moves further and the lifted height of the flame increases more. Moreover, the vortex-ring structure starts to collapse. At 25 ms, the blue flame part in the base flame region disappears but the luminous flame part remains. At 75 ms, the luminous flame area decreases and finally the extinguishment of the diffusion flame is achieved.

Figure 4 shows the typical failure case of the inert-gas vortex-ring extinguishment. The displacement velocity of the vortex ring was 1.33 m/s. Other conditions are as described in Fig. 3. At 0 ms of Fig. 4, the vortex ring reaches the flame base and the flame base is lifting off from the burner rim. At 42 ms, the CO_2 vortex ring loses its vortex structure and almost all of the CO_2 gas has diminished. At the same time, the jet diffusion flame is divided into the luminous flame region and the blue flame region. At between 138 ms and 213 ms, the blue flame region is blown off to the downstream and then stabilized as a lifted flame. Notably, the luminosity of the

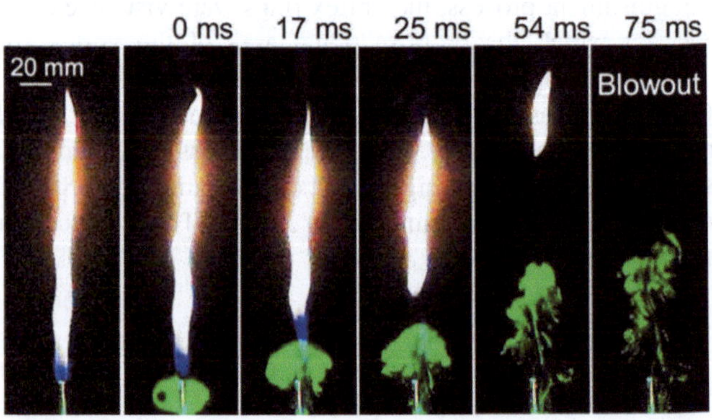

Fig. 3 Blowout process of the jet diffusion flame with CO_2 gas vortex ring at $V = 1.59$ m/s and $Z_b/d_o = 4.3$

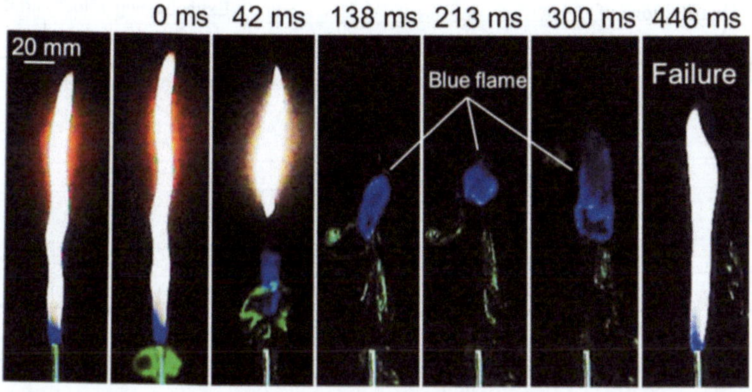

Fig. 4 Failure case of the flame extinguishment with CO_2 gas vortex ring at $V = 1.33$ m/s and $Z_b/d_o = 4.3$

blue flame increases at the edge flame region. Furthermore, from 300 ms, the base region of the blue flame starts to propagate upstream. The flame propagation is caused in that the pre-mixture layer is formed by interdiffusion of the methane gas and the surrounding air in the upstream part of the lifted flame. When the propagation speed of the lifted flame becomes larger than the local incoming flow velocity of premixed flammable gas to the lifted flame base, the flame base can spread toward the upstream [1]. Finally, the jet diffusion flame is restabilized on the burner at 446 ms, and the vortex-ring extinguishment is not achieved.

Thus, in this experiment, the flame base always lifts off due to interaction of the vortex ring regardless of the success or failure of the flame blowout and also the inert-gas species. Therefore, it is said that the behavior of the flame base after the liftoff determines whether the vortex-ring extinguishment is achieved or not. In other words, when the lifted flame is established downstream, the vortex-ring extinguishment is not accomplished.

Distribution of Extinguishment Probability and Extinguishing Velocity Limit

Figure 5 shows the typical probability profile of the vortex-ring extinguishment as a function of the displacement velocity of the vortex ring at the burner rim. The blowout probability distribution of the CO_2 vortex ring is compared with that of the air vortex ring. From the graph, it is seen that the blowout probability increases with increasing the vortex-ring velocity, and the flame-extinguishing region, in which the blowout probability always shows unity, appears for both vortex rings. Moreover, we can define the limitation of the extinguishment region as the extinguishing velocity limit, V_{ex}, which is the lowest velocity of the vortex ring in the extinguishment region. From the limit values, the extinguishment ability of the vortex ring can

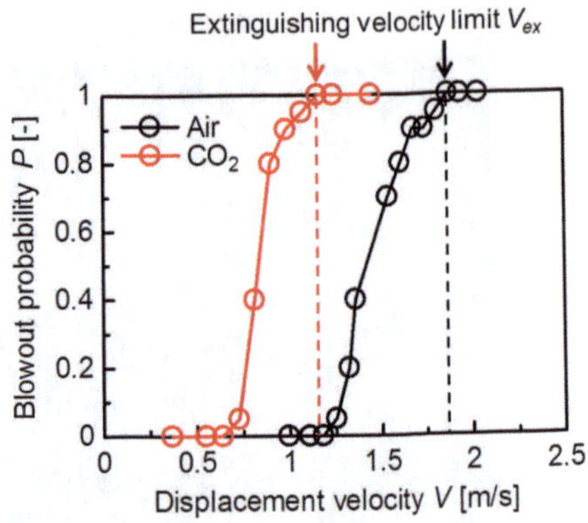

Fig. 5 Distributions of extinguishment probability at $Z_b/d_o = 2.1$

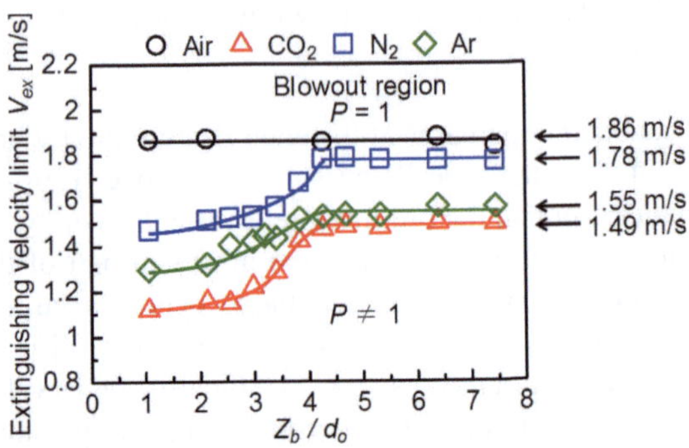

Fig. 6 Distribution of the extinguishing velocity limit, V_{ex}, as a function of nondimensional distance, Z_b/d_o

be evaluated. Figure 5 shows that the CO_2 vortex ring has a lower value of the extinguishing velocity limit than the air vortex ring, that is, the CO_2 vortex ring is superior to the air vortex ring in terms of the ability to extinguish the jet diffusion flame. In case of air, the flame is blown off purely due to the fluid dynamical effects of the vortex ring, such as stretching the flame. On the other hand, in the CO_2 case, the flame extinguishment is caused by both the fluid dynamical effect and the inert-gas effect which dilutes oxygen concentration in the combustion zone and also absorbs heat from the reaction zone [6]. Therefore, it is considered that the difference of the extinguishing velocity limit between inert gas and air indicates the magnitude of the inert-gas extinguishing effect. Next, we will focus on and discuss the extinguishing velocity limit of the inert-gas vortex rings.

Figure 6 shows the relationship between the extinguishing velocity limit, V_{ex}, and the normalized distance by the orifice diameter, Z_b/d_o. For the air vortex ring, the extinguishing velocity limit indicates a constant value (1.86 m/s) and is independent of Z_b/d_o. This means that the air vortex ring having the higher moving velocity at the burner rim than the extinguishing velocity limit always can blow off both attached and lifted flames regardless of the burner position. In other words, when the displacement velocity of the air vortex ring is larger than the extinguishing velocity limit, the incoming flow rate toward the flame base induced by impact of the air vortex ring always overcomes the propagation speed of the lifted flame.

For inert gases, all the vortex rings always show lower values of the extinguishing velocity limit than does the air vortex ring. This may occur because the propagation speeds of the flames lifted by the impact of the inert-gas vortex rings are reduced because of the extinguishing effects of the inert gas, such as oxygen dilution and heat absorption [1, 8]. Moreover, it is also found that the inert-gas vortex ring can retain effective concentration to extinguish the flame even though it travels in air over a distance longer than seven times that of the orifice diameter. The profiles of the extinguishing velocity limits for all inert gases are similar. For Z_b/d_o less than 4.3, the extinguishing velocity limits of the inert-gas vortex rings increase as the traveled distances increase. Therefore, in that range, the inert-gas concentration inside the vortex ring decreases as the vortex ring progresses in air. For Z_b/d_o larger than 4.3, all inert-gas vortex rings show constant values of the extinguishing velocity limits. From the results, it can be said that the vortex ring transports the gaseous extinguishing agent while keeping its concentration over a distance longer than seven times that of the orifice diameter.

For the CO_2, Ar, and N_2 vortex rings, the constant values of V_{ex} are 1.49 m/s, 1.55 m/s, and 1.78 m/s, respectively, as shown in Fig. 6. As a result, the order of the effectiveness of the inert-gas vortex ring in terms of flame extinguishment is $CO_2 > Ar > N_2$. This order, however, disagrees with the extinguishing effectiveness ranking of inert gas determined by the cup-burner test [6]. In case of the cup-burner test, the order is $CO_2 > N_2 > Ar$, and the extinguishing effectiveness is in inverse proportion to the magnitude of heat capacity per unit volume [6].

The variation of the extinguishing effectiveness ranking between the cup-burner test and the vortex-ring extinguishment experiment is thought to be caused by the difference in the transport process of the inert gas to the combustion zone.

Scaling of the Extinguishing Velocity Limit of Inert-Gas Vortex Ring

The ranking order of the extinguishing effectiveness of the inert gases in the vortex-ring extinguishment is $CO_2 > Ar > N_2$ and is considered to depend on the delivery process of the extinguishing gaseous agents to the targeted flame. In order to scale the extinguishing velocity limits of the inert-gas vortex rings, we introduce the Peclet number which represents the ratio between convective transport rate and diffusive transport rate. In the present study, Peclet number is defined as follows:

$$Pe \equiv V_{ex} \times d_o/D \tag{1}$$

where V_{ex} is the value of extinguishing velocity limit, d_o is the orifice diameter which may express the diameter of vortex ring, and D is mass diffusivity of the inert gas in oxygen at temperature of 298 K and pressure of 0.1 MPa. The vertical axis of Fig. 6 is scaled with the Peclet number of the inert-gas vortex ring at extinguishing velocity limit.

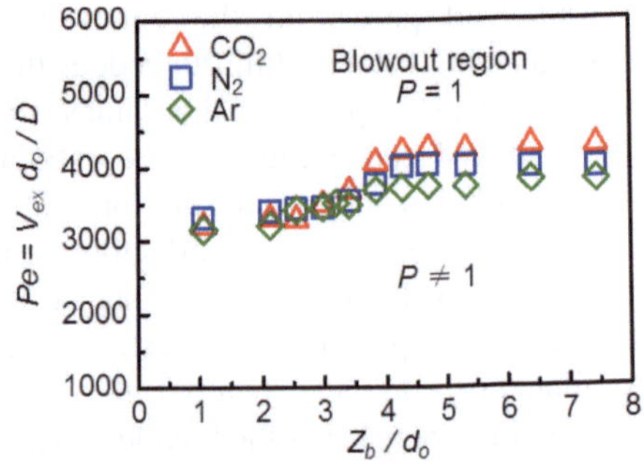

Fig. 7 Peclet number profiles of the vortex rings at each extinguishing velocity limit as a function of Z_b/d_o

As shown in Fig. 7, all the Peclet number profiles are reduced to a single curve. This means that when the inert gas has a lower diffusion rate, that is, the inert gas does not mix faster with the surround air, the vortex ring can maintain the inert-gas extinguishing ability even though the vortex ring travels at slower speed. On the contrary, when the inert gas has a higher diffusion rate, the inert-gas vortex ring has to have a larger displacement velocity and reach the flame before reducing the inert-gas concentration inside the vortex ring below its flame extinguishment limit. As a result, it can be said that the ranking order of the inert-gas extinguishing effectiveness in this study depends on the delivery process of the gaseous extinguishing agent using the vortex ring.

Conclusion

In this paper, we propose the vortex-ring transport method of gaseous fire-extinguishing agents. A vortex ring is easily formed by ejecting various gases out of an orifice with a single application of force so that the vortex travels in a straight line in still air. The vortex ring made of gaseous extinguishing agents could carry the agents effectively over longer distance than jet flows issuing from conventional extinguishers. By using the gaseous agent vortex ring, the targeting delivery of the extinguishing agents to fire area would be developed. The extinguishing-agent vortex ring, however, may lose fire-extinguishing ability as it travels through the atmosphere because the gaseous agent diffuses into the surrounding air.

In order to clarify the extinguishing characteristics of a vortex ring formed with an inert gas, blowout experiments with a methane-air jet diffusion flame have been performed. Nitrogen, carbon dioxide, and argon gases are used as a gaseous extinguishing agent. The extinguishing velocity limit—the lowest displacement velocity of vortex ring at the flame base location able to blow the jet diffusion flame off perfectly—is measured. By changing the distance from the burner to the orifice, the relationship between the extinguishing velocity limit and the traveled distance of vortex ring is clarified, and the extinguishing velocity limits of inert-gas vortex rings are compared with that of air vortex ring.

(continued)

(continued)

As a result, it is found that all inert-gas vortex rings show a lower displacement velocity at the extinguishment limit than the air vortex ring; that is, the inert-gas vortex ring travels in space while keeping its extinguishing effectiveness. All inert-gas vortex rings can keep the flame-extinguishing effectiveness over a distance longer than seven times that of the orifice diameter. Significantly, after the distance larger than four times that of the orifice diameter, the extinguishing velocities indicate constant values independent of traveling distance of vortex ring. Furthermore, the order of the effectiveness of the inert-gas vortex ring in extinguishing is $CO_2 > Ar > N_2$, which is different from the extinguishing effectiveness ranking of inert gas determined by a cup-burner method. Thus, it is found that the ranking order of the inert-gas extinguishing effectiveness is changed depending on the delivery method of gaseous extinguishing agents to flames.

In order to scale the extinguishing velocity limit, the Peclet number which shows the ratio of the convective transport rate due to moving of the vortex ring and the diffusive transport rate of inert gas is introduced. All data of the Peclet numbers at the extinguishing velocity limits lie on a single curve as a function of the normalized traveled distance of the vortex ring. From the results, it can be said that by using the inert gas whose diffusion rate is lower, the inert-gas vortex ring maintains extinguishing ability even though the vortex ring moves at lower displacement velocity. On the contrary, for inert gas having a higher diffusion rate, the vortex ring has to have a larger displacement velocity and reach the flame before reducing the inert-gas concentration in the vortex ring under its flame extinguishment limit.

References

1. Akhmetov, D.G.: Vortex rings, pp. 1–142 (2009)
2. Domon, K., Ishihara, O., Watanabe, S.: Mass transport by a vortex ring. J. Phys. Soc. Jpn. **69**(1), 120–123 (2000)
3. Yagami, H., Uchiyama, T.: Numerical simulation for the transport of solid particles with a vortex ring. Adv. Powder Technol. **22**, 115–123 (2011)
4. Uchiyama, T., Yagami, H.: Numerical simulation for the collision between a vortex ring and solid particles. Powder Technol. **188**, 73–80 (2008)
5. Rajan, K.V., Sanjay, G.: Current status of drug delivery technologies and future directions. Pharm. Technol. On-Line **25**(2), 1–14 (2001)
6. Takahashi, F., Gregory, T.L., Viswanath, R.K.: Extinguishment mechanisms of coflow diffusion flames in a cup-burner apparatus. Proc. Combust. Inst. **31**, 2721–2729 (2007)
7. Ueda, T., Yahagi, Y., Mizomoto, M.: Mie scattering with silicon oil droplets in combustion field. Trans. Jpn. Soc. Mech. Eng. B **57**(541), 395–399 (1991)
8. Qiaoa, L., Gana, Y., Nishiiea, T., Dahmb, W.J.A., Oranc, E.S.: Extinction of premixed methane/air flames in microgravity by diluents: effects of radiation and Lewis number. Combust. Flame **157**(8), 1446–1455 (2010)

Section B
Fire and Explosion - Effect of Gravity on Flame Spread Along a Thin Combustible Solid for Different Sample Orientations in Opposed Flow

Hiroyuki Torikai, Akihiko Ito, and Yuji Kudo

Abstract In order to secure fire safety over the entire period of a manned space mission, gaining a systematic understanding of the effects of gravity on flame spread is important. In this study, opposed-flow flame spread along a thermally thin combustible solid for different sample orientations (−20° downward, horizontal, and +20° upward) was experimentally investigated by changing the ambient oxygen concentration and gravity level. The flame spread rate decreases with decreasing oxygen concentration under normal gravity, and its rate at 18 % oxygen concentration is equivalent to that at 21 % oxygen concentration under microgravity. The downward flame spread rate decreases with an increase in gravity. In contrast, the horizontal and the +20° upward flame spread rates clearly increase as the gravity level increases. The flame spread rate varies remarkably with sample orientation in a supergravity environment. To clarify the effect of gravity on flame spread, the relation between the non-dimensional flame spread rate and the Rayleigh number was examined. The Ra number both for horizontal and upward flame spread increases with increasing gravity, while the Ra number for downward flame spread decreases slightly with a decrease in gravity. The non-dimensional flame spread rate is almost constant under normal and supergravity conditions for Ra numbers less than 10^3 and is equivalent to that under microgravity. When the Ra number is greater than 10^3, the non-dimensional flame spread rate increases with increasing Ra number and is proportional to $Ra^{1/3}$.

H. Torikai • A. Ito (✉)
Graduate School of Science and Technology, Hirosaki University, Hirosaki, Aomori 036-8561, Japan
e-mail: aito@cc.hirosaki-u.ac.jp

Y. Kudo
Department of Mechanical Engineering, Hachinohe Institute of Technology, Hachinohe, Aomori 031-8501, Japan

K. Saito et al. (eds.), *Progress in Scale Modeling, Volume II*,
DOI 10.1007/978-3-319-10308-2_10

Nomenclature

A	The thermal diffusivity of oxidizer
a_s	The thermal diffusivity of paper
g	The gravity level $g = 1$ on Earth
g	Acceleration due to gravity
L	The flame length defined as spreading base flame
Ra	The Rayleigh number
T_f	The averaged temperature in flame zone
T_0	The ambient temperature
U	The averaged opposed airflow velocity
V	The flame spread rate
V_n	The non-dimensional flame spread rate
X_{O2}	The volumetric oxygen concentration

Greek Symbols

β	The thermal coefficient of volume expansion
ν	The kinematic viscosity
θ	The sample orientation angle
δ	The sample thickness

Introduction

In order to promote development in outer space, it is necessary to establish fire safety assessment under a microgravity environment such as in a spacecraft or space station. A fire should be suppressed in the initial stages in order to protect human life. To predict events in the initial stages of a growing fire, it is important to understand the propagation mechanism of a flame that is spreading along a combustible solid. The flame spread rate differs depending on the surrounding conditions, such as the gravity level, the ambient oxygen concentration, and the sample orientation angle. The microgravity environment strongly influences the processes of combustion and consequently influences fire safety itself. Although establishing fire safety is absolutely essential in human-occupied spacecraft and in space missions, fire characteristics under microgravity conditions are not entirely clear. From the results of early tests in a static chamber conducted on the Skylab space station, the observed flame spread rate is slower than the corresponding flame spread rate on the ground [1]. Some research on flame spread over thin sheet combustible materials in the low gravity environments of drop towers, parabolic-trajectory airplanes, the Space Shuttle, and the Mir space station has provided qualitative findings [2–12]. Based on these results, it has been commonly assumed up until now that acceptance testing in normal gravity provides an acceptable fire safety margin.

When considering the launch of a spacecraft, it is necessary to clarify the fire characteristics not only in a microgravity environment but also under supergravity conditions (elevated gravity level) such as during launch or liftoff. Also when considering manned Martian or Lunar missions, it is necessary to clarify the fire characteristics in a partial-gravity environment. The current understanding of the effect of partial or elevated gravity on fire behavior is based on limited experiments. Parabolic-trajectory airplane tests on a downward-spreading flame show that the paper-thin flame spread rate attains its maximum in the partial-gravity range [13].

In order to secure fire safety over the entire period of a manned space mission, gaining a systematic understanding of the effects of gravity on flame spread is important. In a normal-gravity environment, the induced airflow assisted by buoyancy helps to supply oxygen to the flame leading edge, promoting flame spread, while in microgravity, the oxygen supply is limited due to the absence of buoyancy. As a result, for flame spread under microgravity conditions, the flame temperature decreases, resulting in a decrease in the heat transfer rate to the unburned fuel surface. Comparing the flame spread under various gravity conditions with that in various oxygen concentrations, we found a similarity between the effects of gravity and those of restricted oxygen concentrations [14]. Under supergravity, the induced airflow assisted by buoyancy will supply more oxygen to the flame leading edge, promoting flame spread. However, this increase in the induced airflow rate may lead to a decrease in residence time, and this may cause flame blow off. Furthermore, due to an increase in buoyancy, the effect of the sample orientation angle relative to gravity on the flame spread rate may become very marked.

In this study, the flame spread experiments along a thermally thin combustible solid at different sample orientations (−20° downward, horizontal, and +20° upward) were performed by changing the ambient oxygen concentration and gravity level. As the effect of gravity on flame spread is due to the variation in the buoyancy force in the gas phase with gravity level, we investigated the relation between the non-dimensional flame spread rate and the Rayleigh number, which includes the acceleration due to gravity and the length of the base of the spreading flame.

Experimental Apparatus and Method

The microgravity experiments were conducted using the 4.5-s drop tower facility at MGLAB (Micro Gravity Laboratory of Japan). A 4.5-s period of microgravity is obtained by allowing the experiment to free-fall in an evacuated tube through a distance of about 150 m. Figure 1 shows a schematic of our experimental package used for the microgravity test. The package housed the test rack, containing a wind tunnel, a sample holder, a hot-wire igniter, three video cameras, and a gas-control system. A filter paper (Advantec MFS Inc., No.131, thickness $\delta = 0.25$ mm, thermal diffusivity $a_s = 0.0824$ mm^2/s, density 140 g/m^2) was used as the sample paper. Samples were dried in desiccators for a minimum of 24 h to remove moisture. A sample holder (140 mm × 70 mm) was located inside the wind tunnel (7 × 7 cm,

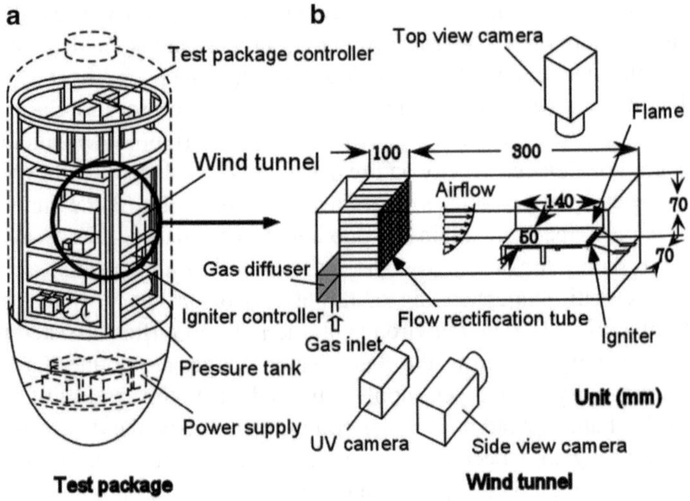

Fig. 1 Experimental apparatus used in the MGLab 4.5-s drop tower test; (**a**) test package, (**b**) wind tunnel and test section

30 cm long, Fig. 1b). The wind tunnel can provide a steady, laminar, forced gas flow. The average gas flow rate was 0.1 m/s. A hot-wire igniter was set up at the end of the sample holder and 1 mm under the sample surface. A top-view camera and two side-view cameras recorded the spreading flame. One side-view camera recorded the visible flame and another recorded an ultraviolet image in order to record the blue flame under microgravity. A rectification section was installed ahead of the sample holder in order to provide a uniform laminar flow over the test section. Compressed dry air was diluted with nitrogen to form a known low-oxygen concentration flow of gas and was introduced from a gas-control system to the wind tunnel. Prior to each flame spread test, the oxygen concentration of the flow gas was measured by the Shimadzu POT-101 O_2 concentration meter (accuracy ±0.5 %). The velocity profile of the flow of gas across the tray was measured by traversing an anemometer (Kanomax Model 6543) 5 mm above the sample surface at several different locations along the tray to establish the average velocity, $U = 0.1$ m/s ±5 % along the center line of the tray, and 5 cm above the fuel surface across the fuel tray. The test rack was contained in a sealed test package filled with dry air at 1 atm. Once a 0.1 m/s gas flow opposing the direction of the flame spread was started, ignition occurred just before the package was dropped. The igniter was engaged until a flame appeared.

The supergravity experiments were conducted using the rotating arm type elevated gravity test facility at JAXA (Japan Aerospace Exploration Agency) located at the Tsukuba Space Center. The test facility has a 7 m long rotating arm and the test rack was attached to the end of the rotating arm. Figure 2 shows the setup of the test rack on the mounting table of the rotation arm. The same test rack was used for the microgravity test and the supergravity test. The elevated gravity created by the spinning of the rotation arm is the resultant of gravity and centrifugal

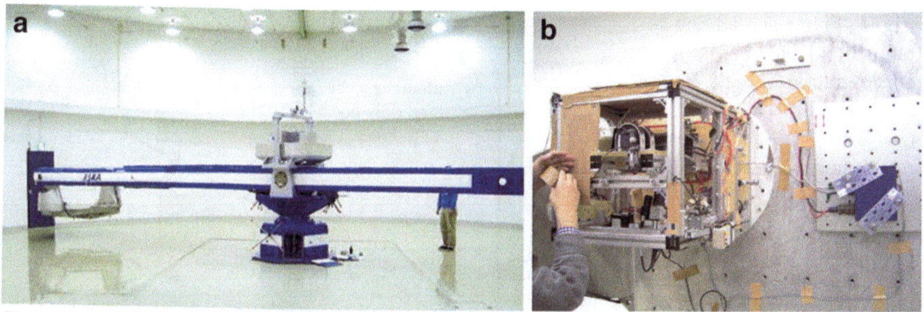

Rotating arm type elevated gravity test facility (JAXA) Setting up of test rack

Fig. 2 Setting up the test rack on a mounting table on the rotation arm (JAXA Tsukuba space center)

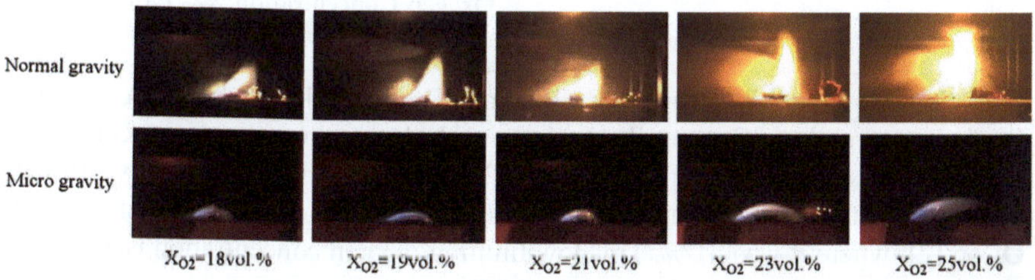

Fig. 3 *Upper line*: five sequential color video images showing a flame spreading over thin filter paper under five different oxygen concentrations in normal gravity. *Bottom line*: five sequential color video images showing a flame spreading over thin filter paper under five different oxygen concentrations in microgravity

force. The wind tunnel was inclined in order to apply the elevated gravity perpendicularly to the sample surface. The gravity level was increased, ranging from normal gravity to 5 g, by controlling the rotation speed of the arm. Flame spread experiments were conducted for three different sample orientations at $\theta = 0°$ (horizontal flame spread), $\theta = -20°$ (downward flame spread), and $\theta = +20°$ (upward flame spread) in 0.1 m/s opposed airflow (21 % oxygen concentration) to investigate the effect of sample orientation on flame spread under supergravity conditions.

Results and Discussion

Flame Spread in Low-Oxygen Gas Flow Under Microgravity

Figure 3 shows horizontal spreading flames at volumetric oxygen concentrations of $X_{O2} = 25$, 23, 21, 19, and 18 % under microgravity and normal gravity. The spreading flame under microgravity takes the shape of a semicircle, and its flame

Fig. 4 Flame spread rate as a function of oxygen concentration in both normal and microgravity. (*Solid circle*): flame spread rate in normal gravity; (*closed rhombus*): flame spread rate in microgravity; (*open square*): downward flame spread rate in normal gravity

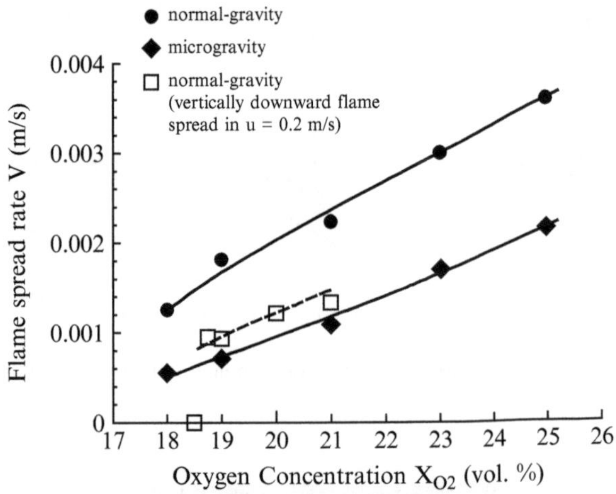

tail is less elongated in the upward direction. The flame length between the flame leading edge and the flame tail decreases with decreasing oxygen concentration. The limiting oxygen concentration was 17.2 % in our microgravity experiment. This is lower than the limiting concentration in a chamber experiment conducted by Olson [2], where it was 21 %. This low limiting oxygen concentration is attributed to the existence of a low velocity opposed flow.

Figure 4 shows the flame spread rate under normal gravity and microgravity for six different oxygen concentrations ranging from 18 to 25 %. The relation of flame spread rate to oxygen concentration is slightly nonlinear near the extinction limit. It is noticeable that the flame spread rate at 21 % oxygen in microgravity is almost identical to that at 18 % oxygen in normal gravity. Also, the flame spread rate in microgravity is almost the same as that for vertical downward spread in normal gravity.

Flame Spread in Airflow Under Supergravity

Figure 5 shows horizontal spreading flames in airflow under supergravity. The flame height increases with increasing gravity level. In this figure, the flame height at 2 and 3 g is lower than the flame height under normal gravity. This can be explained as follows: the airflow at the end of the wind tunnel is influenced by the induced upstream flow due to the rotation of the arm. Therefore, the test rack was covered after these experiments to prevent the effect of this opposed flow caused by the rotation. It is preferable to estimate the effect of the gravity level on flame spread by comparing the flame photograph taken at 1 g with the photograph taken at 5 g. Figure 6 shows the −20° downward and +20° upward spreading flames at 1, 3, and 5 g. The flame height decreases with increasing gravity for downward flame

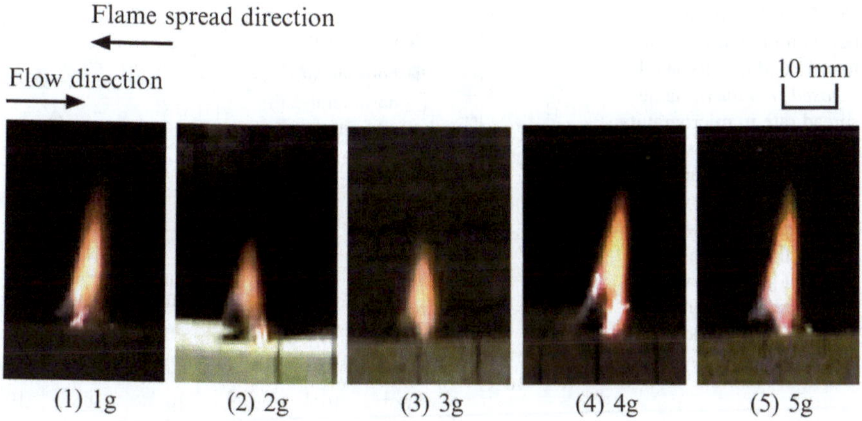

Fig. 5 Six sequential color video images showing horizontal flame spreading over thin filter paper at five different gravity levels in airflow

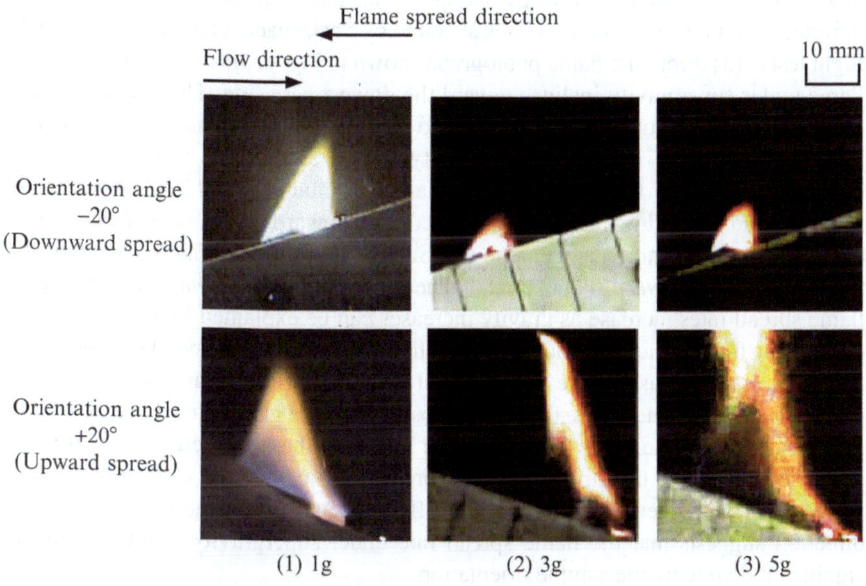

Fig. 6 *Upper line*: three sequential color video images showing downward flame spreading over thin filter paper at three different gravity levels in airflow. *Bottom line*: three sequential color video images showing upward flame spreading over thin filter paper at three different gravity levels in airflow

spread. In contrast, the flame height for upward flame spread increases with increasing gravity.

Figure 7 shows the flame spread rate in microgravity and supergravity. The downward flame spread rate clearly decreases and approaches the limit rate of flame

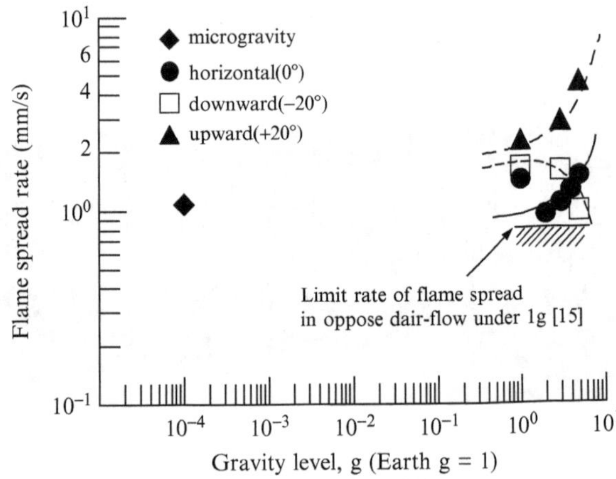

Fig. 7 Relationship between the flame spread rate, V, and gravity level, g. (*Closed rhombus*): flame spread rate in microgravity; (*solid circle*): horizontal flame spread rate at different gravity levels; (*open square*): downward (−20°) flame spread rate at different gravity levels; (*closed triangle*): upward (+20°) flame spread rate at different gravity levels

spreading in opposed airflow of 0.65 m/s [15] as gravity increases. This result is qualitatively in agreement with the experimental flame spread rates for downward burning over a range of gravity levels as found by Sacksteader et al. in their parabolic flight tests [13]. From the flame photograph shown in Fig. 6, the downward-spreading flame under supergravity inclines toward the downstream side. This inclining flame suggests that the induced flow velocity into the flame increases as gravity increases. The increase in the induced flow velocity decreases the heat transfer rate from convection and also the radiation from flame to unburned fuel and decreases the flame spread rate. However, the horizontal and upward flame spread rates under elevated gravity are not in agreement with Sacksteader's result, which was conducted for vertically downward flame spread. The reason why the upward and horizontal flame spread rates increase as gravity increases can be explained by the fact that the concurrent flow induced by buoyancy pushes the spreading flame. This concurrent flow rate may increase as the gravity level increases. The upward spreading flame shown in Fig. 6 inclines more toward the upstream as the gravity level increases. This inclining of the flame toward the upstream increases the heat feedback to unburned fuel. Consequently, the upward flame spread rate increases as the gravity level increases. This tendency for the gravity effect to change depending on the spread direction suggests that the flame spread rate under supergravity is influenced to a significant degree by the sample orientation.

Relation of Non-dimensional Flame Spread Rate to Rayleigh Number

To clarify the effect of gravity level on flame spread based on the buoyancy force in the gas phase, we introduce the Rayleigh number. The Rayleigh number, Ra, is expressed as follows:

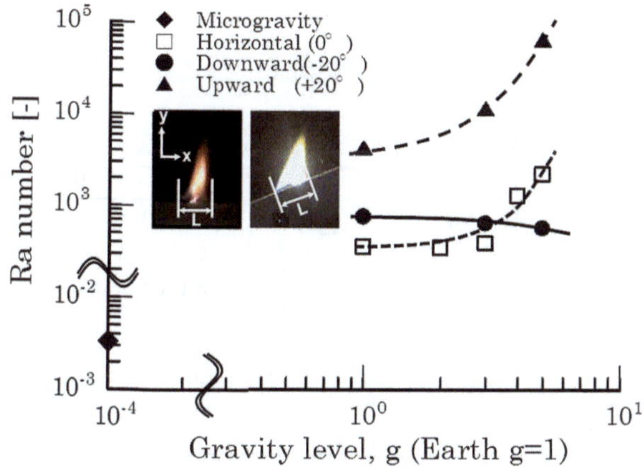

Fig. 8 Relationship between the Rayleigh number, Ra, and gravity level, g. (*Closed rhombus*): Ra number in microgravity; (*open square*): Ra number of horizontally spreading flame at different gravity levels; (*solid circle*): Ra number of downward (−20°) spreading flame at different gravity levels; (*closed triangle*): Ra number of upward (+20°) spreading flame at different gravity levels

$$Ra = \frac{g\beta(T_f - T_0)L^3}{a\nu} \tag{1}$$

where g is acceleration due to gravity, β is thermal coefficient of volume expansion, T_f is average temperature in the flame zone, T_0 is ambient temperature, L is flame length, a is thermal diffusivity of oxidizer, and ν is kinematic viscosity. Figure 8 plots the Rayleigh number as a function of the gravity level for horizontal, downward, and upward spreading flame under microgravity, normal-gravity, and supergravity conditions. The flame length is defined as the length of the spreading flame base as shown in Fig. 8. The Rayleigh number was estimated using the following values: $T_f = 1{,}200$ K, $T_0 = 300$ K, $\beta = ((T_f + T_0)/2)^{-1} = 1/750$, $a = 2.2 \times 10^{-4}$ m^2/s, $\nu = 1.65 \times 10^{-4}$ m^2/s. The Rayleigh number for flame spread under microgravity is on the order of 10^{-3} and under normal and supergravity is on the order of 10^2–10^5. The Ra number for horizontal and upward spreading flame increases as the gravity level increases. In contrast, the Ra number for downward flame spread decreases slightly as gravity increases.

Figure 9 plots the non-dimensional flame spread rate, $V_n = V\delta/a_s$, as a function of the Ra number. The non-dimensional flame spread rate is almost constant in normal and supergravity when the Ra number is less than approximately 10^3 and is similar to that in microgravity. In this region, the effect of buoyancy on flame spread is quite small. The Ra number increases as the gravity level increases for horizontal and upward flame spread. When the Ra number is greater than 10^3, the non-dimensional flame spread rate increases with increasing Ra number and is proportional to $Ra^{1/3}$. Figure 9 suggests that a critical Ra number determines the threshold above which gravity accelerates flame spread.

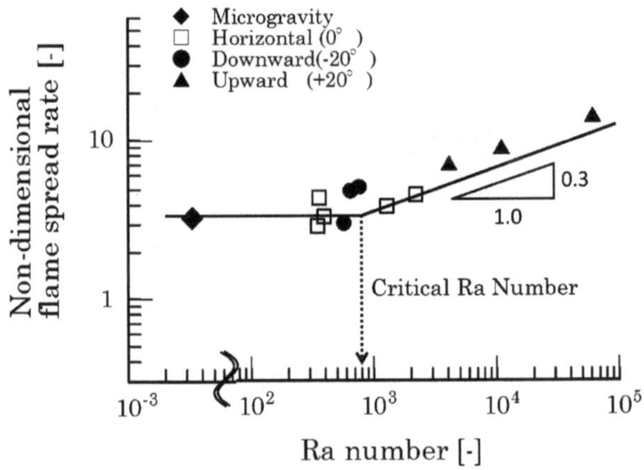

Fig. 9 Relationship between the non-dimensional flame spread rate, V_n, and Rayleigh number, Ra. (*Closed rhombus*): flame spread rate in microgravity; (*solid circle*): horizontal flame spread rate at different gravity levels; (*open square*): downward (−20°) flame spread rate at different gravity levels; (*closed triangle*): the upward (+20°) flame spread rate at different gravity levels

Conclusions

The following summarizes the conclusions of this study:

1. The flame spread rate at an 18 % oxygen concentration under normal gravity is consistent with that in microgravity at 21 % oxygen concentration. The flame spread rate decreases with decreasing oxygen concentration in microgravity, and the limiting oxygen concentration is 17.2 %. This is lower than the limiting concentration obtained in a chamber by Olson [2]. This low limiting oxygen concentration is attributed to the existence of a low-velocity opposed flow.
2. The downward flame spread rate clearly decreases as the gravity level increases; however, the horizontal and the upward flame spread rates increase as gravity increases. This increase is considered to be due to the effect of a concurrent flow. Due to this effect, the upward spreading flame inclines toward the upstream of the flow as the gravity level increases. This inclined flame increases the heat feedback to the unburned fuel surface.
3. The Ra number for horizontal and upward spreading flames increases as gravity increases, while the Ra number for downward flame spread decreases slightly as gravity increases. The non-dimensional flame spread rate is almost constant when the Ra number is smaller than 10^3. In this region, the effect of buoyancy on flame spread is quite small. When the Ra number is greater than 10^3, the non-dimensional flame spread rate increases with increasing Ra number and is proportional to $Ra^{1/3}$, which suggests that a critical Ra number marks the threshold at which gravity affects flame spread or not.

Acknowledgments This study was supported by "Ground-based Research Program for Space Utilization" promoted by Japan Space Forum. We would like to thank T. Nagata, A. Nagadoi, S. Oikawa, T. Okizaki, and Y. Kazehare for their help in experiments and preparing the manuscript.

References

1. Kimzey, J.H.: Proceedings of the 3rd Space Processing Symposium, NASA Space Flight Center, M-74-5, NASA TM-X-70752, **1**, pp. 115–130 (1974)
2. Olson, S.L.: The effect of microgravity on flame spread over a thin fuel, NASA TM-100195 (1987)
3. Friedman, R., Urban, D.L.: Contributions of microgravity test results to the design of spacecraft fire-safety systems, AIAA paper 93–1152 (1993)
4. Shipp, M., Spearpoint, M.: The detection of fires in micro-gravity, fire safety science. In: Proceedings of the 4th International Symposium, IAFSS, pp. 739–750 (1994)
5. Ramachandra, P.A., Altenkirch, R.A., Bhattacharjee, S., Tang, L., Sacksteader, K., Wolverton, M.K.: The behavior of flames spreading over thin solids in microgravity. Combust. Flame **100**, 71–84 (1995)
6. Bryant D., Judd, M.D.: Proceedings of the 10th Anniversary of First ESA Parabolic Flight Campaign Workshop, ESA, pp. 99–101 (1995)
7. Sanchez Tafira, C., Linan, A., Salva, J.A., Conchero, G., Juste, G.L., Esteban, F.: Combustion Experiments During KC-135 Parabolic Flights, ESA SP-1113, pp. 53–64 (1989)
8. West, J., Tang, L., Altenkirch, R.A., Bhattacharjee, S., Sacksteader, K., Delichatsios, M.A.: Quiescent flame spread over thick fuels in microgravity. Proc. Combust. Inst. **26**, 1335–1343 (1996)
9. Kashiwagi, T., MCgrattan, K.B., Olson, S.L., Fujita, O., Kikuchi, M., Ito, K.: Effect of slow wind on localized radiative ignition and transition to flame spread in microgravity. Proc. Combust. Inst. **26**, 1345–1352 (1996)
10. Altenkirch, R.A., Tang, L., Sacksteader, K., Bhattacharjee, S., Delichatsios, M.A.: Inherently unsteady flame spread to extinction over thick fuels in microgravity. Proc. Combust. Inst. **27**, 2515–2524 (1998)
11. Takahashi, S., Kondou, M., Wakai, K., Bhattacharjee, S.: Effect of radiation loss on flame spread over a thin PMMA sheet in microgravity. Proc. Combust. Inst. **29**, 2579–2586 (2002)
12. Bhattacharjee, S., Ayala, R., Wakai, K., Takahashi, S.: Opposed-flow flame spread in microgravity-theoretical prediction of spread rate and flammability map. Proc. Combust. Inst. **30**, 2279–2286 (2005)
13. Sacksteader, K.R., Pettegrew, R.D., T'ien J.S.: Flame spreading over thin fuel samples in partial gravity environments, AIAA paper 98–0567 (1998)
14. Takahashi, K., Kodaira, Y., Kudo, Y., Ito, A., Saito, K.: Effect of oxygen on flame spread over liquids. Proc. Combust. Inst. **31**, 2625–2631 (2007)
15. Sato, K, Suzuki, K., Sakai, Y., Sega, S.: Effects of flame retardant on the behavior and temperature profiles of flames spreading over paper, fire safety science. In: Proceedings of the 4th International Symposium, IAFSS, pp. 503–514 (1994)

Section B
Fire and Explosion - Scale Effects on Consequence Analysis of Accidental Explosions

Ritsu Dobashi, Satoshi Kawamura, Ikuto Nishimura, and Kazunori Kuwana

Abstract Accurate prediction of the blast-wave intensity from a gaseous deflagration is essential for effective consequence analyses of accidental explosions. Most conventional methods are based on the theory of explosives, so adjusting parameters are needed when they are used for evaluating gaseous deflagrations. This review article summarizes new prediction methods developed in [1], where an evaluation equation of the pressure-wave intensity was derived from theoretical analysis of pressure-wave generation by deflagration. The conventional equation is modified by considering the effects of flame instabilities and resultant flame acceleration. The blast pressure evaluated by the modified equation agrees well with the results of large-scale experiments. These results indicate that combustion reaction rate has to be introduced for the prediction of deflagrations. Scaling analyses conducted in [1] are also discussed. The conventional scaling law is significantly improved by introducing the term "combustion reaction rate" represented by the burning velocity. Also a scaling law of new type is discussed. The universal relationship is realized by the new scaling law for deflagrations. These results can be applied to design a scale-model experiment of an explosion hazard and to interpret its result.

Introduction

Risk assessment has become an important tool for effective safety management of chemical plants, hazardous-material sites, and so on. For the risk assessment to be reliable, accurate analysis of the consequences of potential hazards is essential. In a

R. Dobashi (✉) • S. Kawamura • I. Nishimura
Department of Chemical System Engineering, The University of Tokyo, 7-3-1 Hongo, Bunkyo-ku, Tokyo 113-8656, Japan
e-mail: dobashi@chemsys.t.u-tokyo.ac.jp

K. Kuwana
Department of Chemistry and Chemical Engineering, Yamagata University, 4-3-16 Jonan, Yonezawa-shi, Yamagata 992-8510, Japan

K. Saito et al. (eds.), *Progress in Scale Modeling, Volume II*,
DOI 10.1007/978-3-319-10308-2_11

plant or site where flammable gases are handled, accidental gas explosion is a serious potential hazard. Therefore, the consequence analysis of the possible damages caused by an accidental explosion has to be performed. Main damages of accidental explosions are caused by pressure increase, blast wave, fragment scattering, hot gas, and so on. The damage caused by a blast wave spreads quickly and widely and becomes a significant consequence of an accidental explosion. Some predictive models have been already developed for the intensity of blast wave [2, 3]. However, most conventional models are based on the theory of explosives, and thus adjusting parameters are needed when they are used for evaluating gaseous deflagrations. Although explosion phenomena can be simulated by recent CFD techniques, CFD simulation is usually not suitable for the risk assessment of an explosion hazard because such a simulation is computationally expensive and requires detailed information about the explosion conditions.

In [1], more accurate scaling laws were developed for predicting the blast pressure from gaseous deflagration. Following [1], this paper first discusses difficulties of previous models, then explains the new model, and compares the model prediction with available experimental data of large-scale flame propagation [4]. The knowledge presented in this paper can be applied to design a scale-model experiment for an accidental explosion and to interpret its result. Considering the difficulty in conducting a full-scale explosion experiment, scale-model experiment is a powerful tool for assessing the damage caused by an accidental explosion.

Conventional Models

Conventional Models for Evaluating Blast Wave

In this section, conventional models to estimate the blast wave generated by an explosion are reviewed. Several models have been proposed to predict the intensity of the blast wave induced by an explosion. The representative models are TNT equivalency model, TNO multi-energy model, and modified Baker model [2, 3], which are still widely used when assessing the risk of an explosion hazard.

The TNT equivalency model is based on the assumption of equivalence between the flammable material and TNT (trinitrotoluene). The equivalent mass W is calculated by the following equation based on the heat of combustion:

$$W = \frac{\eta M E_C}{E_{\mathrm{TNT}}}, \tag{1}$$

where η is an empirical explosion efficiency, M the mass of flammable material, E_C the heat of combustion of flammable material, and E_{TNT} the heat of combustion of TNT. The explosion efficiency for gaseous deflagration is usually very low (2–15 %). The effects (consequences) of explosion are then evaluated by reference experimental data for TNT of the equivalent mass. In addition, the scaled distance Z is defined using $W^{1/3}$ as

$$Z = \frac{R}{W^{1/3}}, \tag{2}$$

where R is the distance from the explosion point (ignition point).

The TNO multi-energy model provides a detailed procedure for predicting the blast pressure considering multi-energy sources of explosion. In this model, different factors are used for detonation and deflagration, respectively. The scaled distance (called Sachs-scaled distance) is calculated by the following equation, which is based on the same idea of the TNT equivalency model:

$$\overline{R} = \frac{R}{(E/P_0)^{1/3}}, \tag{3}$$

where $\overline{R}$ is Sachs-scaled distance, E the charge combustion energy, and P_0 the ambient pressure.

The modified Baker model also uses Eq. (3). Differing from the TNO multi-energy model, the modified Baker model introduces the effects of reactivity, obstacle density, and geometry.

In these three models, the essential scaling concepts are based on "cubic root scaling" apparent in Eqs. (2) and (3), in which the scaled distances depend on the distance divided by the cubic root of total energy release. It is emphasized here that this concept has been developed for evaluating the detonation of explosives.

The Comparisons with the Results of Large-Scale Experiments

To verify these conventional models, they were applied to large-scale explosion experiments conducted by Japanese research organizations [5–8]. First, the evaluation is made by the TNT equivalency model under the condition of the empirical explosion efficiency η being unity, which corresponds to the basic concept of the conventional models. In Fig. 1, the maximum overpressures measured in the experiments are compared with those evaluated by the TNT equivalency model. The conditions of experiments are listed in Table 1. The flammable materials used in the experiments are combustible gases (hydrogen, city gas, LPG (liquefied petroleum gas), DME (dimethyl ether)) and explosives (pentolite and black powder). The propagation modes (deflagration or detonation) are also indicated in Table 1. The detonation of gas mixtures is initiated by detonating explosives. In the detonation cases in Fig. 1, the evaluation of the TNT equivalency model agrees well with the experimental data (pentolite and DME). However, for the cases of deflagration, the evaluated values overestimate the experimental data (hydrogen, city gas, and DME), while the data for black powder nearly equals the model's

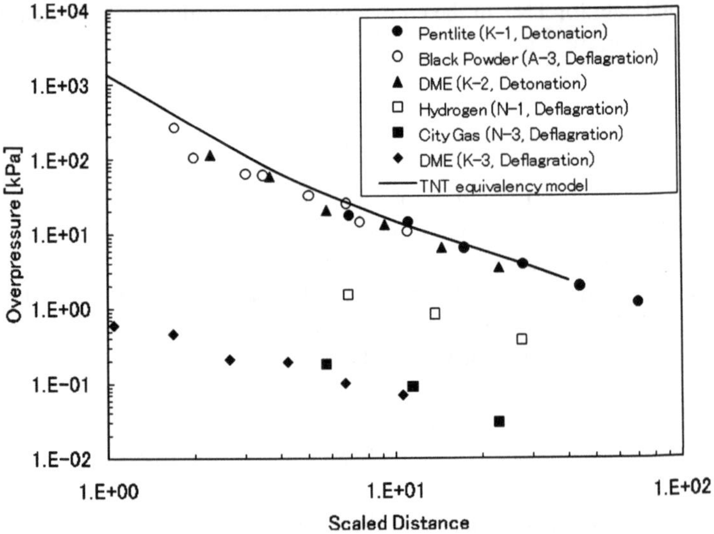

Fig. 1 Comparison between the TNT equivalency model and the experimental data [1, 4]

Table 1 Large-scale experiments [1]

No.	Researcher	Year	Flammable material	Mass (kg)	Volume [m^3]	Vol. %	
K–1	KHK	2003	Pentolite	0.1776	–	–	Detonation
K–2	KHK	2003	DME/air	–	7.74	6.54	Detonation
K–3	KHK	2003	DME/air	–	81	6.54	Deflagration
K–4	KHK	2003	LPG/air	–	81	4.02	Deflagration
N–1	NEDO	2002	Hydrogen/air	–	5.2	30	Deflagration
N–2	NEDO	2002	Hydrogen/air	–	5.2	57	Deflagration
N–3	NEDO	2002	City Gas/air	–	5.2	9.5	Deflagration
J–1	JN IO SH	2007	Hydrogen/air	–	1.4	29.6	Deflagration
A–1	AIST	1997	Black powder	400	–	–	Deflagration
A–2	AIST	2008	Hydrogen/air	–	9.4	30.2	Deflagration
A–3	AIST	2008	Hydrogen/air	–	75	30.55	Deflagration

prediction. Among the data of deflagrations of gas mixtures, the overpressures of city gas and DME are nearly equal and a few hundred times smaller than those evaluated by the TNT equivalency model. The overpressure of hydrogen deflagration is much larger than these gases but a few ten times smaller than the model prediction.

In summary, the TNT equivalency model can accurately predict the maximum overpressure of detonation of gas and explosive and deflagration of explosive. On the other hand, it significantly overestimates the overpressure of the deflagration of a flammable gas. Therefore, the prediction based on the heat release is not suitable

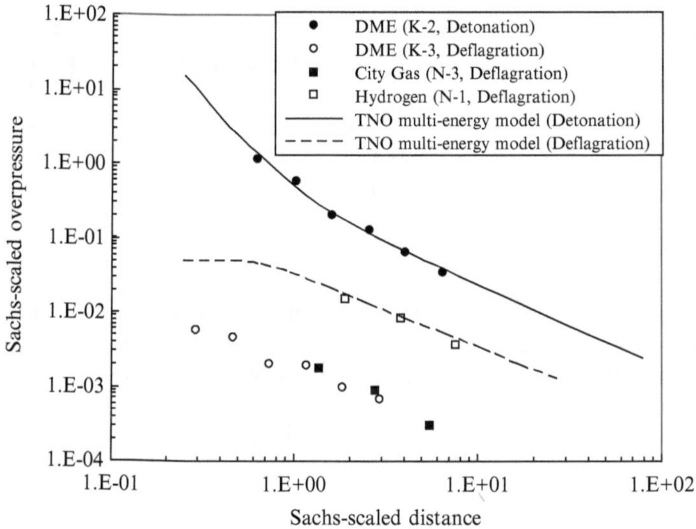

Fig. 2 Comparison between the TNO multi-energy model and the experimental data [1, 4]

for accurately evaluating gaseous deflagration phenomena. The reason for these disagreements is that the heat of combustion is released relatively slowly during the gaseous deflagration; not only the amount of heat release but also the rate of heat release must be considered.

Introduction of the empirical explosion efficiency can adjust the accuracy of evaluation; however, the value of the explosion efficiency cannot be determined systematically without conducting experiments. The evaluations by the TNO multi-energy model are shown in Fig. 2. The data of the deflagration of hydrogen agree well with the evaluated value as deflagration, whereas the data for the deflagration of city gas and DME are still a few tens of times smaller than the evaluated value. In the modified Baker model, reactivity can be considered. However, the reactivity can be set only to high, medium, or low, and the model provides only a rough approximation.

From the survey of the conventional models described above, it is found that the accurate prediction of the consequence by gaseous deflagrations is difficult using these models. In order to predict the consequence of gaseous deflagrations accurately, new prediction methods are needed.

New Model Based on Gaseous Deflagration Analysis

Theoretical Analysis

To accurately predict the blast wave generated from a gaseous deflagration, the generating mechanism was theoretically analyzed [1]. The blast wave is generated

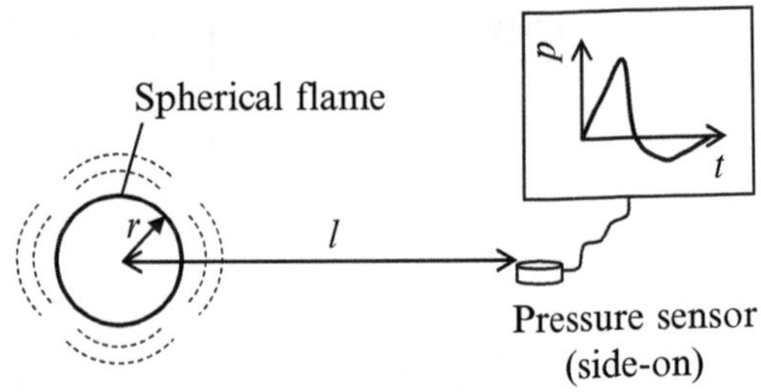

Fig. 3 Model of flame propagation [1]

by the rapid expansion of the burned gas during explosion. The generated pressure wave observed at the position l from the ignition point is obtained by the theory of acoustics as follows under the assumption of isotropic point source:

$$p(t) = \frac{\rho}{4\pi l}(\varepsilon - 1)\left[\frac{\mathrm{d}\dot{Q}}{\mathrm{d}t}\right]_{t-\tau}, \tag{4}$$

where p is pressure, t the time, ρ the density, ε the volumetric expansion ratio, and $\dot{Q}$ the volumetric rate of combustion [9].

When the flammable gas/air mixture is ignited, a flame outwardly propagates from the ignition point. Under the condition that the gas concentration is uniform and that there is no gas flow and obstacle (open space), nearly isotropic flame propagation is achieved, and the shape of the propagating flame is spherical (Fig. 3). Then, the radius r of the spherical flame can be written as follows:

$$r = \varepsilon S t, \tag{5}$$

where S is the burning velocity. The volumetric rate of combustion $\dot{Q}$ can be given by

$$\dot{Q} = SA_f = S \cdot 4\pi r^2, \tag{6}$$

where A_f is the flame front area. Using Eqs. (5) and (6), Eq. (4) is rewritten as

$$p(t) = 2\frac{\rho}{l}\varepsilon^2(\varepsilon - 1)S^3 t \tag{7}$$

From Eq. (7), the observed pressure history at the position of distance l from the explosion point can be obtained.

Consideration of the Flame Front Instabilities

The pressure history measured in a large-scale experiment (A-2) is shown in Fig. 4. In the experiment, hydrogen/air mixture at the stoichiometric concentration

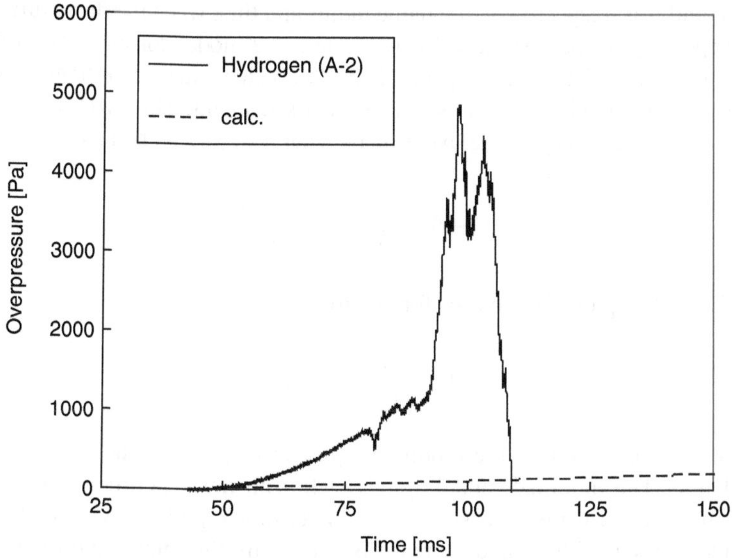

Fig. 4 Measured pressure history and predicted one calculated by Eq. (7) (without the effect of instability), $l=4$ m [1, 4]

(30.2 %) was filled in a vinyl tent house of 9.4 m^2 and ignited at the center of the house. Dashed line in Fig. 4 shows the pressure predicted by Eq. (7). The predicted pressure is much smaller than the measured value. The measured data indicates that the flame propagation velocity in the experiment is much faster than that calculated by Eq. (5). It is supposed that some acceleration mechanism must have influenced flame propagation during the experiment.

In the experiment, the gas mixture is quiescent at the time of ignition and there is no obstacle inside the tent house. Therefore, the acceleration mechanism, before the flame reaches to the tent house, must be some kind of spontaneous instability. Possible mechanisms of spontaneous instability are [9–11]:

- Hydrodynamic instability
- Diffusive-thermal instability

It is known that hydrodynamic instability has more effect when the scale of flame becomes larger. Also diffusive-thermal instability is supposed to have less effect for the stoichiometric hydrogen mixture. Dobashi et al. [1] proposed to modify Eq. (7) by introducing the effects of hydrodynamic instability. Gostintsev et al. [12] derived the following semiempirical formula for the flame propagation with the spontaneous flame instability:

$$r = r^* + c_g \frac{\varepsilon^2 S^2 (t - t^*)^{3/2}}{\sqrt{\alpha}}, \tag{8}$$

where r^* and t^* is, respectively, the flame radius and time when the flame turbulence starts appearing; α the thermal diffusivity; and c_g a model constant. The value of 0.0023 is recommended for c_g in [12]; this value was accordingly used in [1]. According to [12], r^* is a few cm in this experiment. This size is considered negligible in such a large-scale experiment. Then Eq. (8) can be rewritten as

$$r = At^{3/2}\left(A = \frac{c_g \varepsilon^2 S^2}{\sqrt{\alpha}}\right) \tag{9}$$

Using Eq. (9), Eq. (7) can be transformed to

$$p(t) = \frac{21\rho}{4l}\frac{\varepsilon - 1}{\varepsilon}A^3 t^{5/2} \tag{10}$$

The pressure predicted by the modified equation (Eq. 10) is shown in Fig. 5 by dashed line. The modified model agrees well with the experimental data. After $t = 90$ ms, the measured pressure starts increasing rapidly, deviating from the predicted pressure. This rapid increase is caused by the interaction of the flame with the frame pipes of the tent house and the effect of this interaction is not included in Eq. (10). Before such acceleration, the pressure history of blast wave from a gaseous deflagration in an open space can be accurately predicted by the modified theoretical equation (Eq. 10). Kim et al. [13, 14] also confirmed that Eq. (8) accurately predicts the evolution of a spherically expanding flame and that the time history of blast pressure follows Eq. (10).

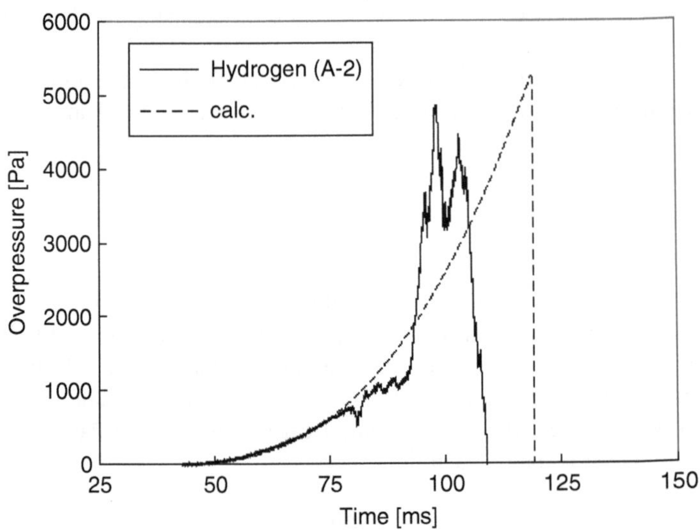

Fig. 5 Measured pressure history and predicted one calculated by Eq. (10) (with the effect of instability), $l = 4$ m [1, 4]

The time exponent of Eq. (8) or (9), 3/2, is associated with the fractal dimension of a wrinkled flame surface, and the value of 3/2 indicates the fractal dimension of 7/3 [12]. Experiments [12] suggest that the flame fractal dimension resulting from hydrodynamic instability is close to 7/3 under typical combustion conditions. Nevertheless, the fractal dimension may depend on the properties of the combustible mixture. In particular, the expansion ratio is expected to influence the fractal dimension because hydrodynamic instability is caused by the volumetric expansion due to combustion [15]. Although a few papers [15–19] discuss the dependence of fractal dimension on expansion ratio, the dependence is not fully understood.

Equation (8) can be used for a CFD simulation. CFD simulation of an accidental gas explosion has to consider a large space, and therefore a turbulent combustion model must be used. Furthermore, the model must account for flame acceleration due to hydrodynamic instability. Use of Eq. (8) or a similar expression based on the fractal concept is a convenient way that can be easily implemented to existing CFD codes [20–23].

The discussion thus far only considers deflagration in an open space. Nishimura et al. [24] analyzed the results of explosion experiments in confined spaces. They found that Eq. (8) is applicable to confined explosions when also considering the change in laminar burning velocity with an increase in the pressure and temperature during explosion.

Scaling Analysis

Modification to the Conventional Models

Scaling analysis was conducted to seek a convenient scaling law for evaluating the blast-wave intensity of a gaseous deflagration [1]. Using the result of the former section, the maximum pressure $P_{\max}$ by a deflagration can be evaluated by the following equation:

$$P_{\max} = \frac{21\rho}{4l} c_g^{4/3} r_0^{5/3} \alpha^{-2/3} \varepsilon^{20/9} (\varepsilon - 1) S^{8/3}, \tag{11}$$

where r_0 is the initial radius of flammable mixture. This equation is derived under the assumption that the maximum pressure appears when the flammable mixture is entirely burned out (at $r = \varepsilon^{1/3} r_0$). From this equation, it is found that the maximum pressure is proportional to $S^{8/3}$. That is, the effect of the combustion reaction rate on the maximum pressure can be represented by $S^{8/3}$. In Eqs. (2) and (3), the scaled distance is determined only by the energy release from the explosion (combustion). However, Eq. (11) implies that the factor of reaction rate must be included, and the maximum pressure is proportional to $S^{8/3}$. Therefore, the scaled distance must be determined by using this factor. As a non-dimensional form, the term of $(S/a_0)^{8/3}$ is introduced to the equation of the scaled distance as

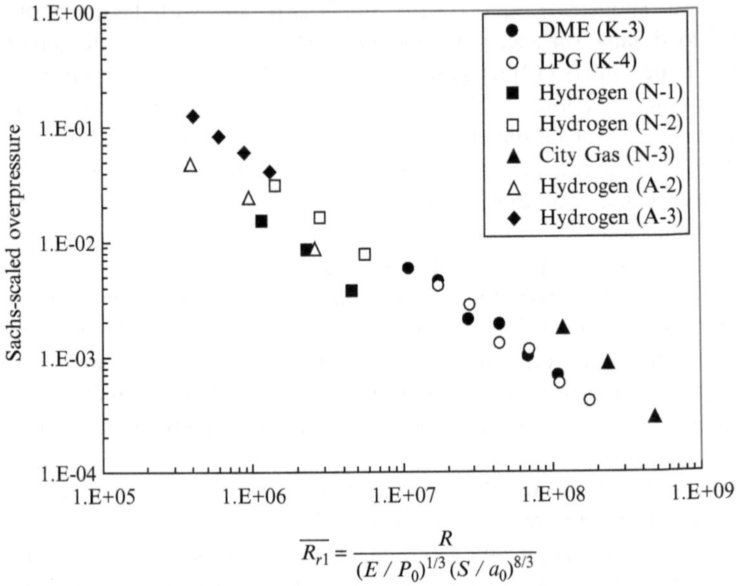

Fig. 6 Experimental data (overpressure) on the modified Sachs-scaled distance [1, 4]

$$\overline{R_{r1}} = \frac{R}{(E/P_0)^{1/3}(S/a_0)^{8/3}}, \tag{12}$$

where $\overline{R_{r1}}$ is the modified scaled distance [replacing the Eq. (3)]. The experimental data in Fig. 1 are rearranged by using this scaled distance (Fig. 6). Figure 6 indicates that all the data (deflagration of flammable gases) can be approximated by one line. The conventional scaling law is significantly improved by this modification.

New Scaling Law

Scaling analysis using Eq. (11) can create a new scaling law for blast-wave intensity from a gaseous deflagration. Assuming that ε and c_g are constant, Eq. (11) can be rearranged to the following scaling relation:

$$\frac{P_{\max}}{\rho\alpha^2/r_0^2} \sim \frac{1}{\frac{l}{r_0(S r_0/\alpha)^{8/3}}}. \tag{13}$$

This is the new scaling law proposed in [1]. The term on the left-hand side of Eq. (13) corresponds to the new scaled pressure, and the denominator of the right-hand side corresponds to the new scaled distance. The data in Fig. 6 are re-plotted in Fig. 7 using this new scaling law. Figure 7 demonstrates that this new scaling law is

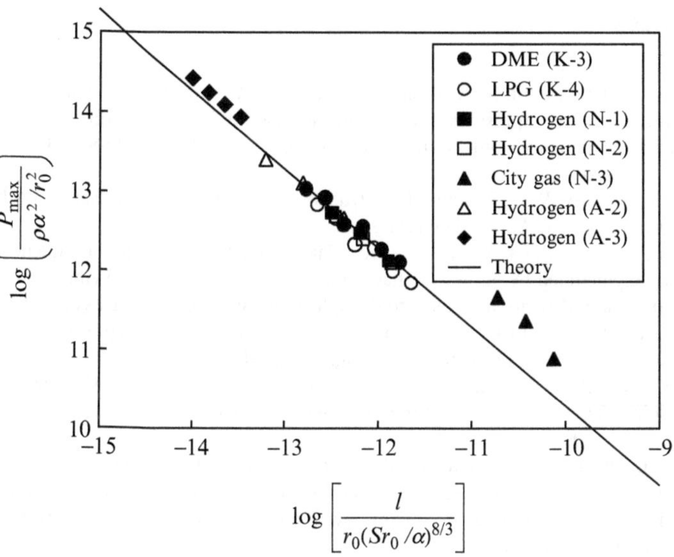

Fig. 7 Relation between scaled pressure and scaled distance using new scaling law [1]

quite relevant. By using the new scaling law, a universal relationship can be realized for the deflagration of all flammable gases used in the experiments. Data of city gas are slightly larger than Eq. (13). Reason for the difference is that r^* cannot be neglected when the radius of the flammable mixture is not so large as in the case of city gas (N-3).

Summary

Conventional prediction methods for blast-wave intensity from a gaseous deflagration are reviewed for the consequence analyses of accidental explosions. New prediction methods developed in [1] are also discussed. The new model, based on theoretical analysis of pressure-wave generation by a deflagration, can predict pressure-wave intensity. The equation considers the effects of flame instabilities and resultant flame acceleration. It is confirmed that the evaluations by this modified equation agree well with the results of large-scale experiments. These results indicate that combustion reaction rate has to be introduced for the prediction of gaseous deflagrations. Scaling analyses based on these results are also reviewed. The conventional scale model was significantly improved by introducing the term "combustion reaction rate" using the burning velocity. Also a new scaling law was proposed. The universal relationship is realized by the new scaling law for deflagrations. This article discusses deflagration mainly in an open space. However, the new model discussed may be used for deflagration in a closed space. It is expected that the knowledge presented in this article is applicable to designing a scale-model experiment for an accidental explosion and for interpreting its result.

References

1. Dobashi, R., Kawamura, S., Kuwana, K., Nakayama, Y.: Consequence analysis of blast wave from accidental gas explosions. Proc. Combust. Inst. **33**, 2295–2301 (2011)
2. American Institute of Chemical Engineers: Guidelines for Chemical Process Quantitative Risk Analysis (2000)
3. Baker, W.E., Cox, P.A., Westine, P.S., Kulesz, J.J., Strehlow, R.A.: Explosion Hazards and Evaluation. Elsevier, Amsterdam (1983)
4. Dobashi, R., Kawamura, S., Kuwana, K., Nakayama, Y.: Study on consequence analysis of blast wave generated by gaseous deflagrations. In: 7th International Symposium on Hazards, Prevention, and Mitigation of Industrial Explosions, St. Petersburg, Russia (2008)
5. LPG Research Laboratory of the High Pressure Gas Institute: The report of basic research on practical application of DME fuel (in Japanese) (2004)
6. NEDO and AIST, International Clean Energy Network Using Hydrogen Conversion (WE-NET), Task 2 Phase II Program: Surveys and researches on safety precautions (in Japanese) (2002)
7. Saitoh, H., Otsuka, T., Mizutani, T., Morimoto, K., Iwasa, T., Shimizu, N., Naruo, Y., Matsui, H., Inatani, Y., Yoshikawa, N.: Field experiments of hydrogen air deflagrations to evaluate appropriate mixture volumes by scaling law. Trans. Jpn. Soc. Mech. Eng. B **73**(728), 1099–1106 (2007) (in Japanese)
8. Iida, M., Matsunaga, T., Nakayama, Y., Tanaka, K.: Blast wave generated by black powder. Sci. Technol. Energy Mater. **58**(6), 271–274 (1997) (in Japanese)
9. Hurle, I.R., Price, R.B., Sugden, T.M., Thomas, A.: Sound emission from open turbulent premixed flames. Proc. R. Soc. Lond. A **303**, 400–427 (1968)
10. Williams, F.A.: Combustion Theory, vol. 2. Westview Press, Boulder (1985)
11. Dobashi, R.: Experimental study on gas explosion behavior in enclosure. J. Loss Prev. Process Ind. **10**(2), 83–89 (1997)
12. Gostintsev, Y.A., Istratov, A.G., Shulenin, Y.V.: Self-similar propagation of a free turbulent flame in mixed gas mixtures. Combust. Explos. Shock Waves **24**, 563–569 (1988)
13. Kim, W.K., Mogi, T., Dobashi, R.: Fundamental study on accidental explosion behavior of hydrogen/air mixtures in an open space. Int. J. Hydrog. Energy **38**, 8024–8029 (2013)
14. Kim, W.K., Mogi, T., Dobashi, R.: Flame acceleration in unconfined hydrogen/air deflagrations by using infrared photography. J. Loss Prev. Process Ind. **26**, 1501–1505 (2013)
15. Blinnikov, S.I., Sasorov, P.V.: Landau-Darrieus instability and the fractal dimension of flame fronts. Phys. Rev. E. **53**(4827–4841) (1996)
16. Wada, Y., Kuwana, K.: Propagation velocity and fractal structure of premixed flame during gas explosion. J. Chem. Eng. Jpn. **45**, 823–828 (2012)
17. Mukaiyama, K., Kuwana, K.: Scale effect of flame instability mechanisms on propagation velocity. J. Loss Prev. Process Ind. **26**, 387–391 (2013)
18. Wada, Y., Kuwana, K.: A numerical method to predict flame fractal dimension during gas explosion. J. Loss Prev. Process Ind. **26**, 392–395 (2013)
19. Mukaiyama, K., Shibayama, S., Kuwana, K.: Fractal structures of hydrodynamically unstable and diffusive-thermally unstable flames. Combust. Flame **160**, 2471–2475 (2013)
20. Molkov, V., Makarov, D., Schneider, H.: LES modelling of an unconfined large-scale hydrogen-air deflagration. J. Phys. D **39**, 4366–4376 (2006)
21. Makarov, D., Molkov, V., Gostintsev, Y.: Comparison between RNG and fractal combustion models for LES of unconfined explosions. Combust. Sci. Technol. **179**, 401–416 (2007)
22. Tomizuka, T., Kuwana, K., Shimizu, K., Mogi, T., Dobashi, R., Koshi, M.: Estimation of turbulent flame speed during DME/air premixed gaseous explosions. J. Loss Prev. Process Ind. **26**, 369–373 (2013)
23. Tomizuka, T., Kuwana, K., Mogi, T., Dobashi, R., Koshi, M.: A study of numerical hazard prediction method of gas explosion. Int. J. Hydrog. Energy **38**, 5176–5180 (2013)
24. Nishimura, I., Mogi, T., Dobashi, R.: Simple method for predicting pressure behavior during gas explosions in confined spaces considering flame instabilities. J. Loss Prev. Process Ind. **26**, 351–354 (2013)

Section B
Fire and Explosion - Preliminary Reduced Scale Experimental Study on Pool Fires in Tunnels

L.W. Pan, S.M. Lo, B.H. Cong, R.K.K. Yuen, Tao Wei, T.S.H. Liang, and G.X. Liao

Abstract Scaling relations for tunnel fires have been extended to include the impact of pool sizes on $\dot{Q}$ of fires. The theoretical analysis deduced that in order to preserve a similar fire and smoke flow, the portion of pool fire sizes D_f/D_m should be proportional to the five-eighth power law of the characteristic length L^*. A dimensionless parameter τ is introduced to eliminate the time diversity. It is defined as the fire-heating time for upper hot layer. Full-scale tunnel fire experiments were conducted with two fire sizes. A series of reduced scale experiments were carried out in the model tunnel (20:1). The experimental results enhanced the reliability of the theoretical models.

L.W. Pan (✉)
State Key Laboratory of Fire Science, University of Science and Technology of China, 96 Jinzhai Road, Hefei 230026, People's Republic of China

Department of Civil and Architectural Engineering, City University of Hong Kong, Tat Chee Avenue, Kowloon, Hong Kong

Public Safety System Integration Engineering Center, East China Research Center of Electronic Engineering, 199 Xiangzhang Avenue, Hefei 230088, People's Republic of China
e-mail: panliwei0813@163.com

S.M. Lo • R.K.K. Yuen
Department of Civil and Architectural Engineering, City University of Hong Kong, Tat Chee Avenue, Kowloon, Hong Kong

B.H. Cong
Shanghai Institute of Disaster Prevention and Relief, Tong Ji University, 1239 Siping Road, Shanghai 200092, People's Republic of China

T. Wei • T.S.H. Liang • G.X. Liao
State Key Laboratory of Fire Science, University of Science and Technology of China, 96 Jinzhai Road, Hefei 230026, People's Republic of China

K. Saito et al. (eds.), *Progress in Scale Modeling, Volume II*,
DOI 10.1007/978-3-319-10308-2_12

Nomenclature

A_f	Area of burning fuel, m^2
D	Pool fire size/side length, m
$L*$	Characteristic length, m
$\dot{Q}$	Heat release rate, MW
$\dot{m}$	Mass burning rate of fuel kg/s
ΔH_c	J/kg of combustible
ΔH_{vap}	J/kg of vaporization of fuel
p	Pressure
$\dot{q}''$	Heat feedback of fire plume kw/m^2
Vwind	Ambient wind velocity

Subscripts

^	Dimensionless symbol
0	Ambient
c	Combustible (heptane)
f	Full scale
i	Products of combustion
m	Reduce model
s	Smoke

Introduction

In consideration of the rapid urbanization in many Asian countries, governments have endeavored to improve the transportation system including the comprehensive subway/road tunnels. Fire in transport tunnels has become a major concern of these governments and infrastructure designers. Tunnel fires often lead to more serious disasters than ordinary fires owing to the confined space and complex traffic volumes. Real-life experiments have been performed to acquire a better understanding of smoke and heat propagation in tunnel fires [1–6]. However, full-scale fire experiments are very time consuming and costly.

The reduced scale study is an alternative for investigating the mechanism of the fire-induced smoke flow. The aim of the scaling method is to produce a smoke flow in the reduced scale tunnel analogous to that in the full-scale tunnel. In the last 15 years, empirical expression and semi-empirical models have been deduced for the physical characteristics of fire in reduced tunnel [7, 8]. Detailed temperature and velocity distributions in the tunnels were recorded [9]. Effects of longitudinal ventilation velocity on the burning rate were reported [10, 11]. In the light of previous scale model studies, the scaling model depends on the relation of heat release rate, but fewer investigations are concerned with the exact production of the heat release rate which is sensitive to the pool size. For previous studies, the portion of pool fire sizes $\frac{D_f}{D_m}$ was equal to the characteristic length L^*,

$$\frac{D_f}{D_m} = \frac{L_f}{L_m} = L^* \tag{1}$$

It follows the relation to $\dot{Q}$ and $\dot{m}$ as,

$$\frac{D_f}{D_m} = L^* = \left(\frac{\dot{m}_f}{\dot{m}_m}\right)^{2/5} = \left(\frac{\dot{Q}_f}{\dot{Q}_m}\right)^{2/5} \tag{2}$$

In small-scale fires, the heat release rate relies on the heating of fuel by the heat feedback of the fire plume $\dot{q}''$, which is dominated by the pool size D. Formula (2) needs to be modified.

This paper aims at exploring the scaling relation of pool size $\frac{D_f}{D_m}$. Heptane pool fires, which offer the advantage of involving only one combustible in the combustion process, were chosen to simulate the fire source.

Theoretical Analysis

Dimensionless Governing Equations

In order to investigate the smoke flow of tunnel fires, two-dimensional forms governing equations are presented as follows:

$$\begin{cases}
\text{Mass :} \\
\frac{\partial \rho}{\partial t} + \frac{\partial(\rho u)}{\partial x} + \frac{\partial(\rho v)}{\partial y} = 0 \\
x - \text{direction momentum :} \\
\rho\left(\frac{\partial u}{\partial t} + u\frac{\partial u}{\partial x} + v\frac{\partial u}{\partial y}\right) = -\frac{\partial p'}{\partial x} + g(\rho_0 - \rho) + \mu\Delta u \\
y - \text{direction momentum :} \\
\rho\left(\frac{\partial v}{\partial t} + u\frac{\partial v}{\partial x} + v\frac{\partial v}{\partial y}\right) = -\frac{\partial p'}{\partial y} + \mu\Delta v \\
\text{Energy :} \\
\rho c_p\left(\frac{\partial T}{\partial t} + u\frac{\partial T}{\partial x} + v\frac{\partial T}{\partial y}\right) = \frac{\partial p}{\partial t} + k\frac{\partial^2 T}{\partial^2 x^2} - 4\kappa\sigma T^4 + \int_0^{4\pi} \kappa I d\omega + \dot{Q}''' \\
\text{RadiativeTransfer Equation :} \\
\frac{\mathrm{d}I}{\mathrm{d}z} = -\kappa\left(I - \frac{\sigma T^4}{\pi}\right) \\
\text{State :} \\
p = \rho R T \\
\text{Energy Boundary Condition :} \\
\dot{m}_c = \frac{D^2}{\Delta h_{vap}}\left[\dot{q}''' + \varepsilon\sigma\left(T^4 - T_0^4\right)\right] \\
p' = p - p_0 \\
\frac{\mathrm{d}p_0}{\mathrm{d}z} = -\rho_0 g
\end{cases} \tag{3}$$

The above equations can be made dimensionless by introducing the following parameters: characteristic dimension L^*, and characteristic velocity V^*. The dimensionless variables denoted by (^) are presented below:

$$\begin{cases} \text{Mass :} \\ \dfrac{\partial \hat{\rho}}{\partial \hat{t}} + \dfrac{\partial (\hat{\rho}\,\hat{u})}{\partial \hat{x}} + \dfrac{\partial (\hat{\rho}\,\hat{v})}{\partial \hat{y}} = 0 \\ x - \text{direction momentum :} \\ \hat{\rho}\left(\dfrac{\partial \hat{u}}{\partial \hat{t}} + u\dfrac{\partial \hat{u}}{\partial \hat{x}} + v\dfrac{\partial \hat{u}}{\partial \hat{y}}\right) = -\dfrac{\partial \hat{p}'}{\partial \hat{x}} + \dfrac{gL^*}{V^{*2}}(1-\hat{\rho}) + \dfrac{\mu}{\rho_0 V^* L^*}\dfrac{\partial^2 \hat{u}}{\partial \hat{x}^2} \\ y - \text{direction momentum :} \\ \hat{\rho}\left(\dfrac{\partial \hat{v}}{\partial \hat{t}} + u\dfrac{\partial \hat{v}}{\partial \hat{x}} + v\dfrac{\partial \hat{v}}{\partial \hat{y}}\right) = -\dfrac{\partial \hat{p}'}{\partial \hat{y}} + \dfrac{\mu}{\rho_0 V^* L^*}\dfrac{\partial \hat{p}'}{\partial \hat{y}^2} \\ \text{Energy :} \\ \hat{\rho}\left(\dfrac{\partial \hat{T}}{\partial \hat{t}} + \hat{u}\dfrac{\partial \hat{T}}{\partial \hat{x}} + \hat{v}\dfrac{\partial \hat{T}}{\partial \hat{y}}\right) = \dfrac{k}{\rho_0 c_p V^* L^*}\dfrac{\partial^2 \hat{T}}{\partial \hat{x}^2} + \dfrac{\kappa L^* \sigma T_0^3}{\rho_0 c_p V^*}\left(\displaystyle\int_0^{4\pi} \hat{I}\,d\omega - 4\hat{T}^4\right) + \hat{\dot{Q}} \\ \text{Energy Boundary Condition :} \\ \hat{\dot{m}}_f = \hat{\dot{q}}'' + (\hat{T}^4 - 1) \end{cases} \tag{4}$$

The following dependent dimensionless variables are defined:

$$\begin{aligned} \hat{\dot{Q}} &= \frac{\dot{Q}''' L^*}{\rho_0 V^* c_p T_0} \\ \hat{\dot{m}}_f &= \frac{\dot{m}_f \Delta h_{vap}}{D^2 \sigma T_0{}^4} \\ \hat{\dot{q}}'' &= \frac{\dot{q}''}{\sigma T_0{}^4} \end{aligned} \tag{5}$$

where the velocity V^* in Eq. (5) is derived as [12, 13],

$$V^* = \left(\frac{g\dot{Q}}{\rho_0 c_p T_0 L^*}\right)^{1/3} \tag{6}$$

Heat Feedback of Fire Plume

Based on the following two assumptions:

(1) The energy to maintain combustion comes primarily from flame convection and radiation heat feedback.
(2) The mass burning rate depends on the heat feedback rate from the flame to the fuel surface, and the mass burning rate controls the total heat release rate and the amount of heat feedback [14–16].

A detailed analysis of the heat feedback rate of the fire plume $\dot{q}''$ was given. As Holboren pointed out [17], the heat feedback rate of the flame to the surface of pool fire was expressed as:

$$\dot{q}'' = a\left[1 - e^{(-bD)}\right] \tag{7}$$

where a, b are the coefficients relating to the maximum heat flux output $\dot{Q}''_{\max}$ and flame height H_f. For a heptane fire, the value of the coefficient a can be deduced from an asymptotic behavior of the experimental data $a = 0.0784\,\text{kg/m}^2\text{s}$ [18]. The value of the coefficient b is $0.8\,\text{m}^{-1}$ [7]. A Taylor expansion of the formula (7) was executed and the third-level item was ignored. Burning rate $\dot{m}_c$ and heat release rate $\dot{Q}$ can be expressed as,

$$\begin{aligned} \dot{Q}_f &\propto \dot{m}_f \propto D^2 \dot{q}_m{}'' \propto D_f{}^4 \\ \dot{Q}_m &\propto \dot{m}_m \propto D^2 \dot{q}_m{}'' \propto D_m{}^4 \end{aligned} \tag{8}$$

The relation of $\frac{D_f}{D_m}$ with $L*$ is

$$\frac{D_f}{D_m} \approx \frac{\dot{Q}_f{}^{1/4}}{\dot{Q}_m{}^{1/4}} \approx \left(L^{*5/2}\right)^{1/4} \approx L^{*0.625} \tag{9}$$

Flow Time Scale

The rates of temperature rise in the full-scale experiments are greater than in the reduced scale experiments. The correlation of temperature is incompatible unless time diversity is eliminated. A dimensionless time τ can be used for representing real experimental time. Classically, the characteristic time t_* is defined as the ratio of the characteristic length to characteristic velocity for natural convection:

$$\tau = \frac{t}{t_*} = \frac{t}{L^*/V^*} \tag{10}$$

In this study, the characteristic time t_* was defined as the heating time of the upper hot smoke layer heated by the freely burning fire [7, 17]:

$$t_* = mc_p T_0 / \dot{Q}_0 \tag{11}$$

The value $\dot{Q}_0$ is the characteristic heat flux to the smoke layer from a freely burning fire. m_{up} is the mass of upper layer smoke:

$$\begin{aligned} &\dot{Q}_0 = \chi \dot{m}_f \Delta h_c \\ &m_{\text{up}} = \rho_s V_s = \rho_s \frac{v_m \sum_{\text{products}} a_i}{a_c M_c} m_c \end{aligned} \tag{12}$$

So, the time scale can be obtained as,

$$\tau = \frac{t}{t_*} = \frac{a_c M_c \chi \dot{m}_f \Delta h_c t}{c_p T_0 \rho_s m_c v_m \sum_{\text{products}} a_i} \tag{13}$$

Experimental Setup

Full-Scale Experiments

Full-scale experiments were performed in a 100 m long tunnel (Fig. 1). The pool was set at cross section A, which was 35 m from the upstream tunnel entrance and 65 m from the downstream tunnel exit. Two tests were carried out with different pool sizes as shown in Table 1.

The temperature of downstream smoke flow was detected by K-type stainless steel sheathed thermocouples with a diameter of 0.25 mm. The thermocouples were fixed 0.5 m below the tunnel ceiling 30 and 40 m downstream of the fire (Fig. 1). The specification of burner employed is given in Table 1. Heptane was used as the fuel.

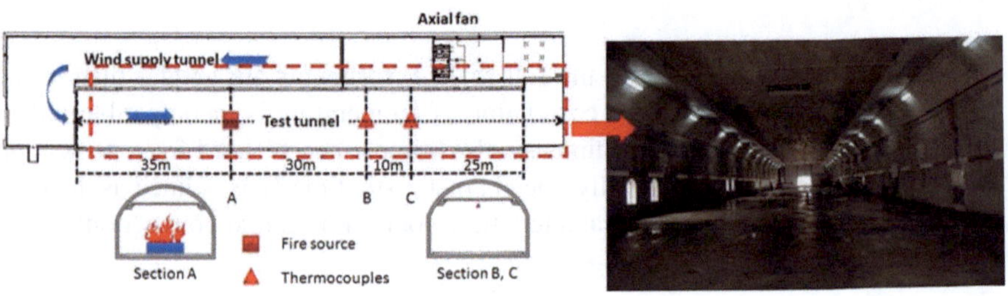

Fig. 1 Interior photograph of the full-scale test tunnel and measuring units

Table 1 List of experiments performed in the full-scale tunnel

Burner no.	Dimension	Location	Notes
Burner 1	1.5 m × 1.5 m	Cross section A, center, 1 m height	$V_{\text{wind}} = 0$, heptane
Burner 2	1.5 m × 3.0 m	Cross section A, center, 1 m height	$V_{\text{wind}} = 0$, heptanes, longer side set along tunnel axis

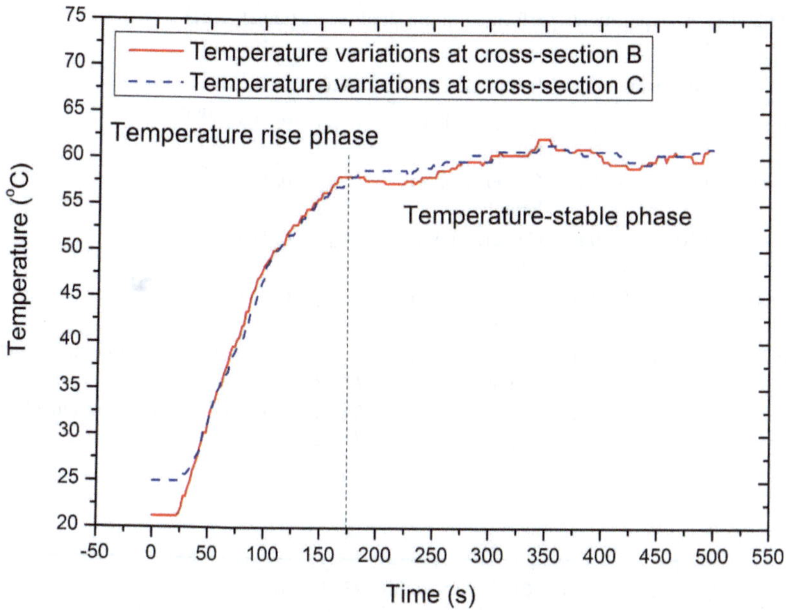

Fig. 2 Diagram and measurement units of the small-scale tunnel (20:1)

Small-Scale Experiments

A reduced scale model tunnel (20:1) was built to validate the reliability of the scaling model (referred by Fig. 2). Heptane pool fires of different sizes were used to simulate the fire source. The pools were set at the same position in each test. The smoke was collected and exhausted by a 1.0 m × 1.0 m × 1.5 m hood where the heat release rate was measured for flow velocity and oxygen concentration. The temperature was measured by thermocouples installed along the tunnel ceiling central line. The mass loss was recorded by an electronic balance. A series of experiments were carried out in this reduced scale tunnel, as shown in Table 2.

Results and Discussion

In full-scale experimental tests, the temperature variations at the ceiling of cross sections B and C were recorded. Figures 3 and 4 show the temperature history. The mass losses of fuel are shown in Figs. 5 and 6.

In test 1, the size of the steel pool is 1.5 m × 1.5 m, with 0.04 m depth heptanes liquid. Temperature results indicate that, 175 s after ignition, the temperature of cross sections B and C reached maximum value and kept steady. In test 2, two square steel pools were combined. The long side was set along the tunnel axis. The dimension was 1.5 m × 3.0 m. In this test, 85 s after ignition, the fire was burned out gradually, making the temperature drop.

Table 2 List of experiments performed in the small-scale tunnel (20:1)

Burner no.	Dimension	Location	Notes
Burner 3	0.23 m × 0.23 m	Center, 0.05 m height	$V_{wind} = 0$, $D \propto L^{*0.625}$
Burner 4	0.15 m × 0.15 m	Center, 0.05 m height	$V_{wind} = 0$, $D \propto L^{*0.75}$
Burner 5	0.10 m × 0.10 m	Center, 0.05 m height	$V_{wind} = 0$, $D \propto L^{*0.9}$
Burner 6	0.08 m × 0.08 m	Center, 0.05 m height	$V_{wind} = 0$, $D \propto L^{*}$
Burner 7	0.23 m × 0.46 m	Center, 0.05 m height	$V_{wind} = 0$, longer side set along tunnel axis, $D \propto L^{*0.625}$
Burner 8	0.08 m × 0.16 m	Center, 0.05 m height	$V_{wind} = 0$, longer side set along tunnel axis, $D \propto L^{*}$

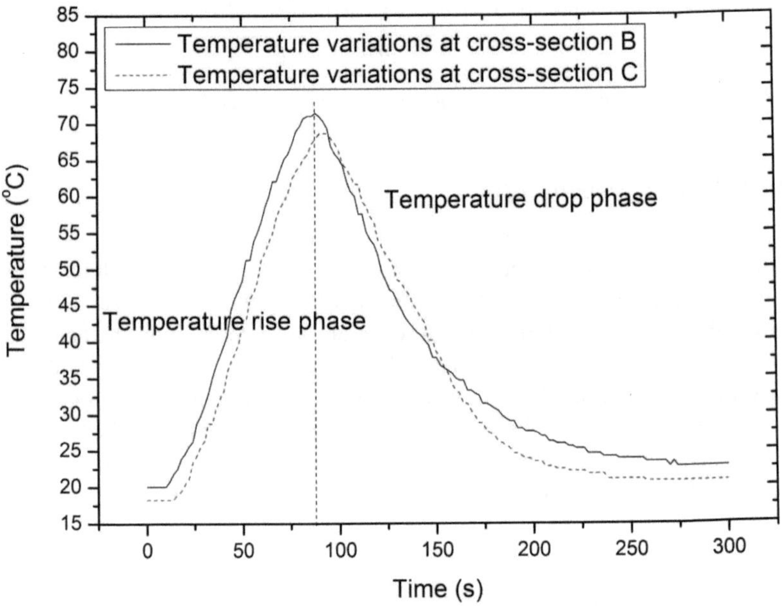

Fig. 3 Temperature variations at the ceiling of cross sections B and C in the full-scale tests (burner 1)

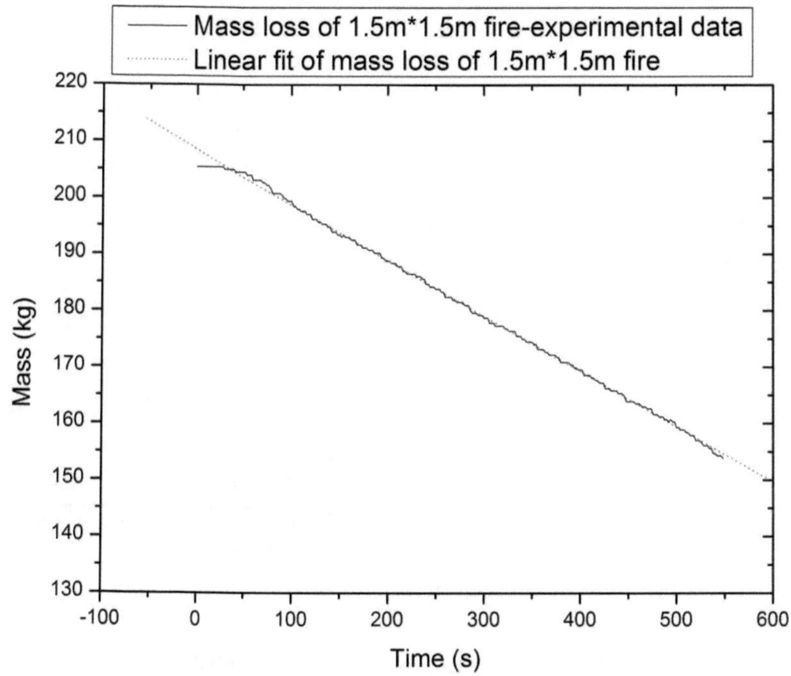

Fig. 4 Temperature variations at the ceiling of cross sections B and C in the full-scale tests (burner 2)

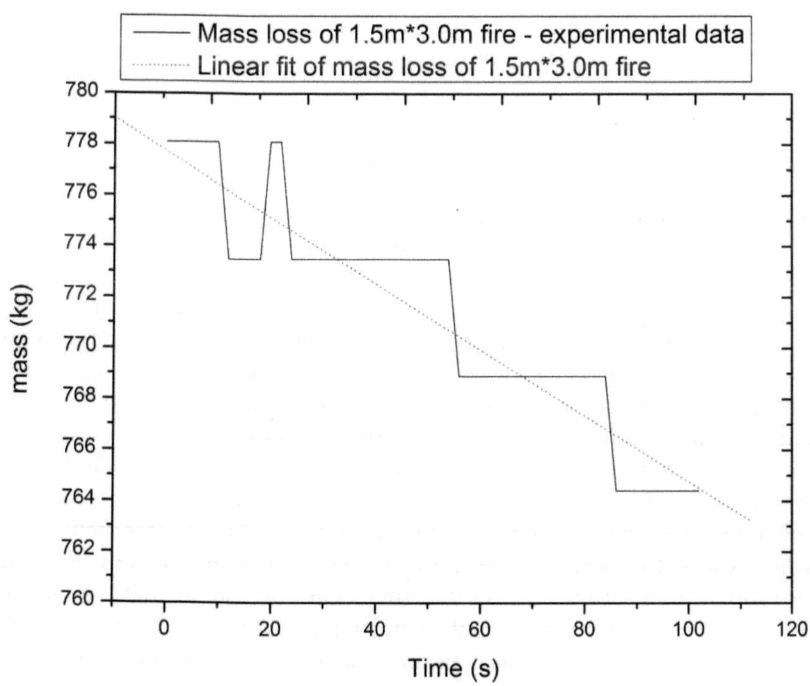

Fig. 5 Mass loss of 1.5 m × 1.5 m fire in the full-scale tests

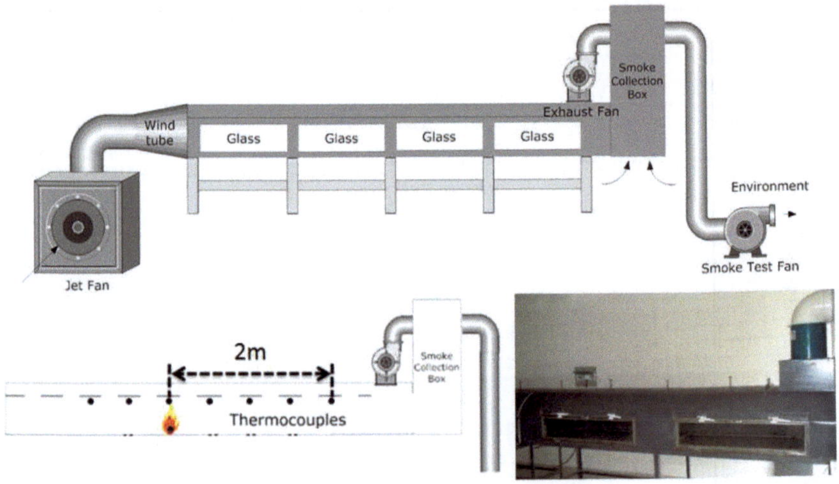

Fig. 6 Mass loss of 1.5 m × 3.0 m fire in the full-scale tests

The mass losses of the fuel for test 1 and 2 were established as follows:

$$\begin{aligned} Y &= 208.469 - 0.09809x \\ Y &= 777.729 - 0.12973x \end{aligned} \tag{14}$$

The mass burning rates are 98.09 and 129.73 g/s for tests 1 and 2, respectively.

Small-scale experimental tests labeled as burner 3–6 were arranged for investigating the new scaling criteria; the dimensionless temperature T^* was set as

$$T^* = \frac{T - T_0}{T_0} \tag{15}$$

Dimensionless time τ was obtained from Eq. (13). Typical results of the correlation of the dimensionless temperatures are shown in Figs. 7, 8, and 9.

Figure 7 shows the temperature correlation at the ceiling of cross section B. Burner 6 (0.08 m × 0.08 m) is derived from previous model; the results show clearly that the temperature variation is inconsistent with the full-scale tests either in the rising trend or maximum temperature value. The results also demonstrate that the temperature rise of burner 3 (0.23 m × 0.23 m) is approximate to the result of burner 1 in the temperature-rise phase. The rise trends of the two curves match very well. In the temperature-stable phase, the maximum difference is within 0.002. The new scaling model is more reliable than the previous model.

Figure 8 shows the temperature correlation at the ceiling of cross section C. The results show that the temperature rise of burner 3 (0.23 m × 0.23 m) is approximated to the result of burner 1 in the temperature-rise phase, and the maximum difference is smaller than 0.001 in the temperature-stable phase. It is obvious that the temperature correlation by the new model at this cross section is better than that at cross section B.

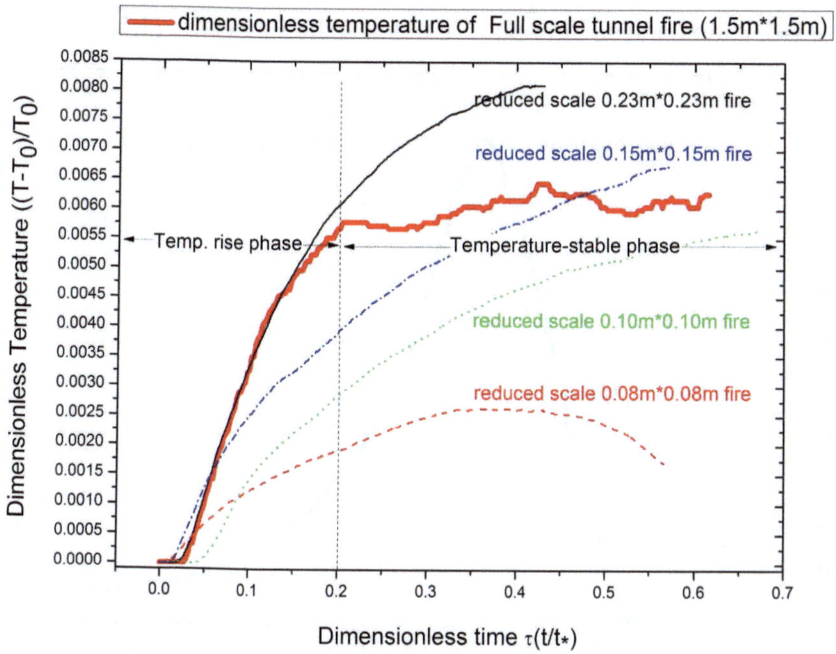

Fig. 7 Temperature correlations of two scale experiments at the ceiling of cross section B

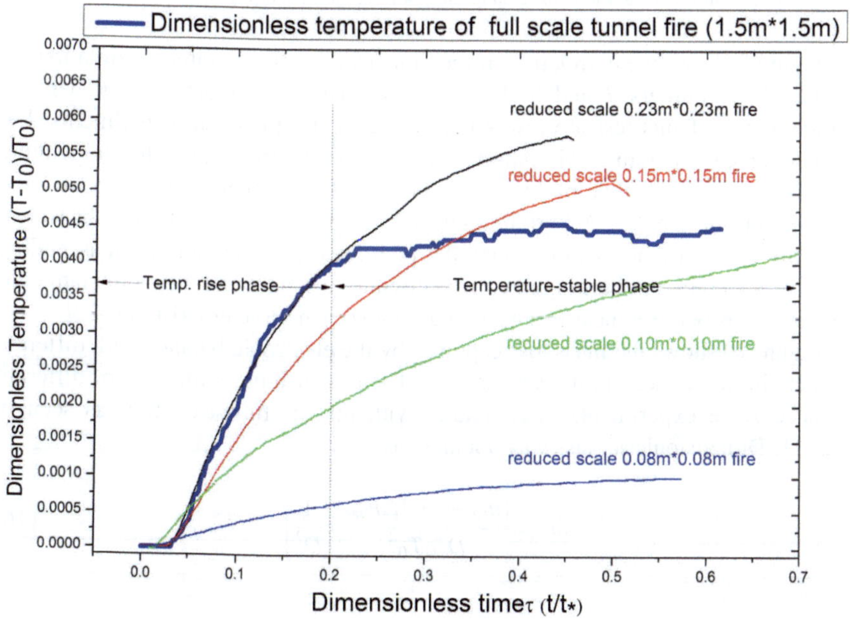

Fig. 8 Temperature correlations of square pool fires at the ceiling of cross section C

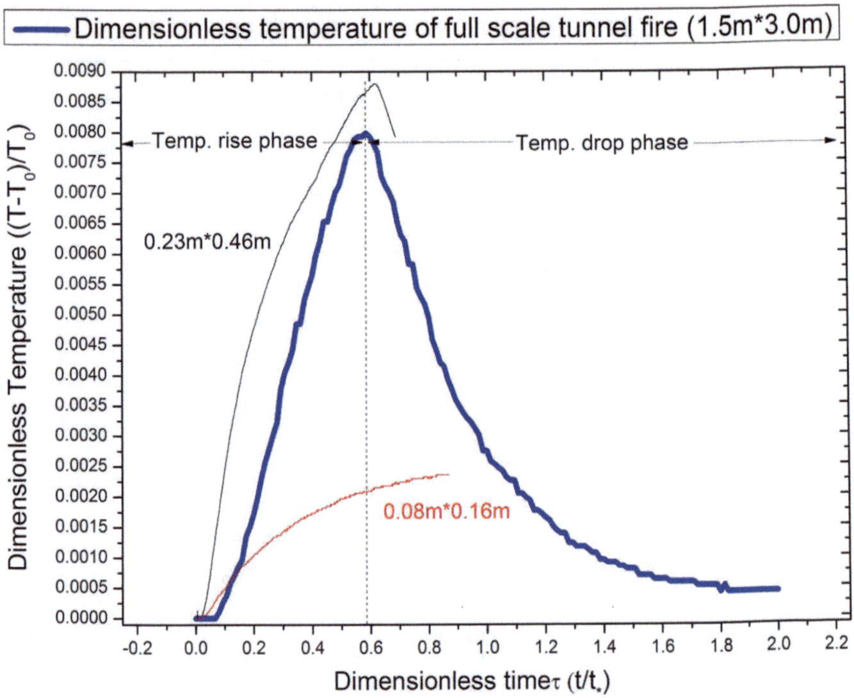

Fig. 9 Temperature correlations of rectangular pool fires at the ceiling of cross section C

Figure 9 shows the correlation of the dimensionless temperature correlation for burner 2 vs. burners 7 and 8. Burner 8 is derived from previous model; the temperatures of this test are lower than those of the full-scale experiments. For burner 7, the temperature correlation is better than for burner 8, but the temperature rise is greater than that of burner 2, both in the temperature-rise phase and temperature-drop phase. According to Figs. 7, 8, and 9, the correlation of the square pool fire is better than that of rectangular pool fires, because there is an ascending order of the quiescent burning rate, from circular to square and then to rectangular. The relations between heat feedback and pool sizes no longer exist [18–22].

Figure 10 shows the mass loss captured by the electronic balance with different burners in the reduced scale experiments. The dimensionless mass $\hat{m}$ of different reduced scale experiments is compared with that of full-scale test, as seen in Fig. 11. Dimensionless mass was obtained as

$$\hat{m} = \frac{(m_0 - \dot{m}_f)\Delta h_{\text{vap}}}{D^2 \sigma {T_0}^4} \left[\frac{1}{D^2}\right] \quad (16)$$

The mass loss of experiment with burner 3 agrees with the full-scale test better than that of the other burners, which accords with the temperature correlation results.

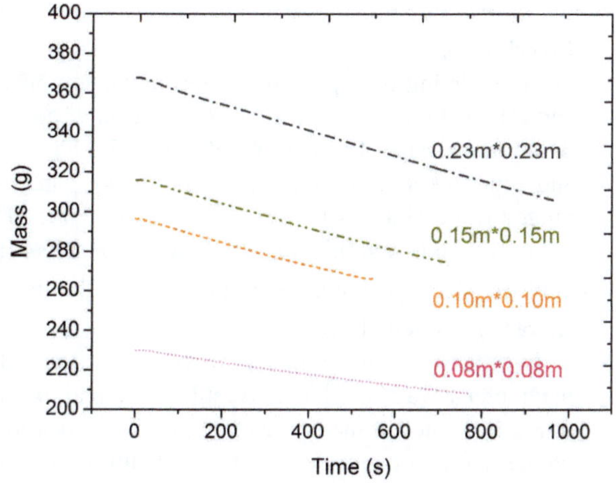

Fig. 10 Mass loss of different burners in small-scale experiments

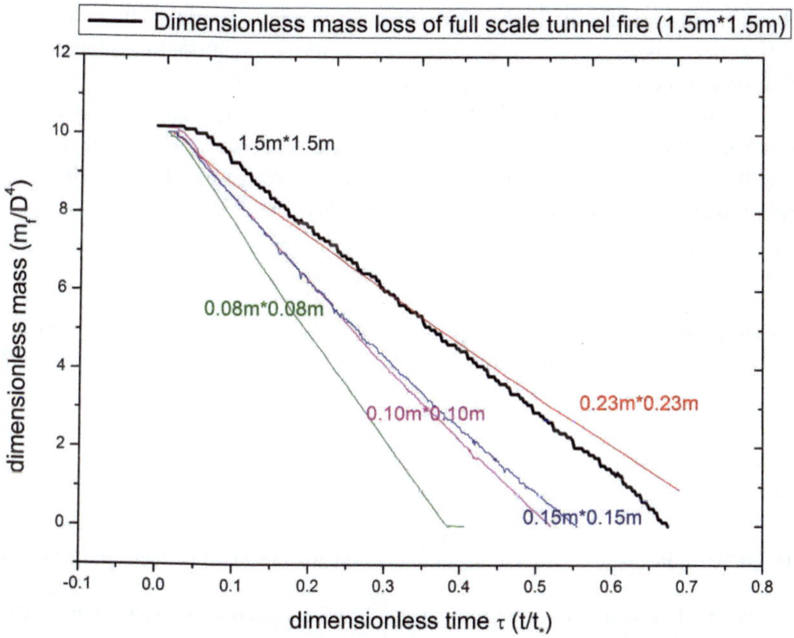

Fig. 11 Mass loss correlations of two scale experiments

Conclusions

Scale modeling is helpful for exploring the complex behavior of tunnel fires and establishing general results. The effects of pool sizes and smoke heating duration on the scale modeling of tunnel fire have been studied theoretically and experimentally in this paper. The square pools in the reduced scale tests ranged from 0.08 m × 0.08 m to 0.23 m × 0.23 m. The rectangular pools are 0.8 m × 0.16 m and 0.23 m × 0.46 m. The results indicate that the portion of pool fire sizes $\frac{D_f}{D_m}$ should be proportional to the five-eighth power law of the characteristic length L^*.

In order to eliminate time diversity and offer better comparability, a new parameter τ was introduced to replace the real experimental time. The scale time τ is defined as the fire-heating time for upper hot smoke layer. It depends on the mass burning rate $\dot{m}_f$ and the combustion products.

When $\tau < 0.2$, it is at the temperature-rise phase. The temperature correlation of the new model matches well with the full-scale experiment. When $\tau > 0.2$, it is at the temperature-stable phase. The temperature correlation of the new model is still better than that of previous model. The maximum difference is within 0.002. The correlation of the square pool fire is better than that of the rectangular pool fire.

The new scale model presents an approximate criterion about the scale relation of the pool sizes between full-scale tests and small-scale experiments. This new model can help to provide more information on the scale relationships. Wind effects and tunnel boundary heat conduction effects will be considered in the further studies.

Acknowledgment This study was supported by the National Natural Science Foundation of China No. 50974110 and the Hong Kong Research Grant Council's Grant No. CityU118708.

References

1. Beard, A.: A theoretical model of major fire spread in a tunnel. Fire. Technol **42**(4), 303–328 (2006)
2. Haack, A.: Fire protection in traffic tunnels: general aspects and results of the EUREKA project. Tunn. Undergr. Space Technol. **13**(4), 377–381 (1998)
3. Haack, A.: Current safety issues in traffic tunnels. Tunn. Undergr. Space Technol. **17**(2), 117–127 (2002)
4. Hu, L.H., Huo, R., Li, Y.Z., et al.: Full-scale burning tests on studying smoke temperature and velocity along a corridor. Tunn. Undergr. Space Technol. **20**(3), 223–229 (2005)
5. Hu, L.H., Huo, R., Peng, W., et al.: On the maximum smoke temperature under the ceiling in tunnel fires. Tunn. Undergr. Space Technol. **21**(6), 650–655 (2006)
6. Lönermark, A., Ingason, H.: Gas temperatures in heavy goods vehicle fires in tunnels. Fire Saf. J. **40**(6), 506–527 (2005)

7. Méret, O., Vauquelin, O.: A model to evaluate tunnel fire characteristics. Fire Saf. J. **34**(4), 393–401 (2000)
8. Oka, Y., Atkinson, G.T.: Control of smoke flow in tunnel fires. Fire Saf. J. **25**(4), 305–322 (1995)
9. Wu, Y., Bakar, M.Z.A.: Control of smoke flow in tunnel fires using longitudinal ventilation systems - a study of the critical velocity. Fire Saf. J. **35**(4), 363–390 (2000)
10. Roh, J.S., Ryou, H.S., Kim, D.H., et al.: Critical velocity and burning rate in pool fire during longitudinal ventilation. Tunn. Undergr. Space Technol. **22**(3), 262–271 (2007)
11. Roh, J.S., Yang, S.S., Ryou, H.S., et al.: An experimental study on the effect of ventilation velocity on burning rate in tunnel fires–heptane pool fire case. Build. Environ. **43**(7), 1225–1231 (2008)
12. Emori, R.I., Saito, K.: A study of scaling laws in pool and crib fires. Combust. Sci. Technol. **31** (5), 217–231 (1983)
13. Quintiere, J.G.: Scaling applications in fire research. Fire Saf. J. **15**(1), 3–29 (1989)
14. Hamins, A., Fischer, S.J., Kashiwagi, T., et al.: Heat feedback to the fuel surface in pool fires. Combust. Sci. Technol. **97**(1), 37–62 (1994)
15. Thomas, P.H.: Dimensional analysis: a magic art in fire research? Fire Saf. J. **34**(2), 111–141 (2000)
16. Wakatsuki, K., Jackson, G.S., Hamins, A., et al.: Effects of fuel absorption on radiative heat transfer in methanol pool fires. Proc. Combust. Inst. **31**(2), 2573–2580 (2007)
17. Holborn, P.G., Bishop, S.R., Drysdale, D.D., et al.: Experimental and theoretical models of flashover. Fire Saf. J. **21**(3), 257–266 (1993)
18. Koseki, H., Yumoto, T.: Air entrainment and thermal radiation from heptane pool fires. Fire. Technol **24**(1), 33–47 (1988)
19. Babrauskas, V.: Estimating large pool fire burning rates. Fire. Technol **19**(4), 251–261 (1983)
20. Koseki, H.: Combustion properties of large liquid pool fires. Fire. Technol **25**(3), 241–255 (1989)
21. Koseki, H., Mulholland, G.W.: The effect of diameter on the burning of crude oil pool fires. Fire. Technol **27**(1), 54–65 (1991)
22. Woods, J.A.R., Fleck, B.A., Kostiuk, L.W.: Effects of transverse air flow on burning rates of rectangular methanol pool fires. Combust. Flame **146**(1–2), 379–390 (2006)

Section B
Fire and Explosion - Scale-Model Experiment of Wind-Generated Fire Whirls

Kazunori Kuwana, Kozo Sekimoto, and Kozo Saito

Abstract This paper reports a series of scale-model experiments of wind-generated fire whirls that occurred in the Hifukusho-ato area shortly after the Great Kanto Earthquake in Tokyo in 1923. The conditions of the scale-model experiments are determined based on scaling consideration. Since it was difficult to make the fuel Froude number (or dimensionless heat release rate) of the scale model equal to that of the prototype, it is proposed to relax the scaling requirement. Then, the Froude number based on the flame height is used to determine the lateral wind velocity of the scale model. The flame height is chosen to be the important length scale because it significantly influences the magnitude of buoyant flow velocity. The scale-model experiments reproduced similar fire whirls to those observed in the Hifukusho-ato area and those reported in our earlier studies.

Introduction

The interaction between a fire-induced buoyant flow and lateral wind may result in the occurrence of a fire whirl. A well-known example is the fire whirls that occurred in Tokyo in 1923 shortly after the Great Kanto Earthquake; the fire whirls killed 38,000 people in less than 15 min. Soma and Saito [1] suggested that the interaction of mass fires with lateral wind was responsible for the generation of the intense fire whirls.

Because of the devastating damage that a fire whirl may cause, a number of studies were conducted to understand the basic nature of fire whirls [2–11]. In most of these studies, however, fire whirls were generated by either a rotating screen or a fixed frame with asymmetrically placed openings. There are, however, several

K. Kuwana (✉)
Department of Chemistry and Chemical Engineering, Yamagata University, 4-3-16 Jonan, Yonezawa-shi, Yamagata 992-8510, Japan
e-mail: kuwana@yz.yamagata-u.ac.jp

K. Sekimoto • K. Saito
Institute of Research for Technology Development (IR4TD), University of Kentucky, Lexington, KY 40506-0503, USA

K. Saito et al. (eds.), *Progress in Scale Modeling, Volume II*,
DOI 10.1007/978-3-319-10308-2_13

studies that discuss the interaction between fire and lateral wind that results in the occurrence of fire whirls [1, 12–16]. This paper also reports the effect of lateral wind on fire-whirl generation, with emphasis on developing the scaling law to conduct the scale-model experiment of large-scale wind-generated fire whirls.

Wind-Tunnel Experiment

Experimental Setup

Figure 1a shows a geographical map of the place, called Hifukusho-ato, where the abovementioned fire whirls occurred in 1923. Fig. 1b shows a schematic diagram of our scale-model experiment [16], where an L-shaped hexane pool fire is used to simulate mass fires surrounding the Hifukusho-ato area. A wind tunnel is used to apply uniform horizontal airflow to the hexane pool fire. The lateral wind velocity of the wind tunnel can be as low as 0.2 m/s. The dimension of the scale model is approximately 1/10,000th that of the prototype. In Kuwana et al. [14, 15], we reported the results of a similar but larger-scale experiment, that is, 1/1000th of the prototype. Scale-model experiments of different scales are expected to provide insights into the scale effect on the generation of fire whirls of this type.

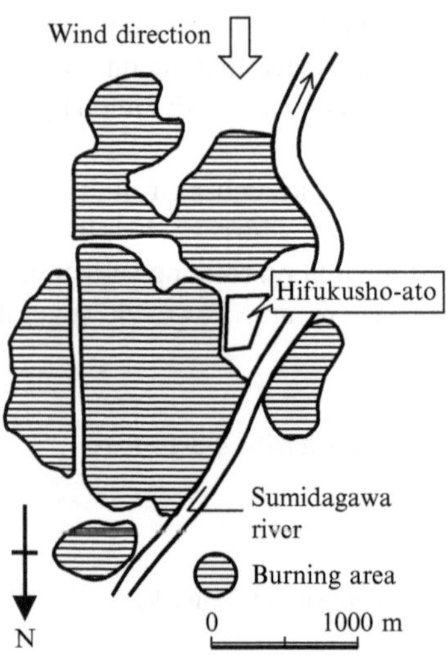

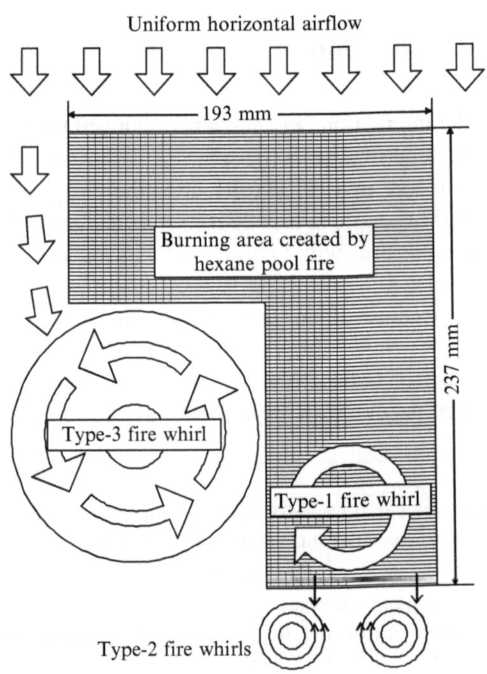

Fig. 1 (**a**) Hifukusho-ato fire whirls and (**b**) schematic of the 1/10,000th scale-model experiment and of fire whirls that were observed

Scaling Consideration

In this section, we discuss scaling laws to reproduce Hifukusho-ato fire whirls using a scale model, in particular, how to determine the lateral wind velocity of the scale-model experiment. The discussion below follows Ref. [16].

In general, to simulate a large-scale, buoyancy-controlled fire phenomenon using a scale model, the fuel Froude numbers (or equivalently dimensionless heat release rates) of the prototype and the scale model are made equal. The fuel Froude number, Fr_f, is defined as follows:

$$\mathrm{Fr}_f = \frac{m^2}{\rho_0^2 g L}, \tag{1}$$

where m is the mass of fuel consumed per unit burning area per unit time, ρ_0 is the air density at the ambient temperature, g is the acceleration of gravity, and L is the horizontal length scale of the burning area. The first scaling requirement is, therefore,

$$\left(\frac{m^2}{\rho_0^2 g L}\right)_{\text{scale model}} = \left(\frac{m^2}{\rho_0^2 g L}\right)_{\text{prototype}}. \tag{2}$$

The lateral wind velocity, U, is then determined to satisfy

$$\left(\frac{U^2}{gL}\right)_{\text{scale model}} = \left(\frac{U^2}{gL}\right)_{\text{prototype}}, \tag{3}$$

which is the second requirement. When both Eqs. (2) and (3) are satisfied, the scale model is expected to simulate the prototype well. Since the flame height, H, of a buoyancy-controlled fire is known to depend strongly on Fr_f as

$$\frac{H}{L} \sim \mathrm{Fr}_f^n, \text{ where } 1/5 < n < 1/3, \tag{4}$$

the scaling requirement Eq. (2) is equivalent, requiring that H/L of the scale model and the prototype be equal. The flame height of the mass fires near the Hifukusho-ato area is estimated to have been about 10 m. The flame height of our 1/10,000th scale-model experiment, consequently, needs to be about 1 mm, which is difficult to achieve if we are to use a hexane pool fire to simulate the mass fires. Therefore, we now propose to relax the scaling requirement Eq. (2).

The question now arises: How do we determine the lateral wind velocity, U, of the scale model? In other words, does Eq. (3) yield reasonable U even under conditions where $(\mathrm{Fr}_f)_{\text{scale model}} \neq (\mathrm{Fr}_f)_{\text{prototype}}$? In this study, the following new scaling law is proposed instead of Eq. (3):

$$\left(\frac{U^2}{gH}\right)_{\text{scale model}} = \left(\frac{U^2}{gH}\right)_{\text{prototype}}. \tag{5}$$

In Eq. (5), the flame height, H, is used as the important length scale instead of L, the horizontal length scale of the burning area. The idea behind Eq. (5) is that U must be on the order of buoyant upward velocity, U_b, which is the important characteristic velocity of the system. U_b can be estimated as $U_b \sim (gH)^{1/2}$; and consequently, the scaling law Eq. (5) is obtained. Combining Eqs. (4) and (5), the scaling law can be written in a different form, i.e.,

$$\frac{U}{(gL)^{1/2}} \sim \mathrm{Fr}_f^{n/2}, \text{ where } 1/5 < n < 1/3. \tag{6}$$

Equation (6) can be interpreted as an equation to estimate the lateral wind velocity that may result in the occurrence of wind-generated fire whirls. In Kuwana et al. [15], we showed that Eq. (6) can be applied to a surprisingly wide range of fire whirls.

The lateral wind velocity around the time when the Hifukusho-ato fire whirls occurred is considered to have been about 4 m/s [1], though the estimation is rather difficult [15]. Hence, $U_{\text{prototype}} \approx 4$ m/s. As mentioned above, $H_{\text{prototype}}$ is estimated to have been on the order of 10 m. The typical flame height of the hexane pool fire used in this study (Fig. 1b) is on the order of 0.1 m. Equation (5), therefore, yields $U_{\text{scale model}} \approx 0.4$ m/s, and so we conducted the scale-model experiments at the lateral wind velocity of 0.4 m/s. The 1/1,000th scale-model experiments reported in Kuwana et al. [14, 15] were conducted at the lateral wind velocity of 1 m/s based on similar scaling considerations. In the next section, we compare the results of the present 1/10,000th scale-model experiment with those of the previous 1/1,000th scale-model experiment to confirm that both scale-model experiments generate similar fire whirls; then, the proposed scaling law Eq. (5) can be validated.

Results and Discussion

Fire Whirls Observed

Figure 1b above also shows schematically fire whirls that were identified from the movie films recorded and from direct observation during the experiments. As in our earlier studies [14, 15], the fire whirls are found to be categorized into three different types:

[Type 1] Fire whirls spinning over the downstream-side burning area creating tall fire columns (Fig. 2)
[Type 2] Fire whirls periodically spinning off from the burning area and traveling to the downstream unburned area (Fig. 3)
[Type 3] Relatively stable spinning of air initially without fire in the unburned area but then entraining fires into its spinning motion from the burning area (Fig. 4)

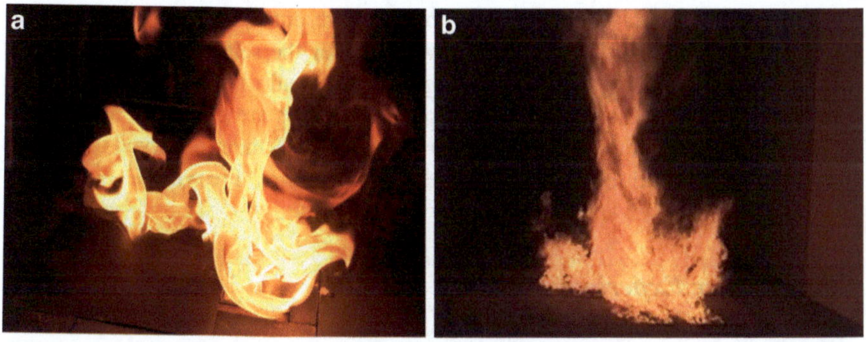

Fig. 2 Type 1 fire whirls. (**a**) 1/10,000 scale model [16] ((©) The Society of Chemical Engineers, Japan 2010, reprinted with permission). (**b**) 1/1000 scale model [15] ((©) Elsevier 2008, reprinted with permission)

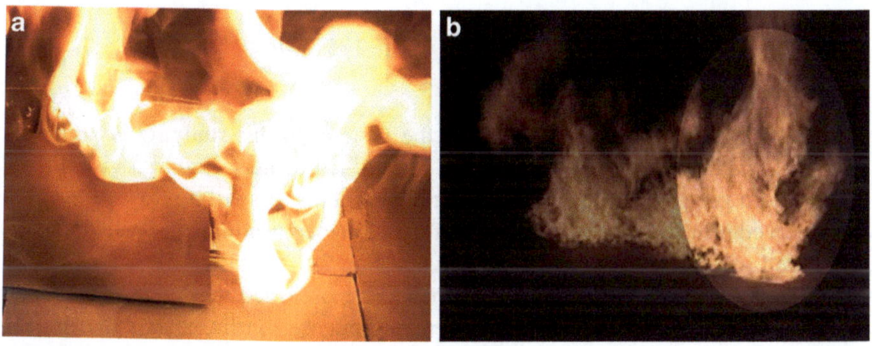

Fig. 3 Type 2 fire whirls. (**a**) 1/10,000 scale model. (**b**) 1/1,000 scale model [15] ((©) Elsevier 2008, reprinted with permission)

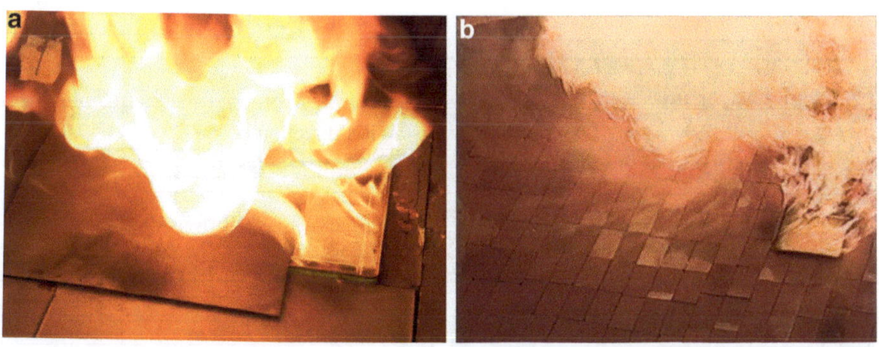

Fig. 4 Type 3 fire whirls. (**a**) 1/10,000 scale model [16] ((©) The Society of Chemical Engineers, Japan 2010, reprinted with permission). (**b**) 1/1,000 scale model [15] ((©) Elsevier 2008, reprinted with permission)

Figures 2, 3, and 4 confirm that both the present 1/10,000th and previous 1/1,000th models generate similar fire whirls under the selected values of lateral wind velocity, validating our scaling consideration and thus scaling law Eq. (5). Among different types of fire whirls, type 3 fire whirls are generated at locations geometrically similar to the Hifukusho-ato area, and their observed behaviors match the description that was given of the Hifukusho-ato fire whirls.

Plate-Heater Experiment

Following the scale-model experiments using the L-shaped hexane pool fire, we conducted similar scale-model experiments using plate heaters. Plate heaters are placed to form an L-shaped heating area, which is similar to the burning area in Fig. 1b. Equation (6), which yields the lateral wind velocity that may result in the occurrence of fire whirls, can be modified for application to plate-heater experiments.

$$\frac{U}{(gL)^{1/2}} \sim \left(\frac{Q^2}{gL^5}\right)^{n/2}, \text{ where } 1/5 < n < 1/3. \tag{7}$$

Here, Q is the total heat release rate of the system. In this study, Q is varied by changing the duty ratio of the power switches of the heaters. Figure 5 visualizes a flow pattern generated by the plate-heater experiments. A swirling flow pattern can be observed near the downstream edge of the L-shaped heating area, which may be compared with the type 1 fire whirl shown in Fig. 2.

Similarly to the hexane pool experiments, we observed that there is a narrow range of lateral wind velocity that can generate intense swirling flow patterns. The velocity range depended on the heat release rate, Q. The critical wind velocity was about 0.2 m/s when the duty ratio was 0.5, while it was about 0.25 m/s at 100 % heating. These results suggest the relationship of $U \sim Q^{0.32}$; its agreement with the proposed scaling law, Eq. (7), is reasonably good.

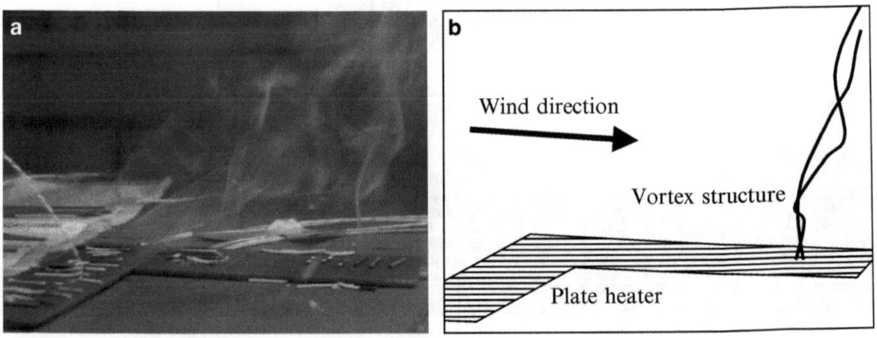

Fig. 5 Swirling motion generated by the scale model using plate heaters. The direction of lateral wind is from *left* to *right*. (**a**) Photo image. (**b**) Diagram

Conclusions

Results of 1/10,000th scale-model experiments are reported. A scaling relationship to determine the lateral wind velocity that can reproduce prototype wind-generated fire whirls using a scale model is proposed. The proposed scaling law is validated by comparing the results of the present scale-model experiment with those of the previous 1/1,000th scale-model experiment.

Acknowledgments This study was supported in part by Japan's Grant-in-Aid for Scientific Research (19860021), in part by the Promotion Program for Fire and Disaster Prevention Technologies, Fire and Disaster Management Agency, Japan, and in part by a general fund from University of Kentucky's Institute of Research for Technology Development.

References

1. Soma, S., Saito, K.: Reconstruction of fire whirls using scale models. Combust. Flame **86**, 269–284 (1991)
2. Emmons, H.W., Ying, S.J.: The fire whirl. Proc. Combust. Inst. **11**, 475–488 (1967)
3. Byram, G.M., Martin, R.E.: The modeling of fire whirlwinds. Forest Sci. **16**, 386–399 (1970)
4. Muraszew, A., Fedele, J.B., Kuby, W.C.: The fire whirl phenomenon. Combust. Flame **34**, 29–45 (1979)
5. Satoh, K., Yang, K.T.: Simulations of swirling fires controlled by channeled self-generated entrainment flows. In: Fire Safety Science–Proceedings of the Fifth International Symposium, International Association for Fire Safety Science, Melbourne, Australia, pp. 201–212 (1997)
6. Battaglia, F., Rehm, R.G., Baum, H.R.: The fluid mechanics of fire whirls: an inviscid model. Phys. Fluids **12**, 2859–2867 (2000)
7. Battaglia, F., McGrattan, K.B., Rehm, R.G., Baum, H.R.: Simulating fire whirls. Combust. Theory Modell. **4**, 123–138 (2000)
8. Snegirev, A.Y., Marsden, J.A., Francis, J., Makhviladze, G.M.: Numerical studies and experimental observations of whirling flames. Int. J. Heat Mass Trans. **47**, 2523–2539 (2004)
9. Hassan, M.I., Kuwana, K., Saito, K., Wang, F.: Flow structure of a fixed-frame type fire whirl. In: Fire Safety Science–Proceedings of the Eighth International Symposium, International Association for Fire Safety Science, Beijing, China, pp. 951–962 (2005)
10. Chuah, K.H., Kushida, G.: The prediction of flame heights and flame shapes of small fire whirls. Proc. Combust. Inst. **31**, 2599–2606 (2007)
11. Chuah, K.H., Kuwana, K., Saito, K.: Modeling a fire whirl generated over a 5-cm diameter methanol pool fire. Combust. Flame **156**, 1828–1833 (2009)
12. Emori, R.I., Saito, K.: Model experiment of hazardous forest fire whirl. Fire. Technol. **18**, 319–327 (1982)
13. Liu, N., Liu, Q., Deng, Z., Satoh, K., Zhu, J.: Burn-out time data analysis on interaction effects among multiple fires in fire arrays. Proc. Combust. Inst. **31**, 2589–2597 (2007)
14. Kuwana, K., Sekimoto, K., Saito, K., Williams, F.A., Hayashi, Y., Masuda, H.: Can we predict the occurrence of extreme fire whirls? AIAA J. **45**, 16–19 (2007)
15. Kuwana, K., Sekimoto, K., Saito, K., Williams, F.A.: Scaling fire whirls. Fire Saf. J. **43**, 252–257 (2008)
16. Kuwana, K., Sekimoto, K., Saito, K.: Wind-tunnel fire-whirl experiments using a scale model. J. Chem. Eng. Jpn. **43**, 952–955 (2010)

Section C
Structures - Seismic Behavior of Batter-Pile Foundation Based on Centrifuge Tests

Hiroyuki Kimata, Takashi Tazoh, Jiho Jang, Yoichi Taji, Yasunobu Sakai, and George Gazetas

Abstract We carried out centrifuge tests to clarify the seismic behavior of batter-pile foundations. A vertical-pile foundation and a batter-pile foundation without the presence of a superstructure were installed parallel to each other in a soil container filled with dry sand and were excited simultaneously. Through a comparison of the acceleration and displacement response of the footing, as well as the bending and axial strain of the piles for the two pile foundations, the seismic behavior of the batter-pile foundation was experimentally investigated.

Introduction

The lateral stiffness of a pile foundation can be increased by adopting batter piles, which is why they are commonly used in landing piers that are subject to large lateral forces. However, batter piles are seldom used for buildings or civil engineering structures even in the case of large lateral forces for the following reasons:

H. Kimata (✉) • Y. Sakai
Civil Engineering Technology Division, Shimizu Corporation, 2-16-1 Kyobashi, Chuo-ku, Tokyo 104-8370, Japan
e-mail: kimata@shimz.co.jp

T. Tazoh
Toyama Prefectural University, 5180 Kurokawa, Imizu-shi, Toyama 939-0398, Japan

J. Jang
Research Institute, GS Engineering & Construction, 417-1, Deokseong-ri, Idong-myeon, Cheoin-gu, Yongin-si, Gyeonggi-do 449-831, South Korea

Y. Taji
Institute of Technology, Shimizu Corporation, 3-4-17 Etchujima, Koto-ku, Tokyo 135-8530, Japan

G. Gazetas
School of Civil Engineering, National Technical University of Athens, Heroon Polytechneiou 9, 15780 Zografou, Greece

K. Saito et al. (eds.), *Progress in Scale Modeling, Volume II*,
DOI 10.1007/978-3-319-10308-2_14

(1) When soil settlement occurs, not only the safety of the pile foundation but also that of the structure as a whole system may be threatened by settlement-induced vertical loads acting on the batter piles.
(2) During an earthquake, the piles in a batter-pile foundation may be subject to excessive axial compression and pullout forces, which are not generated in a vertical-pile foundation.
(3) The strength of concrete piles is reduced by decreasing the compressive force acting on the piles due to rocking motions induced by the adopted batter piles.
(4) Since infinite lateral ground planes cannot be assumed for batter piles, they cannot be expected to have the same horizontal subgrade reaction as that of vertical piles.
(5) In urban areas, the use of batter piles is constrained by the boundary lines of adjacent land.

The 1995 Great Hanshin Earthquake in Japan has increased the demand for pile foundations with high seismic performance, as well as lower cost and easier construction. Batter piles can be used with little additional expense, no special design, and relatively easy construction. Therefore, the seismic behavior of batter piles has recently attracted much research interest, as has research and development related to easy and accurate methods of installing batter piles [1–14].

In this study, we carried out centrifuge shaking table tests to clarify the seismic behavior of batter-pile foundations. A vertical-pile foundation and a batter-pile foundation were installed parallel to each other in a soil container filled with dry sand and were excited simultaneously [12–14]. As our objective was to investigate the fundamental characteristics of the seismic behavior of batter piles, none of the pile foundation models had a superstructure. This study focused on the kinematic interaction of batter piles [10, 15–19]. Through a comparison of the acceleration and displacement response of the footing, as well as the axial and bending strain of the piles for the two pile foundations, the seismic behavior of the batter-pile foundation was experimentally studied.

Centrifuge Tests

The characteristics of soils are significantly sensitive to stress level. In model testing, one of the most important issues, therefore, is the problem of testing models whose stress dependency matches that of the prototype. Since it is very difficult under the gravitational field (1 g model tests), one common approach involves testing under increasing gravitational fields (centrifuge model tests). In this study, we carried out centrifuge model tests.

The most direct and effective way to quantitatively and qualitatively investigate the seismic behavior of batter piles is to compare the seismic behavior between a vertical-pile foundation and a batter-pile foundation under the same input motions. Each test for each model must be carried out under nearly identical conditions with

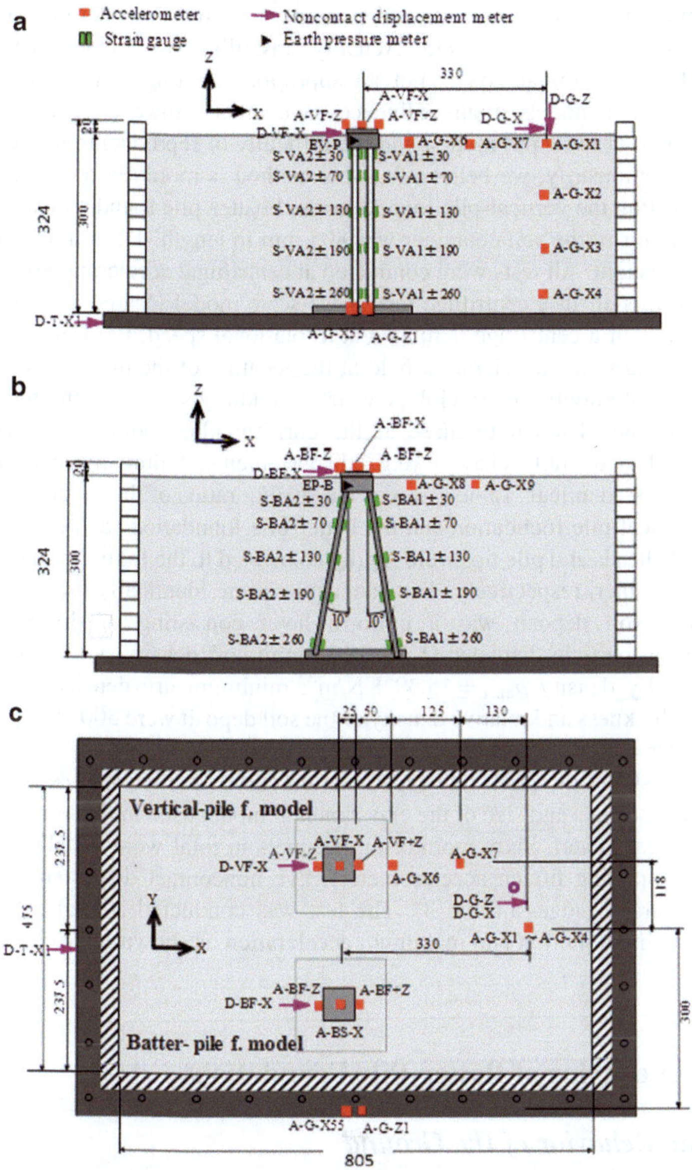

Fig. 1 Longitudinal sections and plan of the 1/30-scale centrifuge model (scale unit: mm). (**a**) Section: vertical-pile foundation model. (**b**) Section: batter-pile foundation model. (**c**) Plan

respect to input motions, soil conditions, and soil behavior. Note, however, that it is impossible to achieve complete similarity between shaking table tests due to the difficulty of reproducing the input motion and nonlinear behavior of the soil. Therefore, a vertical-pile foundation and a batter-pile foundation without the presence of a superstructure were installed parallel to each other in a soil container, as shown in Fig. 1, and were excited simultaneously.

A laminar box was used as the soil container to allow shear deformation of the soil deposit as in the free field. Actually, installing two models that behave differently in a laminar box is not an appropriate testing method because the behavior of the models might influence each other. However, considering the inconsistency of the input motion and the difficulty of reproducing the soil conditions and nonlinearity, we believe that this method is more reasonable than individually testing the vertical-pile foundation and batter-pile foundation separately.

The interior of the soil container was 805 mm in length, 475 mm in width, and 324 mm in height. All tests were conducted at centrifugal acceleration of 30 g on a 1/30-scale model. In a centrifuge test, a $1/\lambda$-scale model located at a distance, r, from the axis of a centrifuge is rotated at a rotational speed, $V=(\lambda/r)^{0.5}$, which is sufficient to raise the acceleration field at the location of the model to λ times the acceleration of gravity. In principle, the stress conditions at any point in the model should then be identical to those at the corresponding point in the full-scale prototype. The overall behavior (e.g., displacements, failure mechanisms, etc.) should also be identical. Table 1 shows the scaling ratios of the models.

The vertical-pile foundation and the batter-pile foundation each had four piles, and the pile heads and pile tips were rigidly connected to the footing and the base of the soil container, respectively. The batter piles were identically inclined at a 10° angle. The soil deposit was a uniform layer consisting of dry silica sand No. 7 (mean particle diameter $D_{50}=0.147$ mm; soil density $\rho_s=26.35$ KN/m^3; maximum dry density $\rho_{\max}=15.39$ KN/m^3; minimum dry density $\rho_{\min}=12.06$ KN/m^3). Thickness and relative density of the soil deposit were 300 mm (prototype: 9 m) and $Dr=60$ %, respectively.

Figure 2 shows the grain size accumulation curve of silica sand No. 7. Table 2 shows the materials and size of the experimental model used in the tests, and Fig. 3 shows the test model. Sixty monitoring channels in total were installed, with the sensors comprising fifteen accelerometers, five noncontact displacement meters, and forty strain gauges (Table 3). The test was conducted a total of 12 times, varying the input motion and maximum acceleration as shown in Table 4.

Seismic Behavior of Batter-Pile Foundations

Dynamic Behavior of the Ground

Figure 4 shows the frequency transfer function calculated by the acceleration records between the soil surface and the input motion of the sweep test. The predominant frequency of the ground is 3.5 Hz in the case of maximum acceleration of input motion at 5 Gal. The predominant frequencies are 3.2–3.3 and 3.0–3.1 Hz, and also the peak acceleration amplification factors decrease corresponding to the increase in maximum acceleration of the input motion to 15 and 30 Gal.

Table 1 Scaling ratios of the test model

Item		Symbol	Unit	Centrifuge model	Prototype	Scale
Sand stratum	Depth	H	m	0.3	9	1/λ
	Density	ρ_t	KN/m^3	15.0	15.0	1
Structure	Width	W	m	0.05	1.5	1/λ
	Height	H	m	0.04	1.2	1/λ
	Mass	M	kg	0.785	21,195	1/λ^3
Footing	Width	W	m	0.05	1.5	1/λ
	Height	H	m	0.03	0.9	1/λ
	Mass	M	kg	0.58875	15,896	1/λ^3
Pile	Diameter	D	m	0.01	0.3	1/λ
	Thickness	t	m	0.0002	0.006	1/λ
	Young's modulus	E	MN/m^2	2.06E + 05	2.06E + 05	1
	Area	A	m^2	6.16E-06	5.54E-03	1/λ^2
	Moment of inertia of area	I	m^4	7.40E-11	5.99E-05	1/λ^4
	Normal stiffness	EA	MN	1.27E + 00	1.14E + 03	1/λ^2
	Bending stiffness	EI	MN-cm^2	1.52E-09	1.23E-03	1/λ^4
Acceleration	Centrifuge	*g*	*g*	30	1	λ
	Earthquake	α	Gal	6,000	200	λ
Other parameters	Displacement	δ	m	1	30	1/λ
	Force	F	N	1	900	1/λ^2
	Stress	τ	KPa	1	1	1
	Strain	γ		1×10^{-6}	1×10^{-6}	1
	Time	t	s	1	30	1/λ
	Frequency	f	Hz	30	1	λ

Figure 5 shows the frequency transfer function between the ground surface and input motion obtained from El Centro record excitation. The predominant frequency of the ground is 3.4 Hz in the case of maximum acceleration of input motion at 50 Gal. The predominant frequencies are 2.8–2.9 and 2.4–2.5 Hz, and also the peak acceleration amplification factors decrease according to the increase in maximum acceleration of the input motion to 100 and 200 Gal. These phenomena were obviously produced by the nonlinearity of the soil.

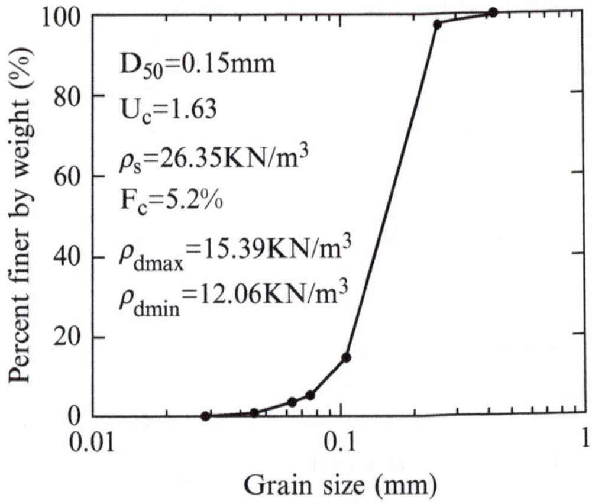

Fig. 2 Grain size accumulation curve of silica sand No. 7 (mean particle diameter $D_{50} = 0.15$ mm, soil density $\rho_s = 26.35$ KN/m^3, maximum dry density $\rho_{max} = 15.39$ KN/m^3, minimum dry density $\rho_{min} = 12.06$ KN/m^3)

Table 2 Materials and dimensions of the test model

Parts	Material and size
Laminar box	805 mm, 475 mm, 324 mm (inner size: length, width, depth)
Soil deposit	Dry sand: silica No. 7 (Dr = 60 %), thickness: 300 mm
Vertical pile	Stainless steel No. of piles: 4 (2 × 2), inclination angle: 0° Length: 270 mm, diameter: 10 mm, thickness: 0.2 mm
Batter pile	Stainless steel No. of piles: 4 (2 × 2), inclination angle: 10° Length: 274 mm, diameter: 10 mm, thickness: 0.2 mm
Footing	Steel Thickness: 30 mm, plan size: 50 mm × 50 mm

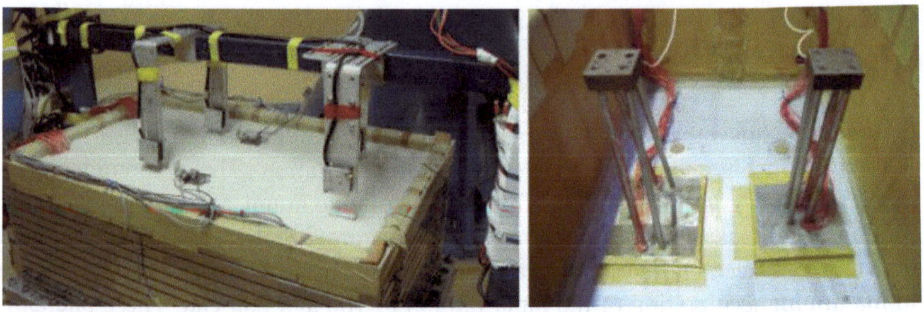

Fig. 3 Test model (the pile foundations have four piles)

Rotational Characteristics of the Footing

Figure 6 shows the relationship between horizontal displacement and rotational angle of the footing based on the data from sinusoidal excitation of 3.5 Hz, in order

Table 3 Installed sensors (60 monitoring channels were installed, with the sensors comprising 15 accelerometers, 5 noncontact displacement meters, and 40 strain gauges)

Transducer	Location	Direction	Number	Subtotal	Total
Accelerometer	Batter pile	X	1	15	60
		Z	2		
	Vertical pile	X	1		
		Z	2		
	Ground	X	6		
	Base	X	1		
	Table control	X	1		
	Centrifugal acc.	Z	1		
Noncontact displacement meter	Batter pile	X	1	5	
	Vertical pile	X	1		
	Ground	X	1		
		Z	1		
	Base	X	1		
Strain gauge	Batter pile	Pile-BA1	10	40	
		Pile-BA2	10		
	Vertical pile	Pile-VA1	10		
		Pile-VA2	10		

Table 4 Test cases

Input motion	Freq. of input motion	Max. acc. of input motion (Gal)	Test case no.
Sweep test motion	1.7–10 Hz	5	1-1
		15	1-2
		30	1-3
Sinusoidal excitation	3.5 Hz	50	2-1
		100	2-2
		200	2-3
Sinusoidal excitation	7.0 Hz	50	3-1
		100	3-2
		200	3-3
El Centro record	El Centro record N-S component	50	4-1
		100	4-2
		200	4-3

to investigate the rotational characteristics of the footing of the vertical-pile foundation and the batter-pile foundation. The rotational angle is calculated by dividing the difference in the vertical displacement based on the data of the accelerometers installed at both sides of the footing by the distance between the two accelerometers.

Fig. 4 Frequency transfer function of the ground surface obtained from sweep test (5, 15, and 30 Gal)

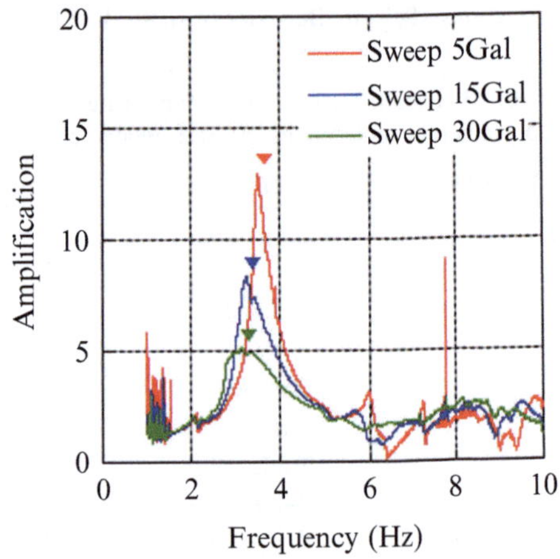

Fig. 5 Frequency transfer function of the ground surface obtained from the El Centro record excitations (50, 100, and 200 Gal)

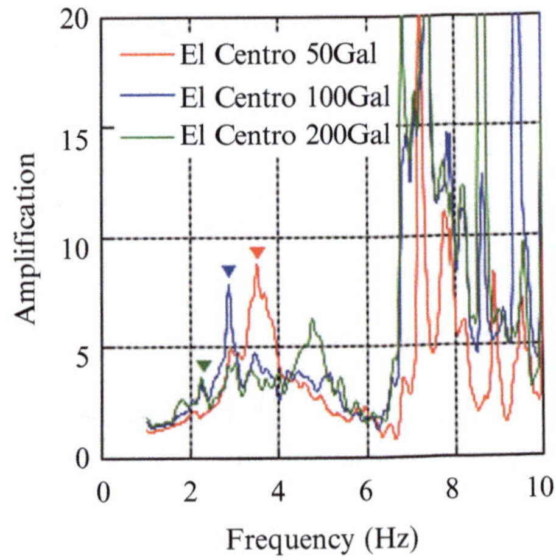

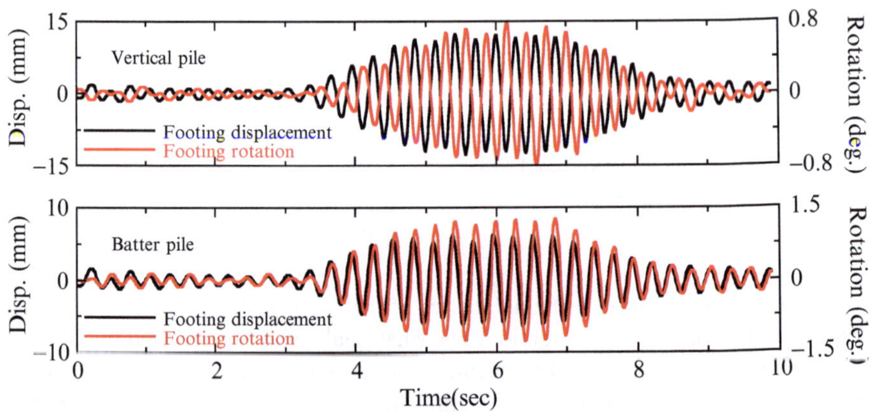

Fig. 6 Comparisons of horizontal displacement and rotational angle of the footings between the vertical-pile foundation and the batter-pile foundation (sinusoidal excitation: 3.5 Hz, 200 Gal)

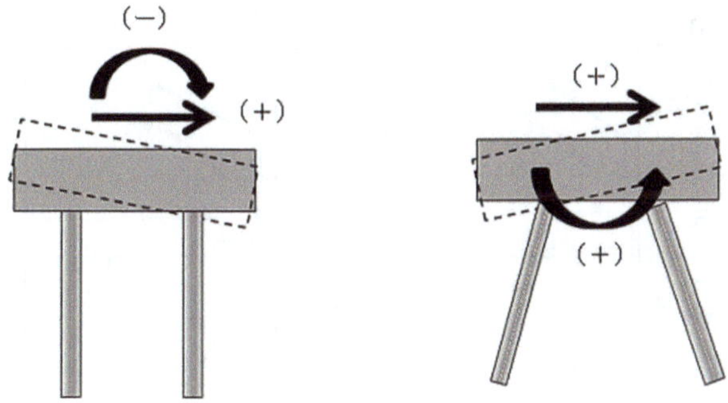

Fig. 7 Kinematic responses of footings

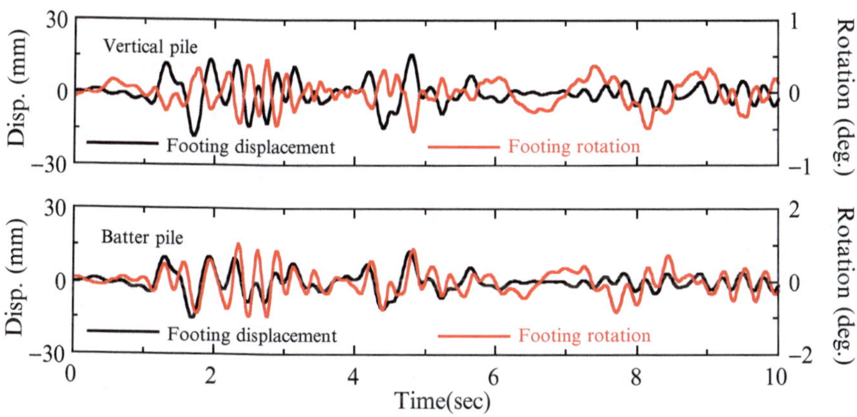

Fig. 8 Comparisons of *horizontal* displacement and rotational angle of the footings between the *vertical*-pile foundation and the batter-pile foundation (El Centro record: 200 Gal)

The fact that there is no phase difference between the sway and rocking motions indicates that the response of the footing to motion to the right is counterclockwise rotation, as shown in Fig. 7. There is no phase difference between the sway and rocking motions of the vertical-pile foundation; on the other hand, anti-phase behavior can be seen in the data for the batter-pile foundation.

Figure 8 shows the data obtained from El Centro record excitation at the maximum acceleration of 200 Gal. The same trend as seen in the case of sinusoidal excitation can also be found in Fig. 8. The phenomena of the opposite phase between the sway and rocking motions of the vertical-pile foundation and the batter-pile foundation can be found in all of the other test data. From Figs. 6 and 8, it can also be seen that the rotation angles of the batter-pile foundation are almost two times larger than those of the vertical-pile foundation. This causes the larger strains of the batter piles compared with those of the vertical piles as shown in the next section.

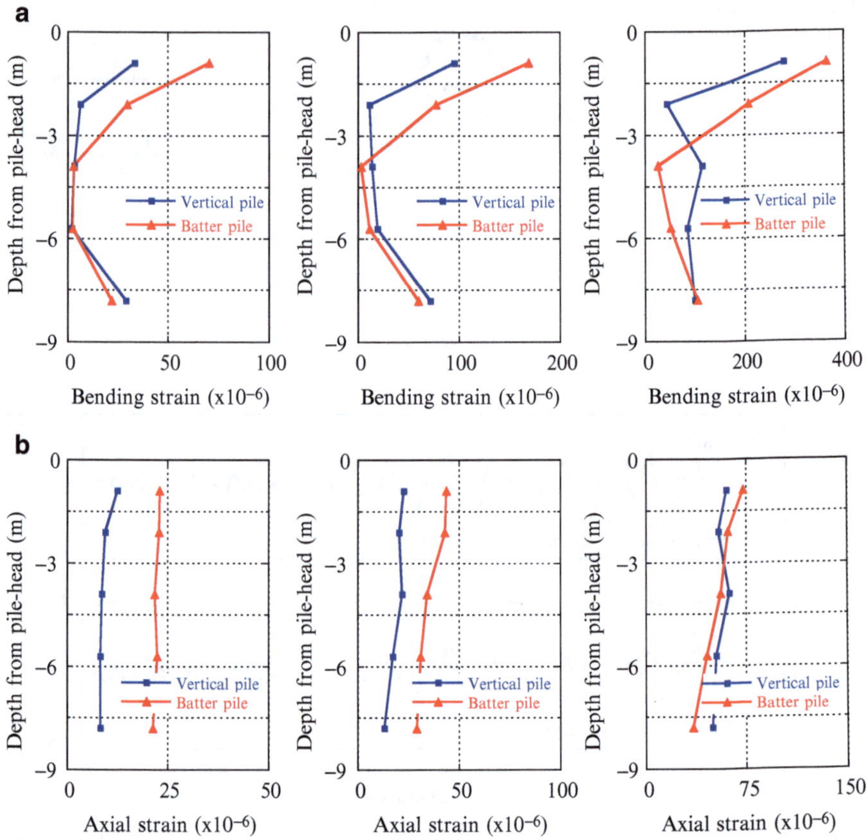

Fig. 9 Bending and axial strain distributions of the vertical-pile foundation (pile-VA1) and the batter-pile foundation (pile-BA1) obtained from the sinusoidal excitation of 3.5 Hz (50, 100, and 200 Gal). (**a**) Bending strain. (**b**) Axial strain

Bending and Axial Strains of the Piles

Figure 9 shows the maximum-value distribution of the bending and axial strains of the piles in the vertical-pile foundation (pile-VA1) and the batter-pile foundation (pile-BA1) obtained from sinusoidal excitation of 3.5 Hz. 3.5 Hz closely corresponds to the predominant frequency of the ground as shown in Fig. 4. The largest values were obtained at the pile heads, and the bending and axial strains of the batter-pile foundation were larger than those of the vertical-pile foundation in all cases, as shown in Fig. 9.

Figure 10 shows the maximum-value distribution of the bending and axial strains of the piles in the vertical-pile foundation (pile-VA1) and the batter-pile foundation (pile-BA1) obtained from El Centro record excitation. The largest

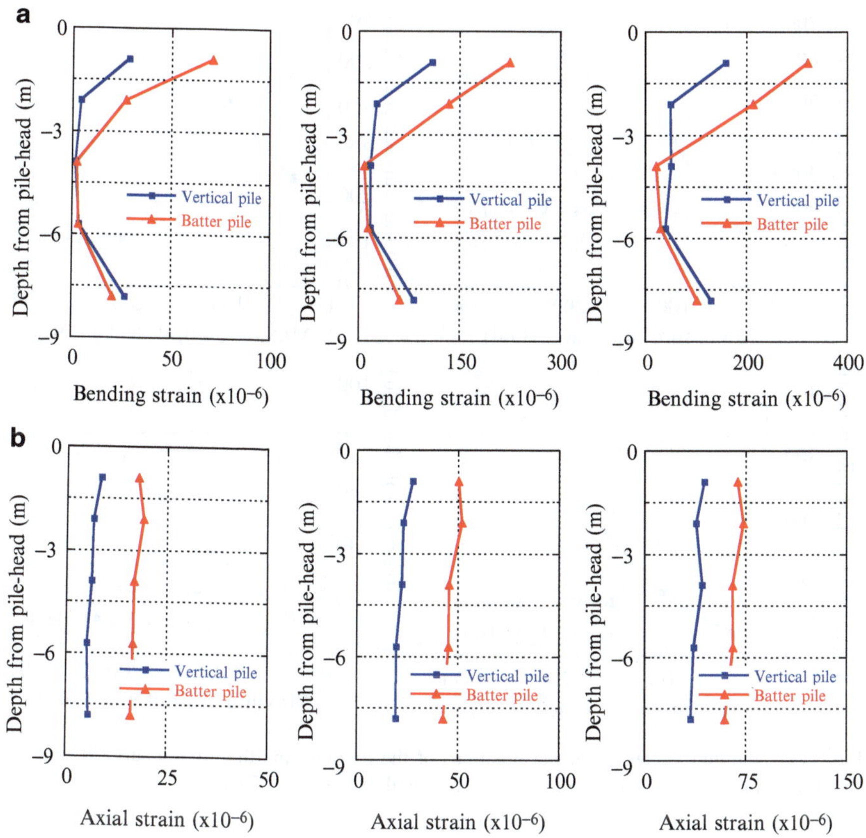

Fig. 10 Bending and axial strains distributions of the *vertical*-pile foundation (pile-VA1) and the batter-pile foundation (pile-BA1) obtained from El Centro record (50, 100, and 200 Gal). (**a**) Bending strain. (**b**) Axial strain

values were obtained at the pile heads, and the bending and axial strains of the batter-pile foundation were larger than those of the vertical-pile foundation, likely due to the sinusoidal excitation.

Maximum Responses of the Footing, the Ground Surface, and the Bending and Axial Strains at the Pile Head

Figures 11 and 12 show the maximum values for acceleration of the footings and the ground surface, and the bending and axial strains at the pile heads corresponding to the increments in maximum acceleration of the input motion. From the figure, it

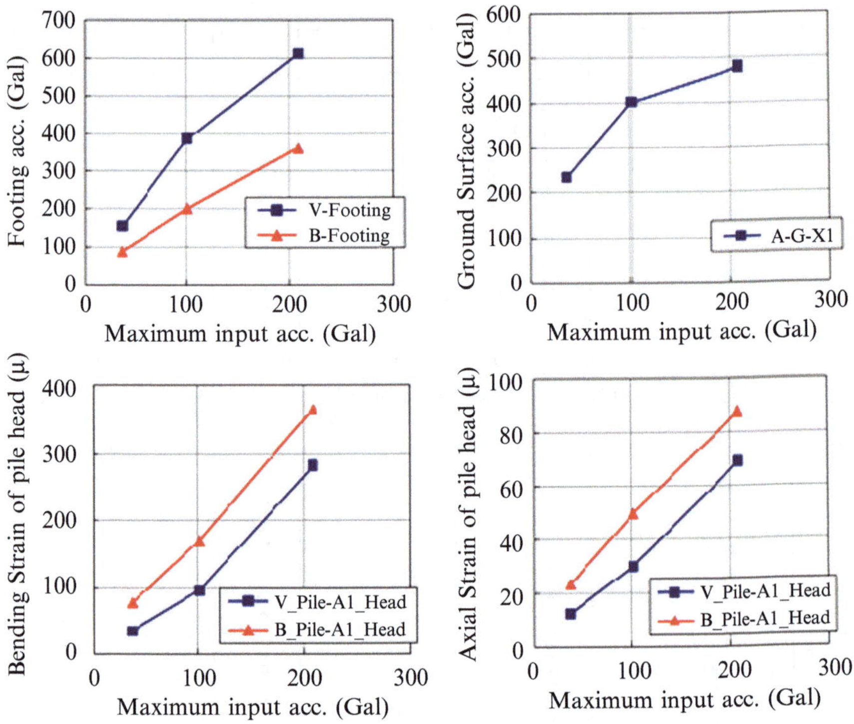

Fig. 11 Maximum values of the accelerations of the footings and the ground surfaces, and the bending and axial strains at the pile heads (sinusoidal excitation: 3.5 Hz)

can be seen that the maximum acceleration of the footing of the vertical-pile foundation is larger than that of the batter-pile foundation and that both the bending and axial pile strains of the batter-pile foundation are larger than those of the vertical-pile foundation in both the sinusoidal and El Centro record excitations.

Aseismicity of Batter Pile

Figures 13, 14, 15, and 16 show the frequency transfer functions of the horizontal acceleration of the footing and input motion and of the bending and axial strains of the piles and input motion, respectively, between the vertical-pile foundation and the batter-pile foundation obtained from the sweep test and El Centro record excitation. In Figs. 13 and 14, the difference between the frequency transfer functions of the two pile foundations, the vertical-pile foundation and the batter-pile foundation, represents the aseismicity of the batter-pile foundation. From these figures, it can be elucidated that the batter-pile foundation has a certain level of aseismicity in almost all of the frequency ranges. From Figs. 15 and 16, it can be

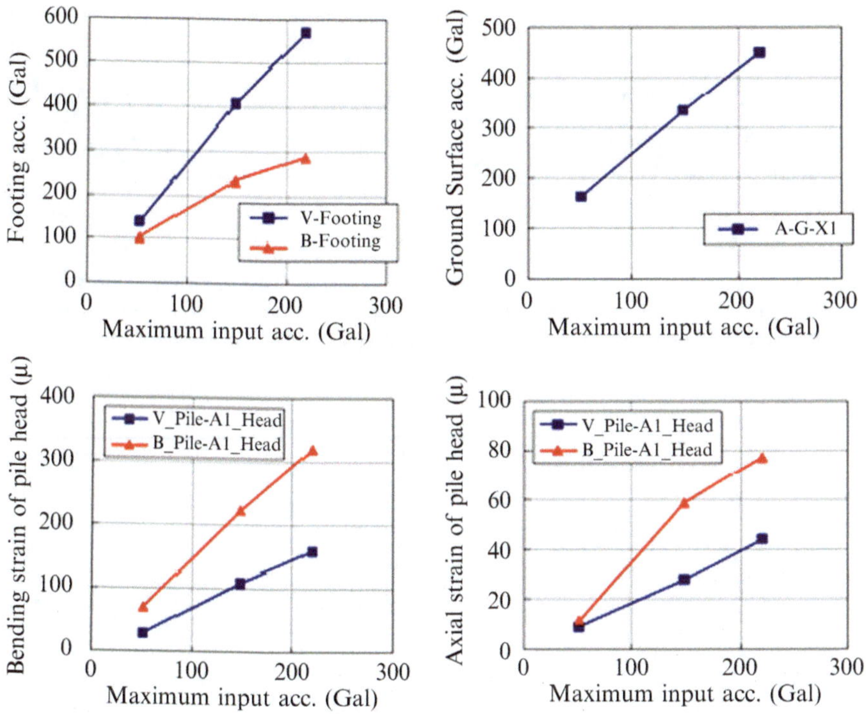

Fig. 12 Maximum values of the accelerations of the footings and the ground surfaces, and the bending and axial strains at the pile heads (El Centro record)

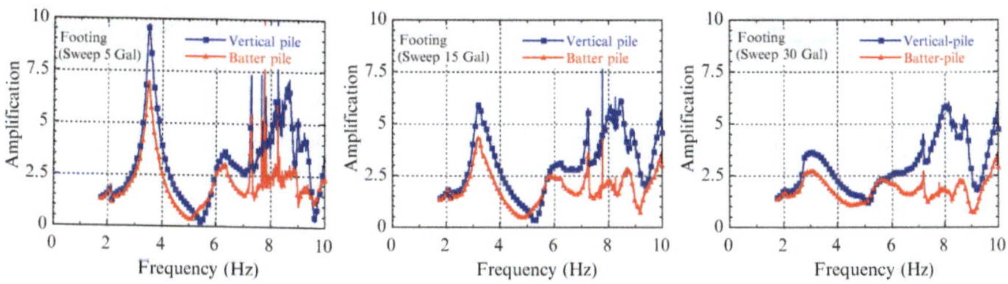

Fig. 13 Aseismicity of the batter-pile foundation: comparison of the frequency transfer function between the *horizontal* acceleration of the footing and input motion in the *vertical*-pile foundation and the batter-pile foundation obtained from sweep tests (5, 15, and 30 Gal)

seen that the bending and axial strains of the batter piles are larger than those of the vertical piles. It is therefore considered that batter piles require large cross-sectional capabilities to compensate for their aseismicity.

Figures 17 and 18 show the transfer function ratios of the vertical-pile foundation to the batter-pile foundation calculated based on the transfer functions in Figs. 14 and 16 which were obtained from El Centro record excitation. The

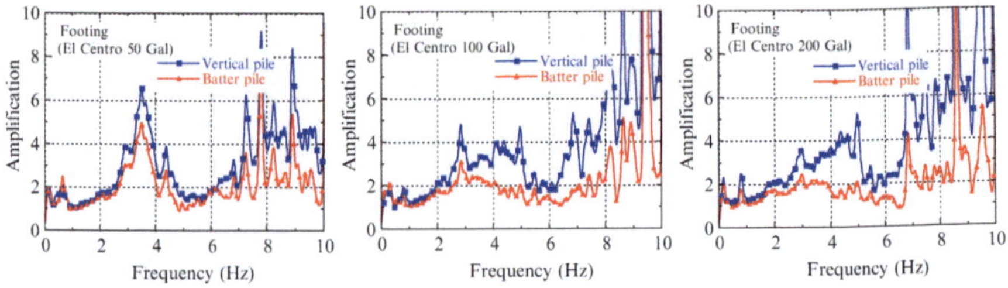

Fig. 14 Aseismicity of the batter-pile foundation: comparison of the frequency transfer function between the *horizontal* acceleration of the footing and input motion in the *vertical*-pile foundation and the batter-pile foundation obtained from El Centro record (50, 100, and 200 Gal)

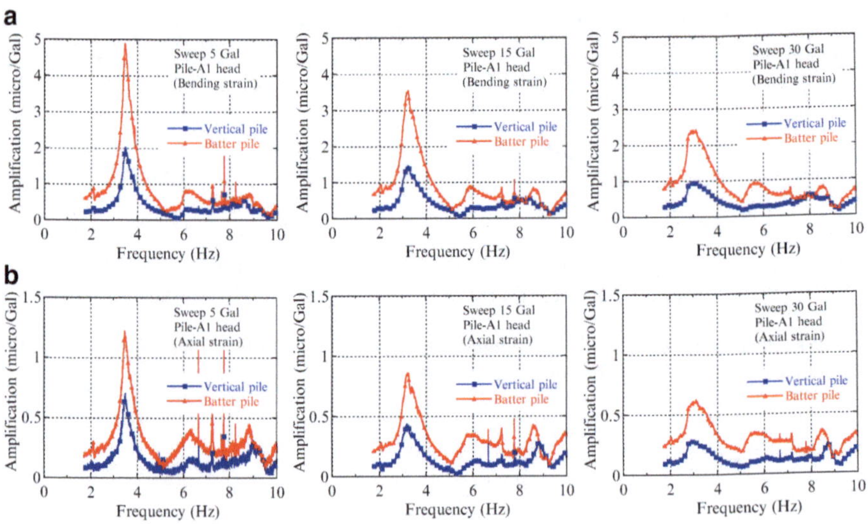

Fig. 15 Comparisons of the frequency transfer functions of the bending and axial strains of the piles and input motion between the vertical-pile foundation and the batter-pile foundation (sweep tests). (**a**) Bending strain. (**b**) Axial strain

magnification of the difference between the lateral line of 1.0 and each curve of the transfer function ratio in Fig. 17 represents the aseismicity of the batter-pile foundation, and those in Fig. 18 represent the compensation for the aseismicity of batter piles.

From the variation of the three curves of 50, 100, and 200 Gal in these figures, it can be seen that the effect of the aseismicity of the batter-pile foundation increased as the maximum acceleration of the input motion increased. However, the bending and axial strains at the pile head of the batter pile appear to approach those of the vertical pile as the maximum acceleration of the input motion increased. It is considered that these results are caused by the soil nonlinearity, which becomes significant as the maximum acceleration of the input motion increases.

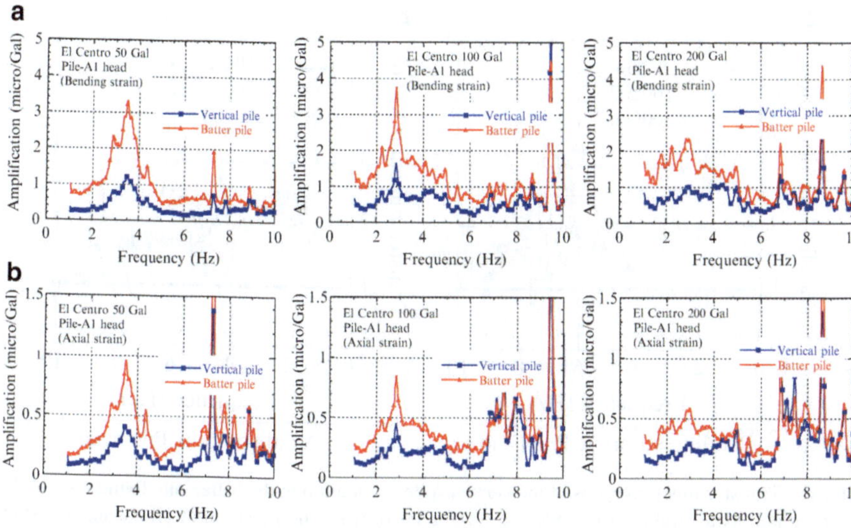

Fig. 16 Comparisons of the frequency transfer functions of the bending and axial strains of the piles and input motion between the *vertical*-pile foundation and the batter-pile foundation (El Centro record). (**a**) Bending strain. (**b**) Axial strain

Fig. 17 Transfer function ratio of the *vertical*-pile foundation to the batter-pile foundation of the horizontal acceleration of the footing

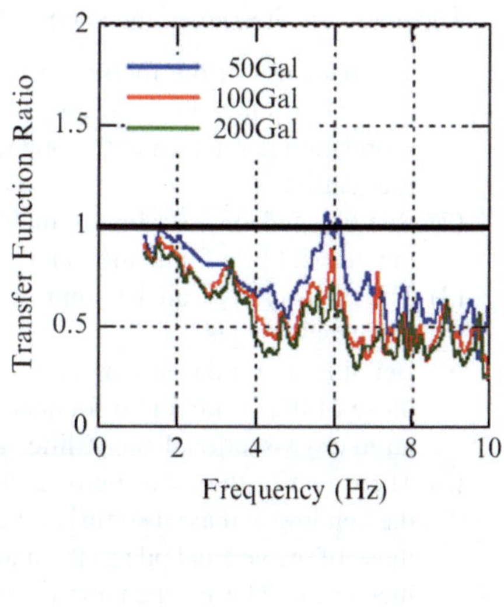

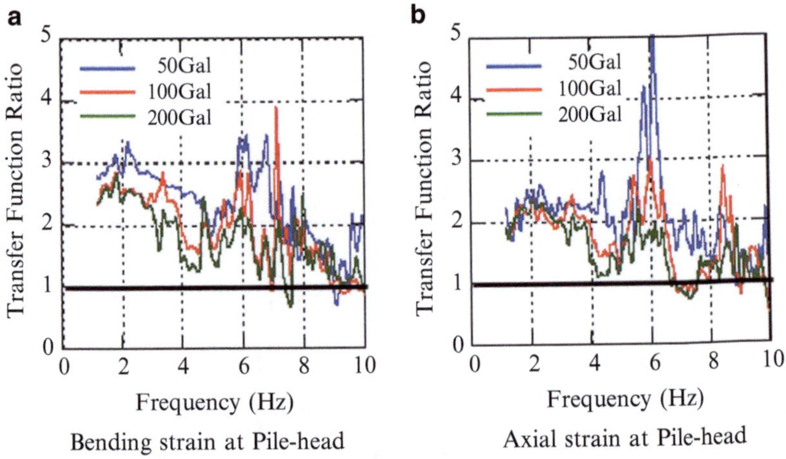

Fig. 18 Transfer function ratios of the *vertical*-pile foundation to the batter-pile foundation of the bending and axial strains of the piles at the pile head. (**a**) Bending strain at pile head. (**b**) Axial strain at pile head

Conclusions

The main conclusions of the study are as follows:

(1) The response of the footing of the vertical-pile foundation to motion to the right is clockwise rotation. On the other hand, that of the batter-pile foundation is rotation in the opposite direction to that of the vertical-pile foundation.
(2) Bending and axial strains attain the largest values at the pile heads in both the vertical-pile foundation and batter-pile foundation.
(3) The aseismicity can be improved by using batter piles in almost all frequency ranges.
(4) Bending and axial strains of the batter-pile foundation are larger than those of the vertical-pile foundation. In other words, batter piles require large cross-sectional capabilities to compensate for their aseismicity.
(5) The effect of the aseismicity of the batter-pile foundation increased and the bending and axial strains at the pile head of the batter pile approached those of the vertical pile as the maximum acceleration of the input motion increased. These phenomena are considered to be related to soil nonlinearity.

Acknowledgment We would like to express our sincerest gratitude to Mr. Katsumi Yoshinari for his invaluable help with the centrifuge tests.

References

1. Bhattacharya, S., Tazoh, T., Jang, J., Sato, M.: A study on the behavior of raked piles in seismically liquefiable soils. In: Proceedings of the 3rd Greece-Japan Workshop on Seismic Design, Observation and Retrofit of Foundations, pp. 1–19 (2009)
2. Deng, N., Kulesza, R., Ostadan, F.: Seismic soil-pile group interaction analysis of a battered pile group. In: 4th International Conference on Earthquake Geotechnical Engineering, Paper No. 1733 (2007)
3. Fscoffier, S., Chazelas, J.-L., Garnier, J.: Centrifuge modelling of raked piles. Bull. Earthq. Eng. **6**(4), 689–704 (2008)
4. Gerolymos, N., Giannakou, A., Anastasopoulos, I., Gazetas, G.: Evidence of beneficial role of inclined piles: observations and summary of numerical analyses. Bull. Earthq. Eng. **6**(4), 705–722 (2008)
5. Giannakou, A., Gerolymos, N., Gazetas, G.: Kinematic and inertial behavior of raked piles. In: Proceedings of the 3rd Greece-Japan Workshop on Seismic Design, Observation and Retrofit of Foundations, 36–50 (2009)
6. Giannakou, A., Gerolymos, N., Gazetas, G.: Kinematic response of groups with inclined piles, and inertial behavior of raked piles. In: 4th International Conference on Earthquake Geotechnical Engineering, Paper No. 1256 (2007)
7. Pinto, P., McVay, M., Hoit, M., Lai, P.: Centrifuge testing of plumb and battered pile groups in sand. Transp. Res. Rec. **1569**, 8–16 (1997)
8. Poulos, H.G.: Raked piles–virtues and drawbacks. J. Geotech. Geoenviron. **132**, 795–803 (2006)
9. Razavi, S.A., Fakher, A., Mirghaderi, S.R.: An insight into the bad reputation of batter piles in seismic performance of wharves. In: 4th International Conference on Earthquake Geotechnical Engineering, Paper No. 1423 (2007)
10. Tazoh, T., Shimizu, K., Wakahara, T.: Seismic observations and analysis of grouped piles. Geotechnical Special Publication No. 11, Dynamic Response of Pile Foundations, ASCE, pp. 1–20 (1987)
11. Tazoh, T., Wakahara, T., Shimizu, K., Matsuzaki, M.: Effective motion of group pile foundations. In: Proceedings of Ninth World Conference on Earthquake Engineering, pp. 587–592 (1988)
12. Tazoh, T., Sato, M., Gazetas, G.: Centrifuge tests on pile-foundation structure systems affected by liquefaction-induced flow due to quay-wall collapse. In: Proceedings of the 1st Greece-Japan Workshop on Seismic Design, Observation and Retrofit of Foundations, Athens, Greece, pp. 79–106 (2005)
13. Tazoh, T., Sato, M., Jang, J., Gazetas, G.: Centrifuge tests on remedial measure using batter piles against liquefaction-induced soil flow after quay wall failure. In: Proceedings of the 2nd Greece-Japan Workshop on Seismic Design, Observation and Retrofit of Foundations, Tokyo, Japan, pp. 431–439 (2007)
14. Tazoh, T., Sato, M., Jang, J., Taji, Y., Gazetas, G.: Kinematic response of batter pile: centrifuge tests. In: Proceedings of the 3rd Greece-Japan Workshop on Seismic Design, Observation and Retrofit of Foundations, pp. 20–35 (2009)
15. Fan, K., Gazetas, G., Kaynia, A., Kausel, E., Ahmad, S.: Kinematic seismic response of single piles and pile groups. J. Geotech. Eng. **117**, 1860–1879 (1991)
16. Mylonakis, G., Nikolau, S., Gazetas, G.: Soil-pile bridge seismic interaction: kinematic and inertial effects. Part I: Soft soil. Earthq. Eng. Struct. Dyn. **26**, 337–359 (1997)
17. Mylonakis, G.: Simplified model for seismic pile bending at soil layer interfaces. Soils Found. **41**(4), 47–58 (2001)
18. Nikolaou, S., Mylonakis, G., Gazetas, G., Tazoh, T.: Kinematic pile bending during earthquakes: analysis and field measurements. Geotechnique **51**(5), 425–440 (2001)
19. Sica, S., Mylonakis, G., Simonelli, A. L.: Kinematic bending of piles: analysis vs. code provisions. In: Proceedings of the 4th International Conference on Earthquake and Geotechnical Engineering (ICEGE), on CD-ROM, Thessaloniki, Greece (2007)

Section C
Structures - Characteristics of the Windmill Structural Fatigue Load in Natural Wind

Hironori Kikugawa, Masaru Sato, and Tomek Ziemba

Abstract In recent years, wind power generation has been gaining attention in many countries around the world as a source of renewable energy. However, windmill-related accidents are occurring frequently in its introductory stages in Japan and, compared to other countries, are leading to a slower increase in installation. Causes of the aforementioned accidents are irregular winds arising from the terrain and specific weather conditions. In this study, experiments using many patterns were conducted in which inflow wind velocity, pitch angle, and rotation speed were changed. The purpose was to observe the bottom load of a wind turbine tower. A model similar to that of an actual wind turbine was produced, and the load, which influenced the bottom, was investigated by using a six-component force transducer. The results showed that each relationship between the pitch angle, load, and wind speed became clear. The relationship between the drag and the lift also became clear by analyzing load data. From the dimensionless data, which was used for the initial load, the increasing tendency of the load, due to the difference of pitch angle, was revealed. In addition, the change of the load following a sudden change of wind velocity became clear. Furthermore, the usefulness of the feathering was also confirmed.

Introduction

The world is faced with many challenges such as global warming and depletion of fossil fuels. To help solve these problems, wind power generation has been attracting much attention as an environmental friendly technology. Therefore, the

H. Kikugawa (✉)
Mechanical Engineering, National Institute of Technology, Oita College, 1666 Maki, Oita, Japan
e-mail: kikugawa@oita-ct.ac.jp

M. Sato
Advanced course student, National Institute of Technology, Oita College, 1666 Maki, Oita, Japan

T. Ziemba
General Education, National Institute of Technology, Oita College, 1666 Maki, Oita, Japan

K. Saito et al. (eds.), *Progress in Scale Modeling, Volume II*,
DOI 10.1007/978-3-319-10308-2_15

amount of installation of wind turbines is growing rapidly around the world. However, the increase in wind turbines is slow in Japan. One of the reasons behind this is breakage and collapse of wind turbines caused by strong gusts of wind and typhoons. Therefore, in Japan, it is necessary to design the strength of the wind turbine structure taking into consideration the influence of strong gusts of wind [1–4]. The performance evaluation test of a wind turbine is a practical field test, which requires a long time, and the present condition leads to poor reproducibility. By carrying out the evaluation test of a wind turbine within a wind tunnel, experiment time can be shortened and cost can also be reduced. Currently, almost every evaluation test is conducted with a steady flow. In order to develop a wind turbine suitable for wind conditions in Japan, it is important to conduct the wind tunnel experiment using a fluctuating flow, which simulates natural wind. Therefore, in this research, the performance test of a wind turbine was conducted by a multiple-fan-type active-controlled turbulent wind tunnel, which simulated natural winds. In this study, the wind turbine model was similar to a real machine, and a load evaluation test of the wind turbine structure in simulated natural wind is conducted. The purpose of this experiment was to investigate the relationship between the rotation speed of blades and wind speed, as well as a change in load due to a difference in pitch angle.

Experimental Apparatus and Experimental Methods

Experimental Setup

Figure 1 shows a schematic of the experimental apparatus. In order to conduct experiments in stable airflow, the wind turbine model was set at a position of 4,000 mm from the inlet of the wind tunnel. The geometric dimensions of the wind turbine model were at 1/214 of a real machine. This model size was decided in consideration of the blockage rate in an experimental wind tunnel. In this research, when changing the wind velocity and the number of blade rotations, the load (FX, FY, FZ, MX, MY, MZ) at the bottom of a wind turbine model tower was investigated. Wind velocity was changed at intervals of 1 m/s to 0 ~ 5 m/s, the rotation speed of the blade was changed at intervals of 100 rpm to 0 ~ 500 rpm, and the data for a total of 36 patterns were collected. The measurement time of one experiment was set at 120 s and carried out three times per pattern in order to increase the accuracy of the data obtained. In addition, the load added to the tower by a different pitch angle was investigated, supposing a rapid wind direction change. In this study, three types of pitch angles were investigated: 90° (which has the largest area receiving wind), 0° (which has the smallest area receiving wind), and 45° (an intermediate condition). Furthermore, the number of loadless free rotations when the wind velocity changed six levels with three types of pitch angles was counted.

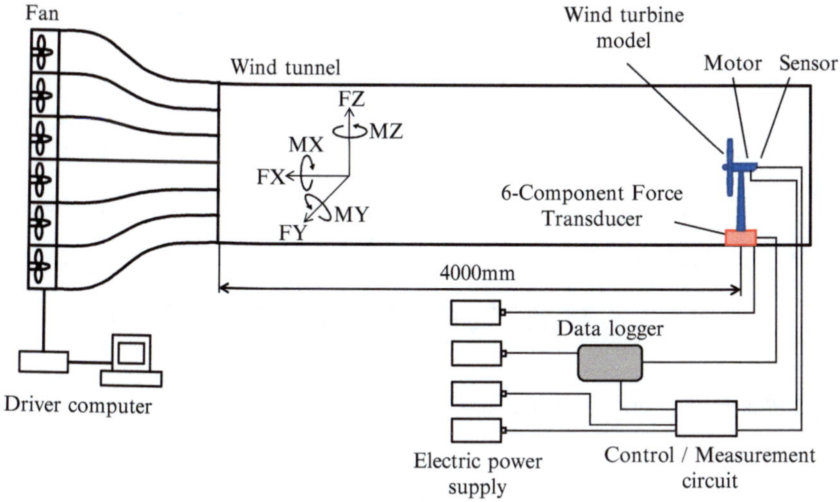

Fig. 1 Schematic of experimental apparatus

Blockage Ratio

An important factor in the wind tunnel experiment is blockage ratio. The blockage ratio is a ratio of the area that received the wind against the model occupied to the cross-sectional area in a wind tunnel. In the experiment, the rate of the area that a model occupies in a wind tunnel becomes larger than actual space. The cross-sectional area of a wind tunnel decreased at a given position of the model. For this reason, around a model, the increasing phenomenon of the flow velocity was generated, compared with a uniform flow. In the experiment, it is advisable that the installation position is away from a wall, because a boundary layer develops on the wall surface. Based on the above reasons, it is generally recommended that the blockage ratio is set to 5 % or less in experiments using wind tunnels [5, 6]. The blockage ratio is shown in Eq. (1). In addition, the relationships between blockage ratio and drag coefficient is shown in Fig. 2:

$$\text{Blockage ratio} = \frac{\text{Wind receiving area of model}}{\text{Cross-sectional area of wind tunnel}} \tag{1}$$

In this experiment, the cross-sectional area of a wind tunnel and the rotation surface of blades are supposed to be circles, and a value is substituted into Eq. (1):

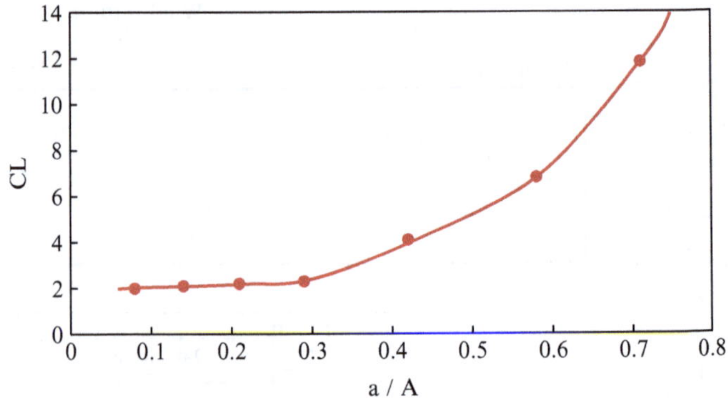

Fig. 2 Relationship between blockage ratio and drag coefficient [7]

$$\begin{aligned}\text{Blockage ratio} &= \frac{\frac{\pi}{4} \times 420^2 + 23 \times 130}{1,020 \times 1,090} \\ &= 0.12730 \\ &= 12.730\,[\%]\end{aligned} \tag{2}$$

From Fig. 2, the drag coefficient has slightly increased between the position of blockage ratio at 0.05 and 0.13. However, since its value of change is very small, the impact of blockage ratio is thus minute.

Scale Modeling

Scale modeling is a test method to more easily, quickly, and conveniently restage phenomena [8]. In the experiment with scale modeling, the essence of the original phenomena must be reproduced. Thus, it is necessary to understand the mechanisms acting on the original phenomena. Dominant physical laws chosen for the scale model of this experiment are as follows:

$$\text{Inertia force of water and air} \quad F_i = \rho l^2 u^2 \tag{3}$$

$$\text{Viscosity of water and air} \quad F_v = \mu l u \tag{4}$$

$$\text{Gravity force of water and air} \quad F_g = \rho g l^3 \tag{5}$$

Dimensionless numbers which are also called pi numbers are derived by Eqs. (3), (4), and (5):

$$\pi_1 = \frac{F_i}{F_v} = \frac{\rho l^2 u^2}{\mu l u} = \frac{\rho l u}{\mu} \quad \text{(Reynolds number)} \tag{6}$$

$$\pi_2 = \frac{F_i}{F_g} = \frac{\rho l^2 u^2}{\rho g l^3} = \frac{u^2}{gl} \quad \text{(Froude number)} \tag{7}$$

$$\pi_3 = \frac{F_i}{F_a} = \frac{u^2 \gamma}{a^2} \quad \text{(Mach number)} \tag{8}$$

Next, for the realization of the scale modeling experiment, physical laws acting secondarily are relaxed. The method is shown below:

$$\pi_1 = \frac{F_i}{F_v} = \frac{\rho l u}{\mu} \xrightarrow{\rho = \rho',\ \mu = \mu'} \pi_1 = lu \longrightarrow \frac{u}{u'} = \frac{l'}{l} \tag{9}$$

$$\pi_2 = \frac{F_i}{F_g} = \frac{u^2}{gl} \xrightarrow{g = g'} \pi_2 = \frac{u^2}{l} \longrightarrow \frac{l}{l'} = \left(\frac{u}{u'}\right)^2 \tag{10}$$

$$\pi_3 = \frac{F_i}{F_a} = \frac{u^2 \gamma}{a^2} \xrightarrow{\gamma = \gamma',\ a = a'} \pi_3 = u^2 \longrightarrow \frac{u^2}{u'^2} = 1 \tag{11}$$

In this case, it is necessary to increase the inflow of wind velocity in inverse proportion to the length in the scale modeling experiment. According to scaling laws, it is 214 times in a 1/214 model of this time. Thus, the negligible influence of the compressibility of air becomes dominant in the prototype. So, it becomes a problem to apply this scaling law. However, if all of the blades are considered to exist in natural wind (turbulence), the blades' surroundings can be considered in that the inertial force is affected, and the influence of viscosity becomes secondary. Therefore, the law of scaling is relaxed:

$$\pi_1 \neq \pi_1' \quad \text{(relaxed)} \tag{12}$$

$$\pi_2 \neq \pi_2' \quad \text{(relaxed)} \tag{13}$$

$$\pi_3 \neq \pi_3' \quad \text{(relaxed)} \tag{14}$$

Hence, in this experiment, relation of resemblance is derived utilizing the circumferential speed ratio of wind turbines [2]. The drag of the hub, nacelle, tower, and rotating surface of the blades is utilized, and it is extended to the moment coefficient applied to the base of wind turbine. The moment coefficient is calculated as the moment coefficient, applied to the base of wind turbine, and derived from the relation of resemblance between the scale model and the actual model:

$$\text{Circumferential speed ratio} \quad \lambda = \frac{\omega R}{V} = \frac{2\pi R n}{V} \tag{15}$$

$$\text{Moment coefficient} \quad C_M = \frac{M_D}{\frac{1}{2}\rho A V^2 L} \tag{16}$$

At this time, the projected area of the wings of the wind direction is utilized as a means to calculate the drag acting on the rotating surface of the blade. In this experiment, the relationship between the scale model and the actual model is standardized by two dimensionless numbers shown in Eqs. (15) and (16) [9].

Results and Discussion

According to each pitch angle, when changing the inflow of wind velocity into a wind turbine model, the changes of load added to a tower bottom are shown in Figs. 3, 4, and 5. From the experimental data, a change of moment load was compared and considered when the rotation speed was fixed to 300 rpm and the wind velocity was changed. A horizontal axis represents circumferential speed ratio, and the vertical axis the value that divided the load Mi by its initial value M0.

From Fig. 3, the MY load is a quadratic curve in a setup of 0° of pitch angles. This supposes that the drag received in the rotation surface of blades became larger with an increase in wind velocity. The MX and the MZ loads have not significantly changed. From Fig. 4, the MY load is increased like a quadratic curve in the negative direction. This is because of the increase of drag, due to the increase in the inflow wind speed. Further, the direction of the MY load is reversed between $\lambda = 6.60$ and $\lambda = 3.30$ of circumferential speed ratio. It is thought that this reason is the influence of the twist angle of the blades. The load MX does not have significant

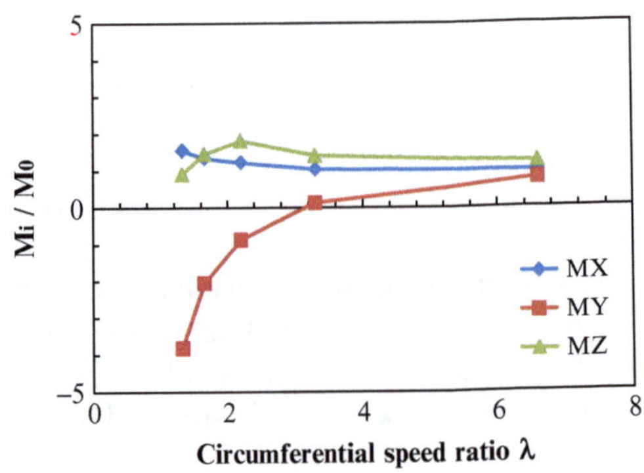

Fig. 3 Moment load and circumferential speed ratio at a pitch angle 0°

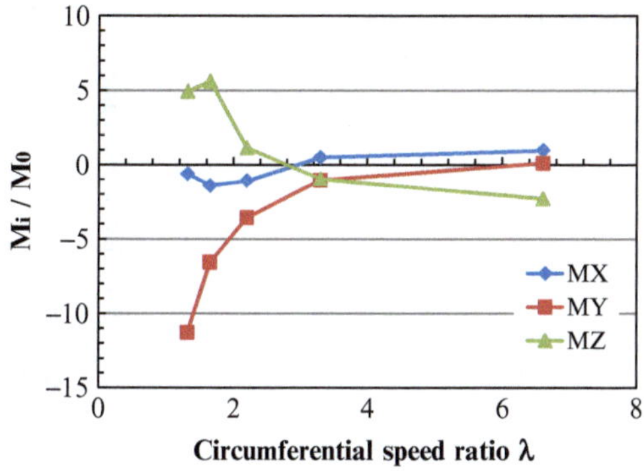

Fig. 4 Moment load and circumferential speed ratio at a pitch angle 45°

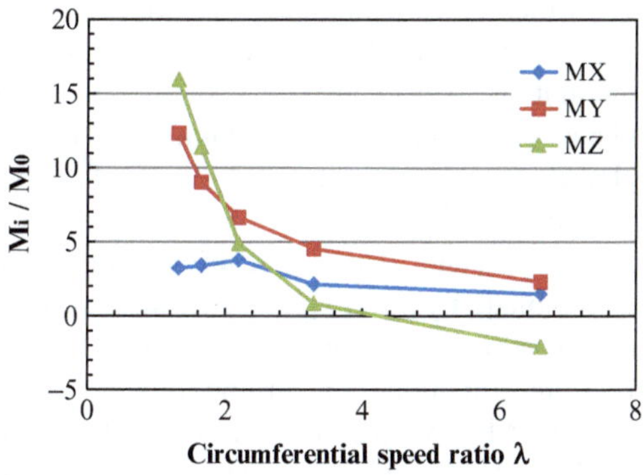

Fig. 5 Moment load and circumferential speed ratio at a pitch angle 90°

change with respect to those in the wind velocity. In Fig. 5, from the data of the MY load, it is thought that the drag of the rotating surface increases with wind velocity.

The three abovementioned conditions were compared. In regards to the MY load generated in the direction of wind, the load under drag became greater as wind velocity increased at every pitch angle. In comparison with windless conditions, and an inflow wind velocity of 5 m/s ($\lambda = 1.32$), about 3.8 times as much load was added in a 0° pitch angle situation. At 45° of pitch angle, about 11.3 times the load was added, and about 12.3 times as much load was added at 90°. These results coincided with the relation of the area size, which receives wind through a rotation surface of blades to a given wind direction. The feathering at a 0° pitch angle makes the rotation surface of the blades smaller, and it is thought to be an effective method of reducing the load on the tower. Moreover, in order to observe the conditions of the load fluctuation in the case of a wind gust, rotation speed was fixed in this experiment. If two points with arbitrary loads of similar conditions are compared, the load under a gust can be assumed. For example, in Fig. 4, when wind velocity changes from 3 to 5 m/s, which is circumferential speed ratio changed from $\lambda = 2.20$ to $\lambda = 1.32$, the load would be about 3.2 times larger.

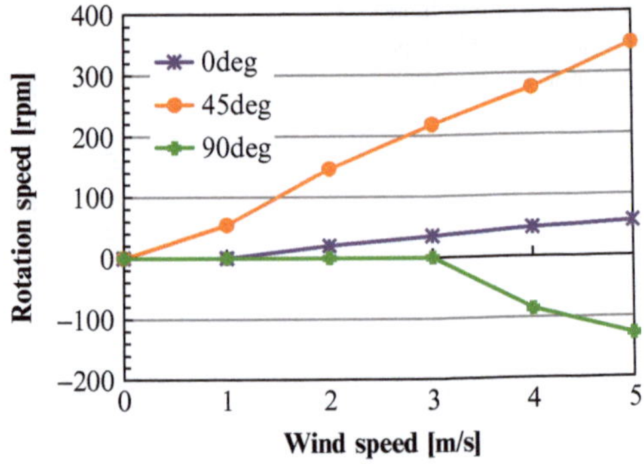

Fig. 6 Unloaded blade rotation speed and wind speed

Next, as a result of changing rotation speeds with fixed wind velocities, it was confirmed that the number of rotations and the load increased. This is because the area that receives wind increased with the increase in rotation speed. However, the amount of load change was smaller than the case in which wind velocity was changed. The change in the number of no-load free rotations of each blade when changing wind velocity is shown in Fig. 6. In the case of 45° of pitch angles, rotation speed also becomes larger with wind velocity, and they are almost proportionately related. This is because of large torque occurring in the rotating direction of the blade by the drag and lift. In 0° of pitch angles, the increase in the rotation speed following the increase in wind velocity was small. This is because the drag of rotating directions of turbine blades increased when the rotation speed was increased by the lift. From these results, it is thought that feathering is an effective method of decreasing rotation speed. Moreover, in the case of 90° of pitch angle, the rotation direction is reversed. It is thought that the reason is a rotation caused by drag due to the torsion angle of the turbine blade.

Finally, the load applied to the wind turbine of the actual model was verified. In Fig. 7, the relationship of the tip speed ratio and drag moment coefficient at each pitch angle is shown. From this figure, at pitch angles of 0° and 45°, the moment drag coefficient goes across the positive and negative. From this result, when the blades are rotated to a certain circumferential speed ratio and over, it is seen that the direction of the load applied to the base of wind turbine is reversed. In addition, there is always a positive moment coefficient in a setup with 90° of pitch angles. It is thought that these reasons are the influence of the lift and the twist angles of the blades. The MY load in conditions of actual operating limits of the actual model was determined from Fig. 7. When the wind velocity and rotation speed conditions were set to 25 m/s and 16.9 rpm, calculation results of the actual model are 750kNm in a setup of 45° of pitch and 4150kNm in a setup of 45°. At this time, the positive direction of the MY load is the direction of backward tipping. Feathering is thought to be the effective method of reducing the moment load of MY applied to the base of the tower.

Fig. 7 Moment coefficient and circumferential speed ratio

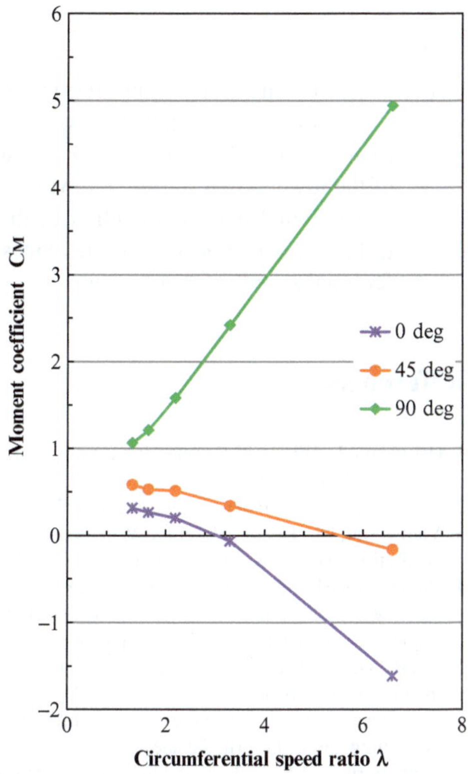

Conclusion

A wind turbine model with a motor was installed in a wind tunnel, and when the wind velocity and rotation speed of a turbine blade were changed, the load at the tower bottom was investigated. In addition, when pitch angle was changed assuming a rapid wind velocity change, the load difference, which influences the tower bottom, was investigated. As a result, the following conclusions were observed:

(1) In the set wind velocity, when rotating a blade with large rotational speed rather than the number of no-load free rotations of a blade, a thrust is generated.
(2) The load fluctuation under the rapid fluctuation of wind velocity in the case of a gust has been verified. The increase of the load in the actual equipment was found from the obtained experimental result.
(3) In feathering, the area that receives a wind direction becomes small. Since the area in which the wind received in rotating directions becomes large, it is hard to rotate a blade under a strong wind.

(continued)

(continued)

(4) Wind velocity and rotation speed both influence the load at the bottom of a windmill tower; however, wind velocity has a greater effect than rotation speed. It has been verified that there is a large load at the bottom of the tower when starting the windmill.
(5) It was found that the direction of the load applied to the base of wind turbine was reversed when the blades are rotated with a certain circumferential speed ratio or higher.

References

1. Ushiyama, I.: Windmill Engineering Introduction, Wind Power Technology and Basic Theory. Morikita (2002) (in Japanese)
2. Japan-Type Wind Power Generation Guidelines: NEDO (2008) (in Japanese)
3. Yamamoto, M., Kondo, H., et al.: Wind characteristics acting on the windmill for wind power generation, No. 52. Kashima Technical Research Institute annual report (2004) (in Japanese)
4. Yamamoto, M., Kondo, H.: Wind tunnel experiments and field observations about wind load on the tower in time of the wind turbine is generating power, No. 36, Issue 1, Japan Wind Engineering magazine (2011) (in Japanese)
5. Wind Tunnel Experiment Guidebook of Building: The Building and Housing Center of Japan, Inc (1995) (in Japanese)
6. Kurokawa, J., et al.: Turbomachinery—Introduction—New revised edition, Japan Industrial Publishing Co. (2007) (in Japanese)
7. Ramamurthy, A.S.: Effect of blockage on steady force coefficients. In: Proceedings of the ASCE, EM4, pp.755–772, Aug 1973
8. Emori, I., Saito, K., et al.: Theory and Application of the Model Experiment. Gihoudo (1973) (in Japanese)
9. Wind Power Generation Equipment Support Structure Design Guidelines, and Commentary: JSCE (2010) (in Japanese)

Part II
Engineering Design Performance Evaluation and Fundamental Understanding Using Scale Models

Summary of Part II

Engineering Design Performance Evaluation and Fundamental Understanding Using Scale Models

Not only in preventing/fighting accidental tragedies such as "unpredictable" natural disasters and "unexpected" structural failures (due to massive load caused by natural disasters), the scale modeling approach can also work to enhance innovative research and development. A typical example would be helping improve the design of engineering devices which enable us to have higher performance/quality systems in the engineering field. Research and development do not exist separately, especially in scale modeling studies. Research-for-development (the hyphens added to indicate the unbreakable link) is mandatory. In this way, scale modeling can cover a wide range of problems related to various engineering fields, e.g., environmental, thermal and fluid, mechanical, chemical, aerospace, safety, etc. In fact, there have been rich articles on this purpose available in past ISSMs, the best of which prior to 2008 we selected the first volume of *Progress in Scale Modeling*. Following that legacy, we felt it would be valuable to again select the most excellent work from ISSM6 and ISSM7 for this volume. Again, editorial committees carefully reviewed all articles submitted to these symposia and selected the 9 best articles contributing to engineering design and its development. These are presented in *Part II: Engineering Design Performance Evaluation and Fundamental Understanding Using Scale Models* in this volume.

For environmental as well as thermo-fluid engineering applications, there is estimation of temperature and flow fields in street canyon studies by introducing the scale model experiment. This work obviously contributes to assess and propose better urban planning/design to make our lives more comfortable. Another example shows the effective use of numerical simulation to evaluate the best performance of louvered fins; the governing scaling law is then extracted. This is one good example of how to utilize the numerical approach to expand the scale modeling approach. For mechanical and chemical engineering applications, two excellent scale model experiments are performed by the group from University of Kentucky; one focuses on chemical looping combustion and the other on an air-based density separator. These are role-model examples in how to design a scale modeling experiment and how to use a scaling law to help tackle real engineering problems. Two articles are selected from aerospace engineering utilizing the scale modeling approach, one an excellent review of the scale-up methodology of a hybrid rocket motor and the other an evaluation of the performance of a locomotion system applicable in an outer space environment (e.g., on other planet where a different gravity field is employed). Since aerospace engineering for outer space is so costly and the environment is as yet so unknown, scale modeling works well to ensure useful evaluation of performance under various conditions. Lastly, three articles are selected from safety engineering. All are related to fire safety; one studies the scale effect on electrical utility cable fires to improve layout and arrangement of electric cables in a nuclear power plant. The other two are focused on large-scale fire fighting: one presents scale modeling for aerial firefighting and the other investigates the scale effect of fire growth in discretely arranged combustibles simulating urban fire spread.

Characteristics of Temperature Fields and Flow Fields in a Heated Street Canyon by Scale Modeling

Hironori Kikugawa, Tomoya Sato, Yuki Yonezawa, Nobumasa Sekishita, and Tomek Ziemba

Abstract The street canyon phenomenon has a negative impact on air quality. Flow, temperature, and the concentration fields of pollutants affect street canyon. Many studies have focused on the concentration and flow fields; model experiments on the temperature field have been less examined. This experiment investigates the effect of model height and placement of flow and temperature fields. Temperature and flow fields are visualized using a smoke-wire method, one of the methods used to visualize a flow field. Visualization images were investigated in comparison to several other images. In turbulent flow conditions, temperature change of the street is verified when the turbulence parameter has changed. As a result of the experiment, data showing a high cooling effect was obtained. Furthermore, a trend was verified for cooling effects under turbulent flow conditions.

Introduction

In recent years, deterioration of air quality in cities, caused by high-rise and high-density urban areas, has become a problem. The street canyon phenomenon, created by lines of high-rise buildings along roadsides, is considered one of the causes of deteriorating air quality. In warm conditions, additional heat is supplied by exhaust heat from air conditioning equipment, reflection of sunlight, and so on, creating an

H. Kikugawa (✉)
Mechanical Engineering, National Institute of Technology, Oita College, 1666 Maki, Oita, Japan
e-mail: kikugawa@oita-ct.ac.jp

T. Sato • Y. Yonezawa
Advanced Course Student, National Institute of Technology, Oita College, 1666 Maki, Oita, Japan

N. Sekishita
Toyohashi University of Technology, Toyohashi, Aichi, Japan

T. Ziemba
General Education, National Institute of Technology, Oita College, 1666 Maki, Oita, Japan

K. Saito et al. (eds.), *Progress in Scale Modeling, Volume II*,
DOI 10.1007/978-3-319-10308-2_16

unpleasant environment. In addition, air pollution is a factor because contaminants are trapped in the street canyon. Therefore, it is necessary to better understand the phenomena taking place in a street canyon. However, street canyon is affected by the flow, temperature, and concentration fields of pollutants [1–4]. Specifically, in turbulent flow conditions, prediction by numerical simulation is difficult. The experiment reported on here investigated the effects of flow and temperature fields, when model height and placement changed. Under turbulent flow conditions, temperature change of streets was verified when turbulence parameters changed.

Experimental Apparatus

Multiple-Fan Wind Tunnel

The turbulence wind tunnel is an actively controlled multiple-fan-type generator with 66 small, independently computer-controlled inlet fans. Each fan duct has separate channels in the air from the inlet to the measurement section, at which point the air streams are mixed to create turbulent flow. The actively controlled fans can generate any natural turbulent wind conditions. The measurement section has a height of 1,015 mm, width of 1,095 mm, and length of 3,600 mm. The wind tunnel apparatus is shown in Fig. 1. To generate wind behavior that mimics real wind conditions, a Kármán-type spectrum was used [5, 6]. This spectrum showed the natural-like wind to set to several parameters in Eq. (1). To create a time series wind velocity fluctuating data for controlling fans of wind tunnel, the inverse FFT method was used.

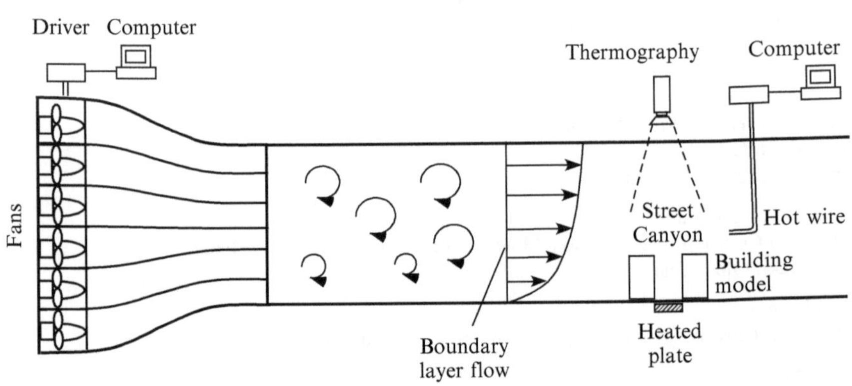

Fig. 1 Multiple-fan-type active controlled turbulence in wind tunnel

$$S_u(f) = 4I_u^2 L_u U \frac{1}{\left\{1 + 70.8\left(\frac{fL_u}{U}\right)^2\right\}^{\frac{5}{6}}} \tag{1}$$

Scale Modeling

Scale modeling uses the laws of nature to make a small replica of a larger system which behaves according to the same key parameters or has the same key properties. The scaling up or down process is carried out by similarity criteria which connect the physical property, characteristics of the fluid flow, and the dimensions of the real system and laboratory model. First of all, it is indispensable to observe and discern the phenomena of the real object carefully, to select the physical law which is governing phenomena.

$$\pi_1 = \frac{F_i}{F_v} = \frac{\rho l^2 u^2}{\mu l u} = \frac{\rho l u}{\mu} \quad \text{(Reynolds number)} \tag{2}$$

$$\pi_2 = \frac{F_i}{F_g} = \frac{\rho l^2 u^2}{\rho g l^3} = \frac{u^2}{gl} \quad \text{(Froude number)} \tag{3}$$

The π_1 to π_2 are nondimensional numbers for similarity laws of air flow, which shows the Reynolds number and the Froude number, respectively. In reality, it is impossible to set up the law of similarity such that it is simultaneously satisfied by all of π-numbers in a scale modeling experiment. Thus, the relaxation need for similarity law, a procedure by elimination of other physical laws, employing the analytical method and separation of phenomena. By substituting the known parameters in this formula, phenomenon in model experiments are reproduced. In this experiment, actual wind speed and building height was assumed to be 5 m/s and 10 m, respectively. The height of the model was assumed to be 0.1 m. If you specify the number π_1 Reynolds number, the ratio of the characteristic length is the inverse of the ratio of the wind speed. Therefore, the use of the number π_1 Reynolds number is difficult. If you specify the number π_2 to Froude number, the ratio of the characteristic length is the square root of the ratio of the wind speed. In the urban scale model that was assumed in this experiment, a Froude number π_2 is specified and indicates a possible experimental wind speeds in the wind tunnel.

$$\boxed{\pi_1 = \frac{F_i}{F_v} = \frac{\rho l u}{\mu}} \xrightarrow{\ \boxed{\rho = \rho'}\ ,\ \boxed{\mu = \mu'}\ } \boxed{\pi_1 = lu} \longrightarrow \boxed{\frac{u}{u'} = \frac{l'}{l}} \tag{4}$$

Table 1 Configurations of three-dimensional models

Case	Upstream model	Downstream model	Arrangement
Case 1	1:1:1	1:1:1	Linear
Case 2	2:1:1	2:1:1	Linear
Case 3	1:1:1	2:1:1	Linear
Case 4	2:1:1	1:1:1	Linear
Case 5	Alternating (symmetrical)		Linear
Case 6	Alternating (asymmetrical)		Linear
Case 7	1:1:1	1:1:1	Staggered
Case 8	2:1:1	2:1:1	Staggered
Case 9	1:1:1	2:1:1	Staggered
Case 10	2:1:1	1:1:1	Staggered
Case 11	Alternating (left diagonal)		Staggered
Case 12	Alternating (right diagonal)		Staggered

$$\pi_2 = \frac{F_i}{F_g} = \frac{u^2}{gl} \xrightarrow{\;g = g'\;} \pi_2 = \frac{u^2}{l} \longrightarrow \frac{u}{u'} = \left(\frac{l'}{l}\right)^2 \qquad (5)$$

Street Canyon Model

The street canyon model comprising a heated street and model buildings was constructed. To heat the street model, electrical current was introduced to a thin metal plate set directly beneath the street. The building models were made of plastic blocks in three-dimensional configurations. Two normalized sizes were used: 1:1:1 and 2:1:1, where the model size is normalized to the street canyon width. The building models were arranged in several configurations along the heated street, as shown in Table 1. The building models had various combinations of upstream and downstream building heights in both linear and staggered arrangements. Thermography was used to observe the temperature distribution of the heated street. In addition, the flow field in the canopy was observed using the smoke-wire method [7]. The wind tunnel was used to test under steady flow and fluctuating turbulent flow conditions (Fig. 2).

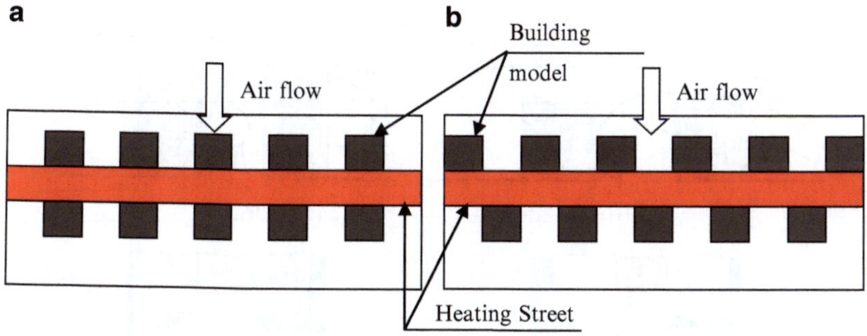

Fig. 2 Building model placement pattern. (**a**) Linear placement. (**b**) Staggered

Experimental Results and Discussion

Steady Experiment

The temperature distributions in low-turbulence steady flow are shown in Fig. 3. Case 1 configuration, per Table 1 (i.e., WS /Hb = 1 with 1:1:1 upstream and downstream building models in a linear arrangement), in which the areas in the middle of the street and in front of the downstream buildings were well cooled due to the frontal eddy generated at separation flow reattachment, is shown in Fig. 3a. Case 2 configuration (i.e., 2:1:1 upstream and downstream building models in a linear arrangement), in which the street areas in line with the building gaps were significantly cooled due to flow acceleration, is shown in Fig. 3b. However, the areas just behind the upstream buildings were not well cooled due to flow stagnation. Cavity flow was also observed in the street by flow visualization.

The temperature distributions of case 3 and case 4 models are shown in Fig. 4. Per Table 1, case 3 configuration has taller downstream buildings, whereas case 4 configuration has taller upstream buildings. Horseshoe vortices formed in the gaps between the upstream buildings in both configurations. In case 3, the separation flow from the upstream buildings collided with the downstream buildings and formed a frontal eddy, resulting in effective cooling of the street surface in front of the downstream buildings. Because the upstream buildings in case 4 are taller, the air flowing between the buildings was accelerated and a lee eddy was formed over the street.

The temperature distributions of case 5 and case 6 model configurations, both of which feature a linear arrangement of alternating building heights, are shown in Fig. 5. Case 5 model is arranged symmetrically, where both upstream and downstream buildings have the same height. Case 6 model is arranged asymmetrically, where the upstream and downstream buildings have different heights. In case 5, the street areas in line with the building gaps were well cooled; however, the areas just

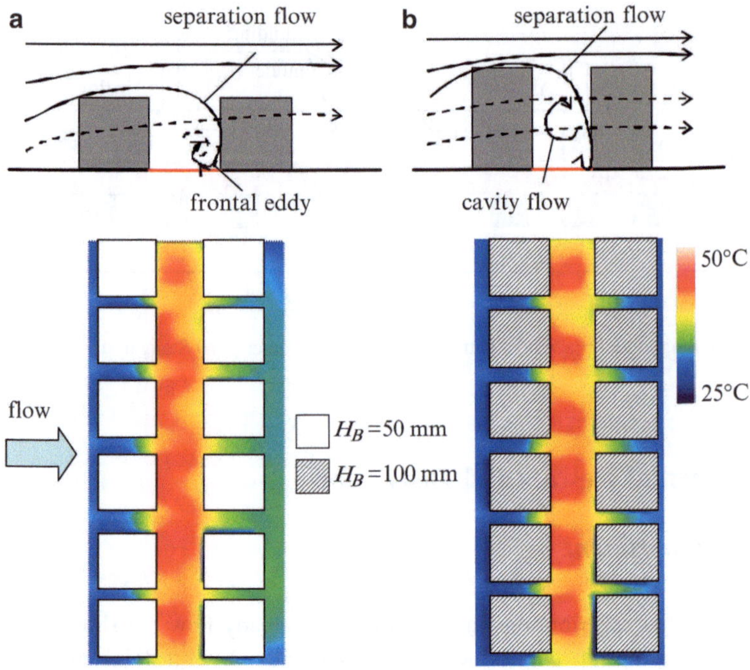

Fig. 3 Temperature distributions of case 1 and case 2 models. (**a**) Case 1 (**b**) Case 2

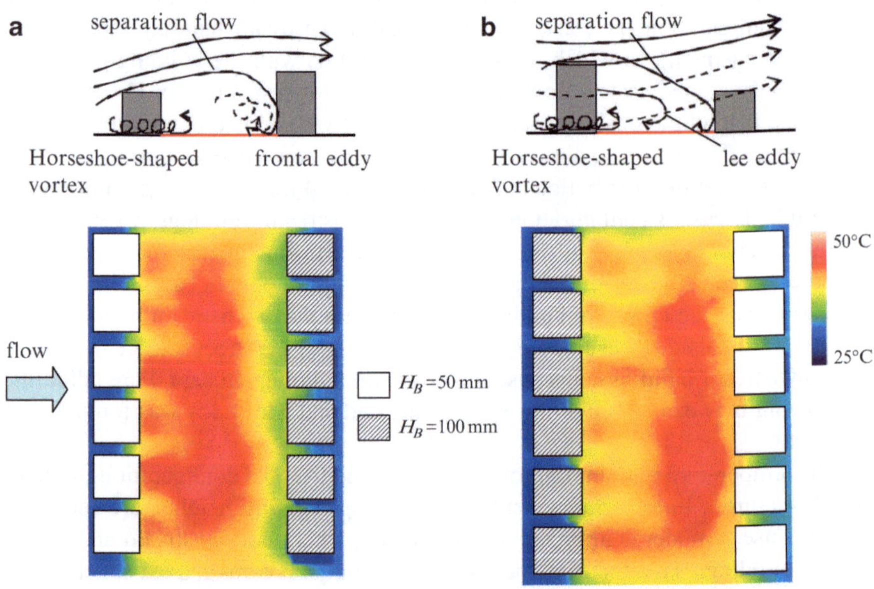

Fig. 4 Temperature distributions of case 3 and case 4 models (**a**) Case 3 (**b**) Case 4

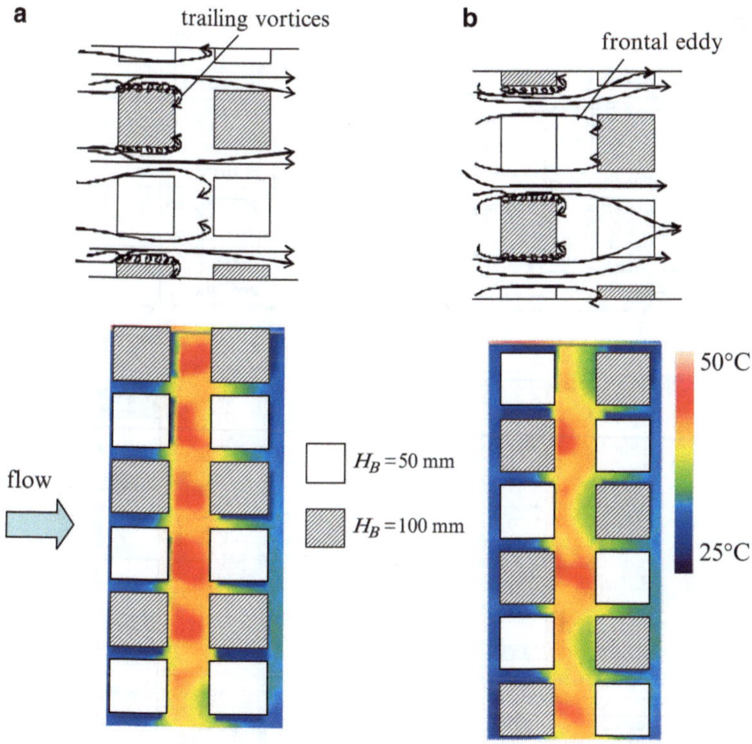

Fig. 5 Temperature distributions of case 5 and case 6 models (**a**) Case 5 (**b**) Case 6

behind the upstream buildings were not well cooled due to flow stagnation, and the separation flow from the upstream model skimmed over the downstream model. Case 6 exhibited a considerable cooling effect over the street in front of the downstream model, specifically when the air flow collided with the taller downstream buildings and formed a frontal eddy.

The temperature distributions of case 4 and case 10 model configurations are shown in Fig. 6. Both configurations feature taller upstream buildings, but case 4 featured a linear arrangement, whereas case 10 featured a staggered arrangement. The areas behind the upstream buildings nearest to the gaps were well cooled by the trailing vortices that were generated from the upstream side walls. In case 10, the trailing vortices collided with the downstream buildings and formed a horizontal frontal eddy. Therefore, the cooling effect of case 10 model configuration is slightly higher than that of case 4. However, if the upstream model were taller than the downstream model, it is clear that the cooling effect would decrease due to flow stagnation behind the upstream model.

The data in Figs. 3, 4, 5, and 6 reveal a significant correlation between the formation of an eddy and the cooling effect on the street; how the eddy forms on the street surface determines the cooling effect.

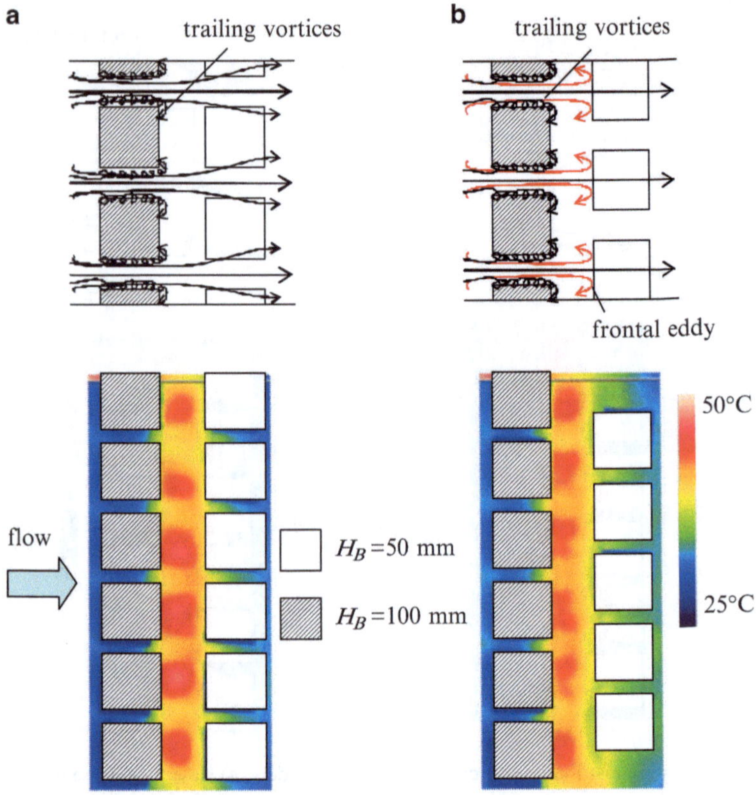

Fig. 6 Temperature distributions of case 4 and case 10 models (**a**) Case 4 (**b**) Case 10

Turbulence Experiment

The temperature fluctuation in turbulent flow produced by actively controlled multiple fans is shown in Fig. 7. It is clear that the cooling effect of the turbulent flow is greater than steady flow; how well the street is cooled depends on the fluctuation of the turbulence and the downdraft. It is also evident that a turbulent eddy contributes to the cooling of a heated street surface. The average temperature is significantly lower in the turbulence flow compared to the steady flow. The maximum and minimum temperature fluctuation in turbulent flow depends on the wind velocity. Because a significant velocity fluctuation exists when the actively controlled multiple fans are employed, the street temperature varies compared to that of under steady flow, and the temperature distribution never reaches steady state.

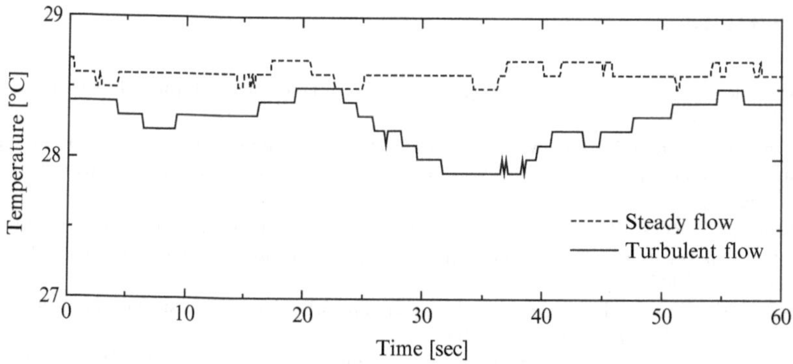

Fig. 7 Changes in the street in temperature due to turbulence

Conclusion

The purpose of this research was to build the database of street canyon phenomena in a turbulent flow wind tunnel. Visualization experiments for temperature fields were conducted under turbulent and steady flow. The following conclusions were obtained by measuring the mean temperature on the surface of a street.

(1) If a Froude number is specified, the wind speed values are reproducible in our wind tunnel.
(2) The arc-like cooling domain is formed by the downward flow ahead of the downstream model. In the case where the upstream model is lower than the downstream model, a downward flow is generated in staggered arrangement.
(3) If the upstream model is higher than the downstream model, flow through the model gap by lower wind path area is increased. Therefore, an increased cooling effect is obtained for street canyon surface.
(4) Cooling effect by accelerating flow shows a temperature fall near the upstream models. The cooling effect by downstream has small influence near model gap.
(5) In turbulent flow conditions, temperature change at the street canyon surface is increased by turbulence intensity.

References

1. Oke, T.R.: Street design and urban canopy layer climate. Energy Build. **11**, 103–113 (1988) (in Japanese)
2. Uehara, K., Yamao, Y., Oikawa, S., Mochida, A.: Wind-tunnel experiments on improving the natural ventilation of a street-canyon. J. Jpn. Soc. Atmos. Environ. **42**, 301–309 (2007) (in Japanese)
3. Gromke, C., Ruck, B.: Influence of trees on the dispersion of pollutants in an urban street canyon-experimental investigation of the flow and concentration field. Atmos. Environ. **41**, 3278–3302 (2007)
4. Khan, I.M., Simons, R.R., Grass, A.J.: Upstream turbulence effect on pollution dispersion. Environ. Fluid Mech. **5**, 393–413 (2005)
5. Simiu, E., Scanlan, R.H.: Wind Effects on Structures, 2nd edn, pp. 48–65. Wiley, New York (1986)
6. Ishii, K.: Summary of visualization technology of flow. Kyushu University Research Institute for Applied Mechanics, Technical staff Technical Report, vol. 8, pp. 142–165 (2007)
7. Von Kármán, T.: Progress in the statistical theory of turbulence. Proc. Natl. Acad. Sci. U.S.A. **34**, 530–539 (1948)

A Study of the Transition from Natural Convection to Force Convection in Plain and Louvered Fins with Scaling Simulations

Keng Hoo Chuah and Woh Peng Fun

Abstract The configuration and performance of louvered fins for compact automotive radiators is a subject matter widely studied [1–7]. The effect of buoyancy on the performance of these fins, however, is assumed to be negligible because fanned radiators normally operate at higher Reynolds numbers. This paper explores the possibility of radiators operating with acceptable performance at lower Reynolds numbers and presents a series of scaling simulations to measure the effect of buoyancy force on the performance of plain and louvered fins. The simulations are generated from a finite element model of a single fin with attached base surfaces and periodic boundary conditions. Transient calculations are performed until the model reaches steady or quasi-steady state, at which point the effect of buoyancy force is quantified for Reynolds numbers between 45 and 2,179 and Froude numbers between 0.4 and 18.9 by intentionally varying the inlet velocity and the gravity constant. At higher Reynolds numbers, the result is consistent with the published reports in the literature, validating the model to an extent. At lower Reynolds numbers, the result shows that the Stanton number scales with Froude numbers, indicating that natural convection is a significant force of the heat transferred. Natural convection is observed coming from the bottom base surface of the fin rather than the fin itself. As the fin is vertical while the base surface horizontal, this effect depends on the orientation of the radiator. On the amount of heat transferred, Nusselt number has a local saddle point at the transition between natural convection and force convection ($Fr = 3 \sim 7$). The performance at this point is lower because airflow pattern for each mode of convection differs and, at the transition, neither mode of convection dominates the airflow.

Nomenclature

A Heat transfer surface area
A_c Inlet cross-sectional area

K.H. Chuah (✉) • W.P. Fun
School of Science and Engineering, Malaysia University of Science and Technology, GL33, Block C, Kelana Square, 17 Jalan SS7/26, 47301 Kelana Jaya, Petaling Jaya, Selangor, Malaysia
e-mail: khchuah@must.edu.my

K. Saito et al. (eds.), *Progress in Scale Modeling, Volume II*,
DOI 10.1007/978-3-319-10308-2_17

C_p	Specific heat
Eu	Euler number
F_p	Fin pitch
Fr	Froude number
Gr	Grashof number
g	Gravity constant
h	Convective heat transfer coefficient
k	Thermal conductivity
L_p	Louver pitch
Nu	Nusselt number
Re	Reynolds number
St	Stanton number
ΔT	Temperature difference
u_i	Inlet velocity
x	Distance from the inlet
ΔP	Pressure drop
Θ	Louver angle
ρ	Density

Introduction

The goal of a compact automotive radiator is to transfer excess heat from the engine to the environment efficiently. The compact nature of the radiator limits its size and performance. Fins on the radiator occupy the most space but are necessary for increasing the rate of heat transfer. As air conducts less heat than fluid from the engine, air dominates the thermal resistance of a radiator. A small improvement in the fins at the air side of the radiator can thus lead to significant improvement in overall performance.

The use of louvered fins on the radiator creates small discontinuous surfaces that remove thick boundary layers over the entire fin, thereby increasing the heat transferred [2]. In addition, properly angled louvers can create favorable flow direction [3–5] and favorable fluid separations and vortex sheets [6, 7]; both of these factors contributed to better performance. For these reasons, louvered fins are classified according to louver pitch, louver angle, fin pitch, and number of louvers, as shown in Fig. 1a. Favorable flow direction, namely, louvered directed flow, as well as favorable flow separations, however, is only possible at higher Reynolds numbers. The small openings on the louvered fins are pressure resistant to airflow at lower Reynolds numbers, creating duct-directed flow and thus neutralizing the benefits of louvered fins [3].

Studies of louvered fins in the literature ignored the buoyancy force because buoyancy force is small. Bellows [1] estimated the strength of buoyancy force based on the ratio of Gr and Re^2, that is, if $Gr/Re^2 = gL_p/u^2$ is greater than the order of unity. The use of L_p as the characteristic length for the estimation, however, may

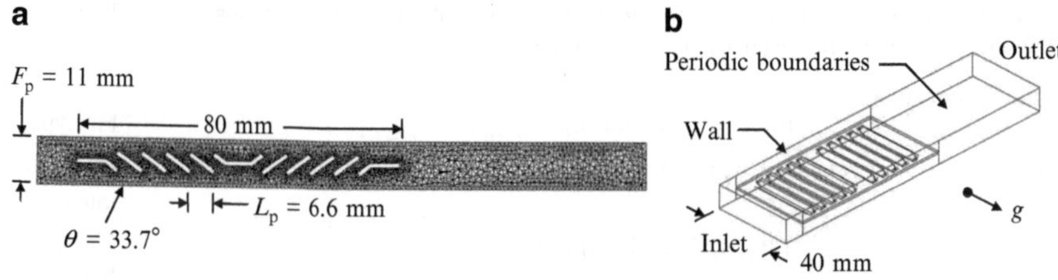

Fig. 1 Drawings of the finite element model: (**a**) fin cross section (**b**) isometric projection

not always be appropriate. Depending on the orientation of the fins, the contribution of buoyancy force is larger from the fin base, but the fin base is seldom included in studies. For studies that include the fin base, the fin length or the hydraulic diameter may be characteristic.

In this paper, the effect of buoyancy on louvered fins from natural convection to the onset of transition to force convection and beyond is investigated. The investigation includes the fin base that enhances the buoyancy force. Besides louvered fins, plain fins are also studied for comparisons. The method of investigation is a series of 3D, incompressible flow simulations conducted with finite element method (FEM). The choice of using FEM is a matter of convenience. From the simulations, the results are analyzed and plotted using the scaling laws. The results as well as the discussion and conclusions are reported below.

Numerical Method

Figure 1a is a cross section of the simulation model showing the details of the louvered fin. The model is three dimensional and the cross section is a mesh of tetrahedral elements around a solid fin. The total number of elements is 940,661. The mesh is relatively coarse for detailed analysis, such as for analyzing the shear layers or the vortex sheets, but is sufficiently fine for the purpose of a scaling analysis based on total pressure drops, average heat transfer rates, and average velocities. The coarseness of the mesh reduces the simulation time, such that many simulation cases are completed.

Table 1 is a list of all the simulation conditions, and it includes a definition for *Re*. By varying the airflow velocity at the inlet and the gravity constant, we collect temperature profiles, velocity profiles, and pressure profiles as given by the respective simulations. The reason to vary the gravity is to modify *Fr*, specifically the strength of buoyancy force, thereby uncovering the influence of *Fr* on these profiles.

Figure 1b is an isometric projection of the model. In the model, the fin is standing vertically against two horizontal walls; both walls have a fixed temperature higher than ambient. The temperature difference, ΔT between the walls and the ambient, is 65 K. The fin is made of aluminum and the space around it contains air. The model

Table 1 Simulation conditions: the inlet velocity, u_i, starts from 10 cm/s; the louver pitch, L_p, is 0.66 cm; the kinematic viscosity, ν, is that of air

Gravity constant, $g = 9.81$ m/s^2	Reynolds number, $Re = u_i L_p / \nu$	Fin type
0 g	136,272,409	Plain, louvered
½ g	136,182,227,272,318,363,409	Plain, louvered
1 g	45,91,136,182,227,272,318,363,409,454, 499,545,681,817,1,090,1,362,1,634,1,907,2,179	Plain, louvered
2 g	136,272,409,545,681	Plain, louvered

has an inlet, an outlet, and four periodic boundaries. The inlet supplies ambient air at a fixed rate, and the outlet is a zero-gradient Neumann boundary condition. The periodic boundaries simulate multiple fins stacked together, vertically and horizontally. Considering the size of a radiator and the number of fins stacked in between, the error from the radiator boundaries is assumed to be a negligible factor. In the FEM simulations, heat conducts from the walls to the fin as well as to the surrounding air. The air is assumed to be laminar at low Reynolds numbers ($Re < 600$), while large eddy simulation (LES) is used for the rest. A comparison between the laminar simulation and the LES simulation at $Re = 681$ shows the general results to be very similar. The average heat transfer rate for the LES simulation is higher than that for the laminar simulation by 6.7 % for plain fin and 3.2 % for louvered fin.

Results and Discussion

General Observations

Natural convection is observed coming from the bottom fin base rather than the fin itself. This is clear from low to moderate Reynolds numbers. Figures 2 and 3 are visualizations of the temperature profile as observed from the side (periodic boundary) of the louvered fin and the plain fin, respectively. For both fins, buoyancy force tends to push the air upward while the air is moving toward the right. For the louvered fin, the effects of louver-directed flow are observed as the air is moving across the side periodic boundary, carrying the higher temperature air. There is a clear interaction between louver-directed flow and buoyancy, in addition to more heat transferred.

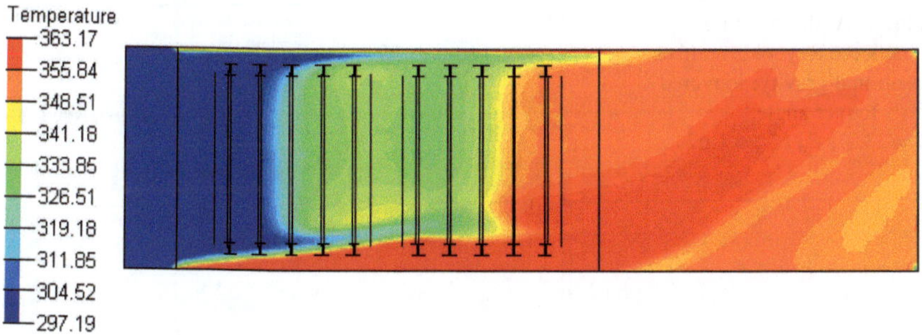

Fig. 2 A visualization of the temperature profile: the inlet airflow speed is 0.9 m/s; $Re = 409$; $g = 9.81$ m/s^2; the temperatures are in Kelvin; the *black lines* in the figures are outlines of the fin

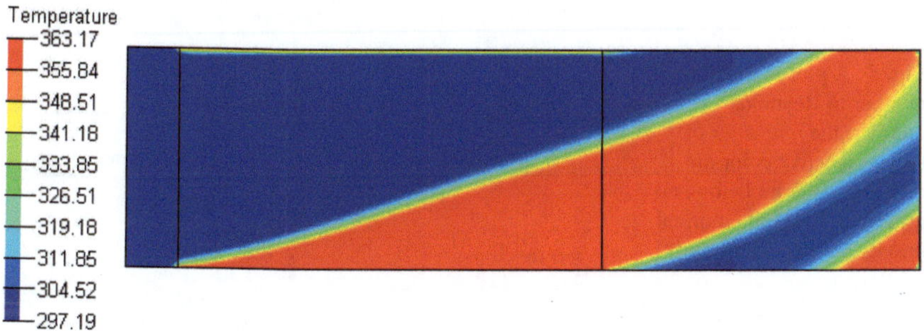

Fig. 3 A visualization of the temperature profile: the inlet airflow speed is 0.9 m/s; $Re = 409$; $g = 9.81$ m/s^2; the temperatures are in Kelvin; the *black lines* in the figures are outlines of the fin

Pressure Drops

Euler number $Eu = \Delta P \,/\, \rho u_{\mathrm{i}}^2$ is a measure of pressure losses in the flow. As shown in Fig. 4, louvered fin has consistently higher pressure resistance than plain fin. The results are comparable to those reported by DeJong and Jacobi [2], as noted in the graph.

Heat Transfer Rates

Stanton number, $St = \Delta T/LMTD$ is a measure of heat convection, where LMTD is the log mean temperature difference as defined for the heat exchangers. Figure 5 is a plot of St to Re for all the cases listed in Table 1. St = 0.4 is the maximum rate based on the limit on ΔT between the wall and the ambient. It can be seen that except for the case where there is no gravity, St is the same between the louvered fin and the plain fin at $Re = 136$, an indication that the mechanism for heat transfer is that of a simple duct-directed flow. As Re does not account for gravity changes, the lines for

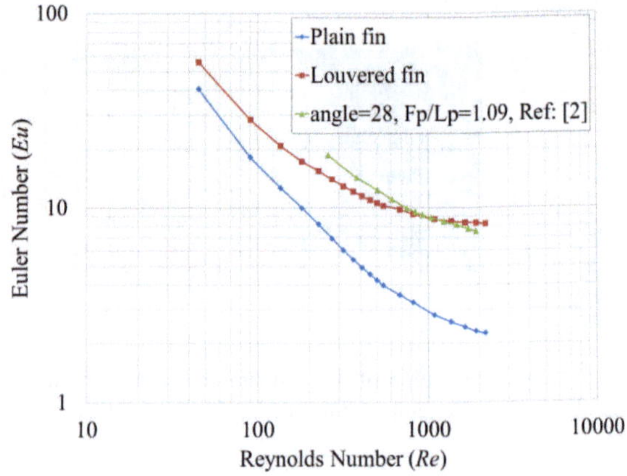

Fig. 4 A plot of Euler number to Reynolds number at normal gravity: louvered fin's louver angle, $\theta = 33.7°$; $F_p / L_p = 1.67$

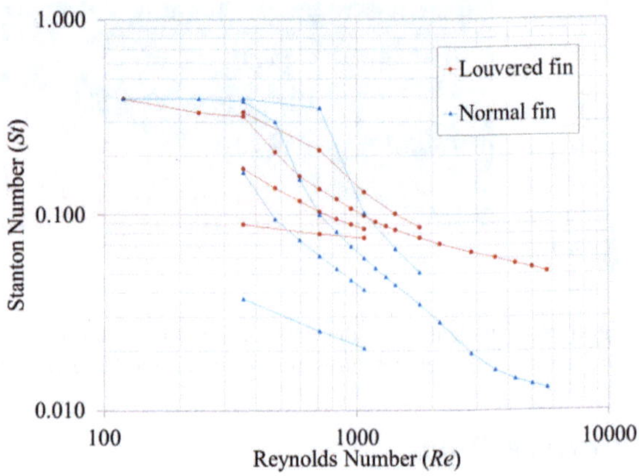

Fig. 5 A plot of Stanton number to Reynolds number: the *four lines* of their respective color are respectively from below: no gravity, half gravity, normal gravity, and double gravity

their respective fins spread over different St. Nevertheless, the lines are converging toward specific values as Re increases. The convergence indicates that the buoyancy force is being overcome.

To account for the gravity parameter, a plot of St to Froude number, Fr is shown in Fig. 6. $Fr^2 = u_i^2/gL_p$ is a measure of inertia force over the buoyancy force. The plot shows that the lines have collapsed to one and that the buoyancy effect is thus completely captured. By scaling Fr, three regions are observed, where at $Fr < 2$, St is limited by a maximum value due to the temperature difference; at $Fr = 3 \sim 7$, the inertia force and the buoyancy force are in the middle of transition; and at $Fr > 10$, the inertia force or force convection is becoming strong.

There is an indication in Fig. 6 that the midpoints of transition between louvered fin and plain fin are different, where the midpoint for the plain fin occurs at a higher Froude number. To show the difference, Fig. 7 is a plot of Nusselt number, $Nu = hL_p/k$ to Fr, where $h = StC_p u_i A_c/A$ is the convective heat transfer coefficient. The plot has a number of lines with a local saddle point at the transition between natural convection and force convection, where the midpoint transition $Fr = 3$ for

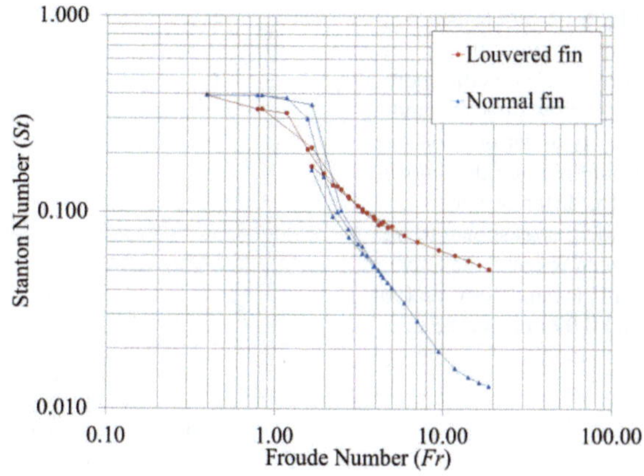

Fig. 6 A plot of Stanton number to Froude number: the *three lines* of their respective color are respectively from below at $Fr = 2$: half gravity, normal gravity, and double gravity

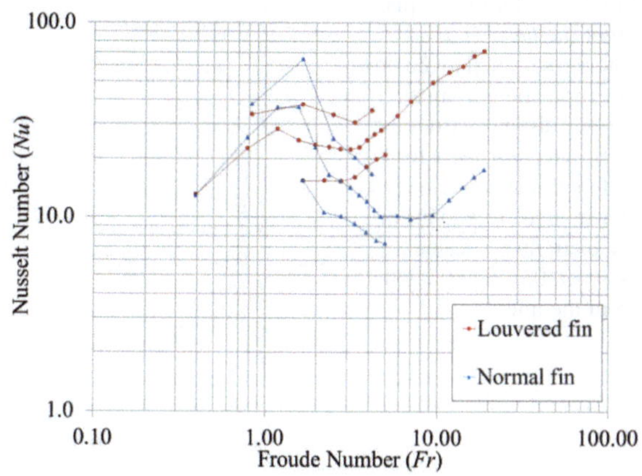

Fig. 7 A plot of Nusselt number to Froude number: the *three lines* of their respective color are respectively from below at $Fr = 2$: half gravity, normal gravity, and double gravity

louvered fin and the midpoint transition $Fr = 7$ for plain fin. The performance at these points is lower because airflow pattern for each force differs. The buoyancy force is naturally an upward force, while the inertia force is sideward. At the transition, neither of these forces dominates the airflow, resulting less heat transferred. The louvered fin transition occurs earlier due to louver-directed flow, as this is the only difference between the two types of fins.

Flow Direction

Louver-directed flow contributes to a higher heat transfer. From the velocity profiles, data are extracted to show the louver-directed flows in the small openings of the fins, as shown in Fig. 8. As a comparison, the lack of louver-directed flows is shown in Fig. 9. As the airflow velocity increases, more air is louver directed, consistent with reports from the literature.

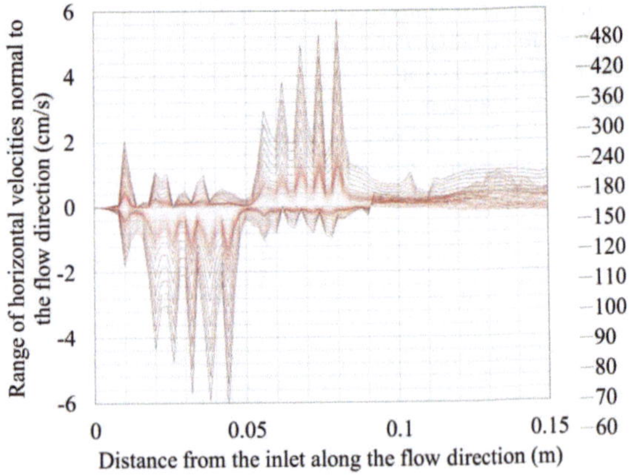

Fig. 8 A plot for the louvered fin of the *horizontal* velocities normal to the flow direction and gravity at various distances from the inlet: the *coloring* is for different inlet velocities as shown in the legend in cm/s; the louvered fin is located between 0.01 and 0.09 m from the inlet

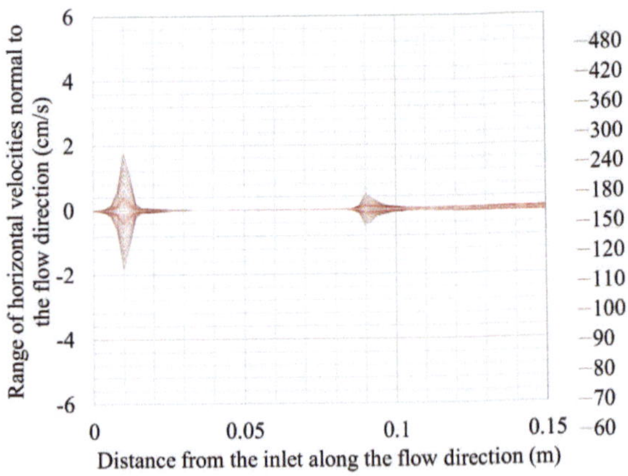

Fig. 9 A plot for the plain fin of the *horizontal* velocities normal to the flow direction and gravity at various distances from the inlet: the *coloring* is for different inlet velocities as shown in the legend in cm/s; the plain fin is located between 0.01 and 0.09 m from the inlet

Conclusions

This paper shows that the buoyancy force has certain effects on the performance of radiator fins. At worst, it lowers the performance at low Reynolds numbers. From the study of the transition, the performance at low Reynolds number can be improved by an optimization of the airflow patterns between buoyancy-supported flow and louver-supported flow. The lower performance at the transition point requires further studies based on different orientations of the fins.

Acknowledgements This study is supported by APM Engineering and Research Sdn. Bhd. (APMER). The authors thank APMER engineers for the valuable technical discussions and their help in this research.

References

1. Bellows, K.D.: Flow visualization of louvered-fin heat exchangers. M.S. Thesis, Department of Mechanical and Industrial Engineering, University of Dlinois at Urbana Champaign (1996)
2. DeJong, N.C., Jacobi, A.M.: Localized flow and heat transfer interactions in louvered-fin arrays. Int. J. Heat Mass Transfer **46**, 443–455 (2003)
3. Lyman, A.: Spatially resolved heat transfer studies in louvered fins for compact heat exchangers. MSME Thesis, Virginia Tech (2000)
4. Beauvais, F.: An aerodynamic look at automotive radiators. SAE Technical Paper 650470 (1965)
5. Davenport, C.J.: Heat transfer and fluid flow in louvered triangular ducts. Ph.D. Thesis, CNAA, Lanchester Polytechnic (1980)
6. Tafti, D.K., Wang, G., Lin, W.: Flow transition in a multilouvered fin array. Int. J. Heat Mass Transfer **43**, 901–919 (2000)
7. Zhang, X., Tafti, D.K.: Effect of fin pitch on flow and heat transfer in multilouvered fins. Air Conditioning and Refrigeration Center, University of Illinois (1999)

Scale and Numerical Modeling of an Air-Based Density Separator

Tathagata Ghosh, Mohammad Rezaee, Rick Q. Honaker, and Kozo Saito

Abstract The development of an air-based density separator for effectively treating finer-size fraction (6 mm × 1 mm) of run-of-mine coal is essential for the industry. Based on the derived scaling laws, the full-scale model of a laboratory air-based density separator and a prototype model scaled down by the factor of ½ in the X, Y, and Z directions were simulated using numerical methods. The air velocity contour plots, vector plots, and turbulent intensity values were used to validate the scaling laws. The absolute percent difference of air velocity between the two models was 3.83 %. In addition to this finding, the air velocity contour and vector plot and the turbulent intensity contours in two models showed modest variation, thereby indicating the validity of the scaling laws. The publication details the development of the full-scale and the prototype numerical models and the correlation of the derived scaling laws with the numerical and the experimental results.

Nomenclature

d	Particle diameter (m)
F_g	Gravitational force (N)
F_i	Inertia force (N)
F_υ	Viscous force (N)
F_μ	Friction force (N)
g	Gravitational acceleration (m/s^2)
l	Characteristics length of the table (m)
t	Time (s)
U	Horizontal component of the table velocity due to the vibration (m/s)
V	Vertical component of the table velocity due to the vibration (m/s)
V_{air}	Velocity of air (m/s)

T. Ghosh • M. Rezaee • R.Q. Honaker
Department of Mining Engineering, University of Kentucky, Lexington, KY, USA

K. Saito (✉)
Department of Mechanical Engineering and IR4TD, University of Kentucky, Lexington, KY, USA
e-mail: ksaito@uky.edu

K. Saito et al. (eds.), *Progress in Scale Modeling, Volume II*,
DOI 10.1007/978-3-319-10308-2_18

μ Dynamic viscosity (Pa.s)
μ' Friction coefficient (N/m^2)
ρ Particle density (kg/m^3)

Subscripts

1 Full-scale model
2 Prototype model

Introduction

Dry separation technologies have a long history of successful employment in the US mineral industry. However, their use declined from a peak of treating around 25 million tons in 1965 to less than 4 million tons annual production in 1991 due to environmental regulations and dust suppression requirements [1]. The advent of efficient wet processing units and demand for higher-quality minerals facilitated the decline. However, scarcity of water, downstream process water decontamination issues, and stricter environmental regulations concerning effluent discharge augmented by increasing capital expenditure on treating slurry impoundments have resulted in a significant increase in industry interest in effective dry cleaning solutions [2–6]. Dry cleaning is making a strong comeback in the twenty-first century.

The effective top particle size for most of the dry mineral separators is around 75 mm and the effective size ratio for which a good separation is achieved is between 2:1 and 4:1. It is noted that this effective particle size range is much smaller than most wet density-based separators due to the difference between the density of water and air [7]. The reported probable error (Ep) values vary in the range of 0.12 up to 0.3 due to the treated particle size distribution. These values indicate that the dry separation is inferior in separation efficiency in comparison with the wet coarse cleaning units. However, dry beneficiation devices have some advantages over wet separation such as lower capital cost; lower operating costs due to reduced need for pumping and aqueous chemicals; smaller plant footprint; lower product moisture values resulting in higher calorific value per ton; no slurry disposal which reduces disposal costs and rehabilitation demands as well as limiting environmental impacts of waste slurry disposal; no dewatering units such as thickeners, filters, and driers; fines generated in dry separation utilized as an ideal fuel for fluidized bed combustion units; and some dry processing units that may be more efficient regarding the recovery and achievement of a target product ash than wet separation, depending on the process [8]. The beneficiation of thermal coals in India with the help of a pilot scale FGX unit demonstrated the capability to effectively treat coarse particles rejecting almost 80 % ash while recovering almost 90 % of the combustible material [9].

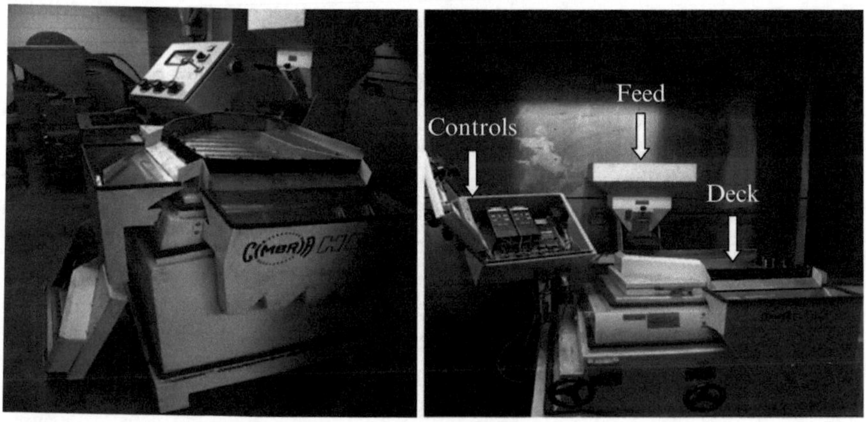

Fig. 1 Laboratory scale air-based density separator setup

Several modifications of processing technologies used during the peak years of dry beneficiation have been successfully developed and commercialized. Such technologies are Accelerator [10], Rotary Breaker [11], Allair Jig [12, 13], FGX unit [14, 15], Air Dense Medium Fluidized Bed Separator [16], AKAFLOW [17], Tribo-Electrostatic Separator [18], MagMill [19], and Optical and X-Ray Electronic Sorting Technology [20].

The modern air-based density separators are effective in treating run-of-mine material in the size range of 75 mm down to 6 mm. However, separation performance declines considerably in the finer-size range of 6 mm × 1 mm. Therefore, there is a demand to modify a dry density-based separator for treating particles having a size less than 6 mm. A study on a modified laboratory scale Bratney Company air table (Fig. 1) showed that dry separation of 6 mm × 1 mm size fraction is feasible [21]. This publication focuses on better understanding of the mechanism of the air-based density separator and modifications needed to be applied to its design to effectively treat finer fraction of feed with less than 6 mm in size using the art of scale modeling [22–24] and numerical modeling techniques.

Process Description

The dry cleaning system employs the separation principles of a fluidized autogenous medium and a shaking table concentrator. The feed to the system is introduced into a hopper from which the underflow is controlled using a vibrating feeder. The separation process generates multiple products of varying grades. Industrial and pilot scale units are equipped with dust collection systems to clean the recycled air and to remove the dust from air being emitted into the atmosphere. Major components of the table consist of a deck, vibrator, air blower fan, and table deck angle adjustment mechanism. A blower provides air that passes through holes on the deck surface at a rate sufficient to transport and fluidize the light particles. An upward movement of air through the table deck suspends the light particles, while the heavy particles remain on the deck and are driven by table vibration in a direction opposite to the light-suspended particles. Upon the introduction of the feed into the separation chamber, a particle bed of certain thickness is formed on the deck. The particles near the bottom of the bed directly contact the vibrating deck and move from the feed end toward the front discharge end under the influence of the vibration-induced inertia force. Upon striking an artificial lip along the edge of the deck, the heavier particles lose momentum and accumulate to build the bed, while the lighter particles travel along the width of the table toward the opposite end. Light particles are lifted up and travel along the deck mesh surface at a higher elevation than the higher-density particles before discharging over the lip along the product side. As such, light particles create the upper layer of the bed and are collected along the side of the table. Particles of sufficient density are able to penetrate the autogenous medium formed due to the fluidized bed of particles and settle on the deck surface. These heavy particles are forced by both vibration and the continuous influx of new feed material to transport toward the front end of the table where the final refuse is collected (Fig. 2).

The fundamental physical forces involved in this air-based density separator are vibration force in the form of inertia (F_i), gravity force acting on particles (F_g), friction force between the particles and the deck (F_μ), force of air acting on particles in the form of inertia (F_i), pressure (F_p) or viscous forces (F_υ), and particle–particle interaction in the form of collision or friction.

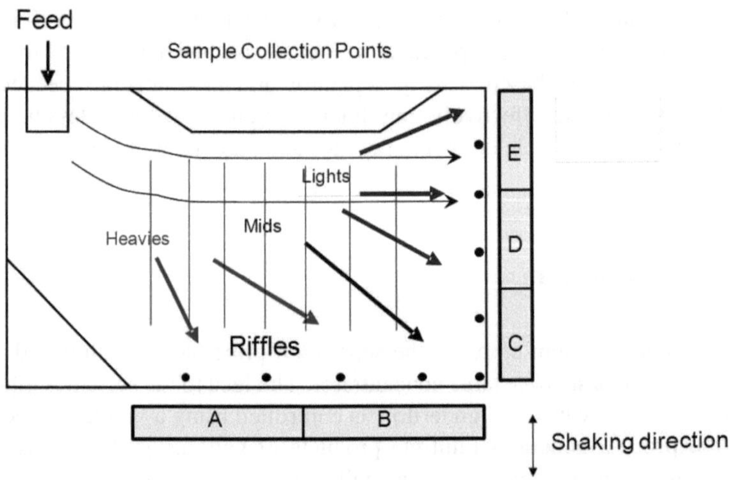

Fig. 2 Schematic view of the laboratory table with sample collection points

Scaling Laws

Scale modeling is based on scaling laws, and if they are correct, they describe the actual system; thus the prototype model can be used for the design and optimization of the full-scale model. If otherwise, then the assumptions made to derive scaling laws should be reconsidered.

To derive the scaling laws, the reference X, Y plane (horizontal and transverse coordinates) are located on the deck; the particle–particle interactions are ignored to simplify the model; the forces of air acting on particles are assumed to be mostly viscous forces (due to low Reynolds number of air flow passing on to the deck prior to particle injection); and the velocity of air is considered as a constant, which is equal to the value needed to lift up the coarsest particle to facilitate its movement along the deck.

Z direction

Deck motion has an angle relative to the horizontal plane, so the inertia force caused by the vibration of the deck acts on both X and Z directions. Therefore, the main forces in the vertical Z direction are gravity force (F_g), inertia force caused by the vibration (F_i), and the viscous force caused by the air acting on the particles (F_υ), as defined in Eqs. 1, 2, and 3:

$$F_i = \frac{\rho d^3 l}{t^2} \tag{1}$$

$$F_g = \rho d^3 g \tag{2}$$

$$F_\upsilon = \mu \frac{d^2 V_{air}}{d} \tag{3}$$

The governing pi numbers are Froude numbers, the ratio of inertia force over gravity, and Reynolds number over Froude number, the ratio of gravity force over viscous force (Eqs. 4 and 7):

$$\pi_1 = \frac{F_i}{F_g} = \frac{\frac{\rho d^3 l}{t^2}}{\rho d^3 g} = \frac{l}{t^2 g} = \frac{v}{tg} = \frac{v^2}{\lg} \tag{4}$$

Therefore,

$$\frac{v_2}{v_1} \approx \sqrt{\frac{l_2}{l_1}} \tag{5}$$

$$\text{and} \quad \frac{t_2}{t_1} \approx \sqrt{\frac{l_2}{l_1}}. \tag{6}$$

The vertical velocity of the vibration and time are proportional to the square root of the length characteristic of the table:

$$\pi_2 = \frac{F_g}{F_v} = \frac{\rho d^3 g}{\mu \frac{d^2 V_{\text{air}}}{d}} \tag{7}$$

Therefore,

$$V_{\text{air}} = \frac{\rho d^2 g}{\mu} = \text{const.} \tag{8}$$

X Direction

In the horizontal plane, particle movement along the X direction is governed by the frictional force between the particles and the deck and the inertia force caused by the vibration of the table. Therefore, the pi number is (Eq. 9)

$$\pi_1 = \frac{F_i}{F_\mu} = \frac{\frac{\rho d^3 l}{t^2}}{\mu' d^2} = \frac{\rho d l}{\mu' t^2} = \frac{\rho d u^2}{\mu' l} \tag{9}$$

Therefore,

$$\frac{u_2}{u_1} \approx \sqrt{\frac{l_2}{l_1}} \tag{10}$$

$$\text{and} \quad \frac{t_2}{t_1} \approx \sqrt{\frac{l_2}{l_1}} \tag{11}$$

So the horizontal velocity of the vibration and time are proportional to the square root of the table length.

Y Direction

In the Y direction, particle movement is governed by the viscous force caused by the air acting on the particles (F_υ) and the gravity force. Therefore, the pi number can also be described by Eq. 7.

Computational Fluid Dynamic Simulation

A numerical simulation was performed to simulate the operation of the full-scale air-based density separator and the prototype model with size reduction ratio of ½ in all three axes, and the results were then compared to evaluate the validity of the scaling laws.

The commercial CFD software ANSYS Fluent was used for the simulation. A preprocessing model was developed using GAMBIT (Fig. 3). The vibration of the separator pan was accounted for by using dynamically moving grids, while the feeder tube, contained in a different sub-domain, is kept stationary. The airflow pattern of the blower was simulated using Fluent's Eulerian model. The simulated geometrical model includes the feeding tube, the vibrating pan with the air supply, and a portion of the surrounding area. The geometry of the simulated air-based density separator offers no symmetries or periodicity to be exploited as simplification. As a result, the full 3-D simulation was mandatory which increased the size of the computational model and the inherent cost of the simulations. In addition, due to the pan vibration, the modeling approach has to be transient with a time step restricted by some fraction of the vibration period. To perform these demanding calculations in a timely manner, parallel calculation using a server with 30 cores running at 2.2 GHz each and 128 GB of random access memory was utilized.

A user-defined function (UDF) was developed and utilized to account for the amplitude of the table vibration occurring in a plane which is positioned at 44° angle with the horizontal axis. The mesh structure contained 151,679 nodes and a total of 794,319 elements. The mesh was divided into three separate volumes (Fig. 4), the top, the bottom, and the screen volume. The screen volume was set up as a porous media. The modified Darcy law formulations were utilized to account for the pressure drop and define the porous media resistance factors. The K-Epsilon model was used to define the turbulence regime around the screen volume. Table 1 describes the parameter values used to define the boundary conditions. Simulations were performed to characterize the airflow patterns for

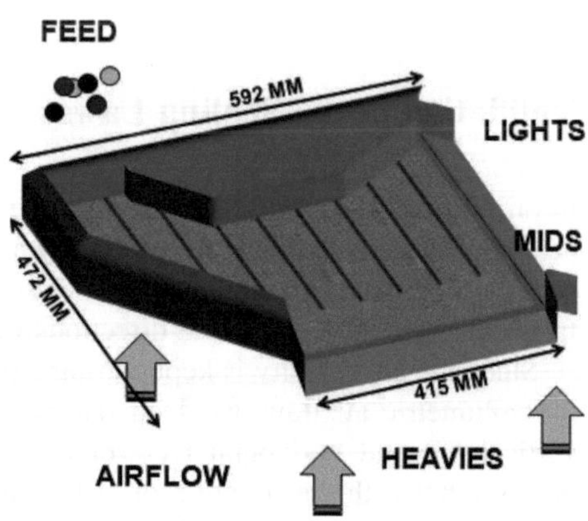

Fig. 3 Schematic of the full-scale air-based density separator table geometry and simplified operating principle

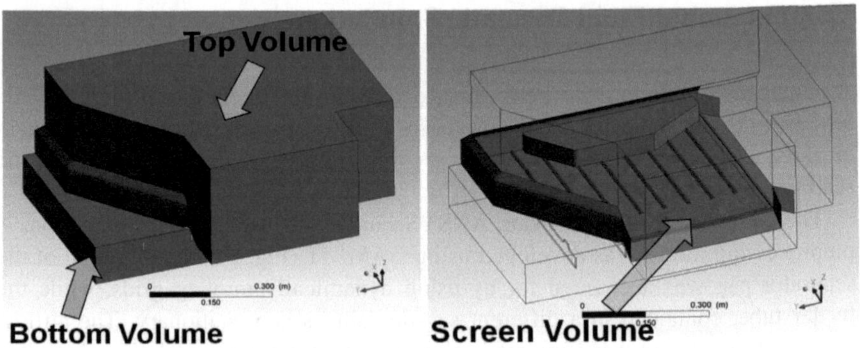

Fig. 4 Volume mesh setup for the air-based density separator CFD model

Table 1 Boundary parameter values (full-scale model vs. prototype)

Model	Airflow rate (kg/s)	Longitudinal angle	Transverse angle	Table frequency (Hz)	Vibration amplitude (mm)
Full-scale model	0.78	3°	8°	35	7
Prototype model	0.196	3°	8°	35	3.5

1 s simulation of the full-scale model and 0.7 s simulation for the scaled prototype model.

The time-step size was 2×10^{-5} s with 10–30 iterations per time step. One second flow-time simulation translates to approximately 500,000 iterations. Since the iterative solver scheme solves two equations for K-Epsilon, 1 s flow-time simulation was completed in around 20 days. The scaled prototype was run for 0.7 s real time simulation.

Validation of the Scaling Laws

In order to validate the scaling laws and evaluate the modifications needed in the design of the full-scale model of the dry air-based density separator to effectively treat particles having a size range of 1–6 mm, a prototype model scaled down by the factor of ½ in the X, Y, and Z directions was simulated.

Since the air velocity is kept constant and the area of the table decreases by ¼, the volumetric air flow and feed rate were reduced by a factor of ¼. Since the vertical (Z) and horizontal (X) velocity components of the table vibration are proportional with the square root of the characteristic length of the table, and the

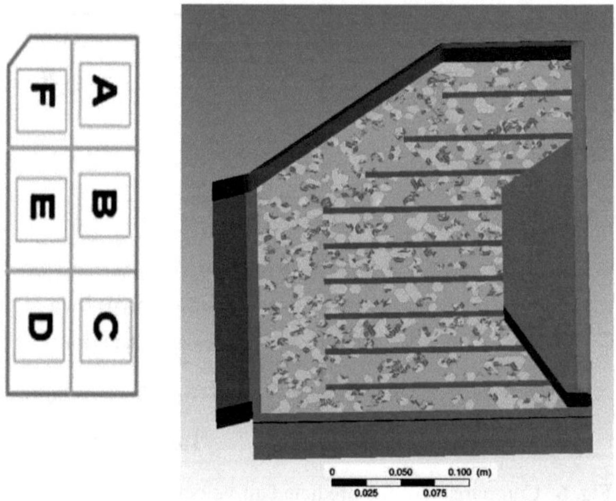

Fig. 5 Top view of the table deck displaying the dividing sections (A, B, C, D, E, and F)

Table 2 Average air velocity data measured at each section on top of the table deck (full-scale model vs. prototype vs. experimental data)

Sections	Full-scale model	Prototype model	Absolute % diff.	Experimental
A	2.52	2.69	6.75	2.32
B	3.03	3.05	0.66	2.50
C	3.22	3.29	2.17	2.64
D	2.93	3.09	5.46	3.04
E	2.69	2.74	1.86	2.19
F	2.62	2.46	6.11	2.19
Average	2.84	2.89	3.83	2.48

scaling ratio of the prototype model is ½ of the full-scale model, these parameters were reduced by a factor of 0.7 (square root of ½). Since the time is proportional with the square root of length characteristic, the time period for the prototype model should be 0.7 times that of the time period of the full-scale model. Hence, the full-scale model was run for 1 s, while the prototype was simulated for 0.7 s. In order to better compare the results, the area of the table is divided to the six sections (Fig. 5).

Results of the air velocity profile on top of the deck in 1 s of the full-scale model, 0.7 s in prototype model and in the experiment are presented in Table 2 and Fig. 6. The average absolute percent difference of the air velocity on the top of the deck in two models is 3.83 % which shows that variance between the model data is trivial and thus can validate the scaling laws. Slightly higher variance is recorded with respect to the experimental data. Nevertheless, in section D of the table, the experimental data matches the simulated values. The experimental measurements have been recorded on various sections of the table, and an average dataset was prepared while the simulated dataset is imported from a post-processor using

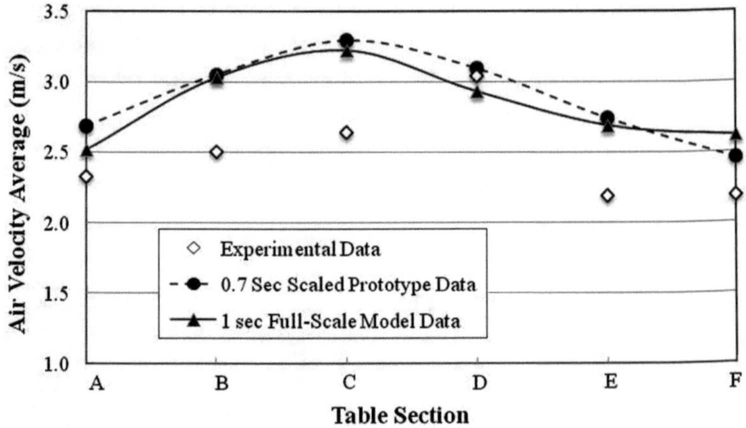

Fig. 6 Comparison of the predicted air velocity values occurring directly above the table surface in defined cross-sectional areas using the full-scale and the prototype models with the average values obtained from direct measurements at several points in each area

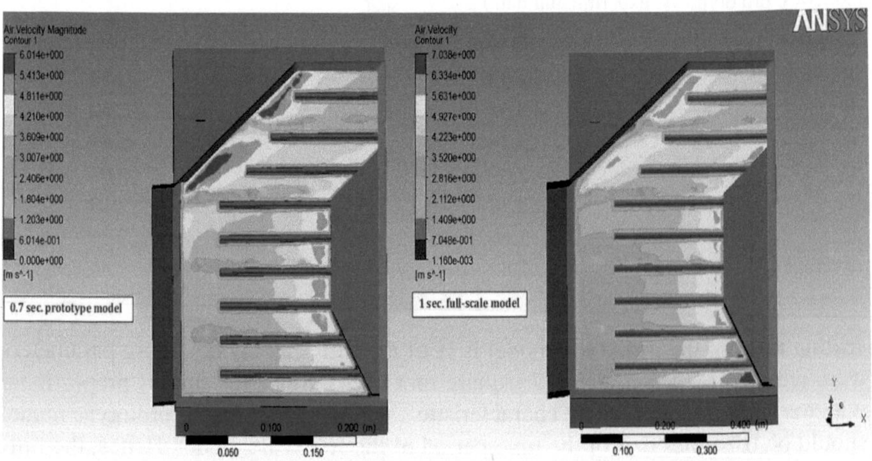

Fig. 7 Air velocity contour plot (0.7 s prototype vs. 1 s full-scale model)

average sectional domain analysis. However, the difference between the average simulated data for each section and the experimental data is less than 15 %; hence, the numerical results are validated.

The air velocity magnitude contours on the top of the deck of the full-scale model in 1 s and prototype model in 0.7 s are demonstrated in Fig. 7. The velocity vectors for 1 s simulated time for the full-scale model and 0.7 s for the prototype model are depicted in Fig. 8. The air velocity contour plot and the vector plot show modest variation between the two models, validating the scaling laws.

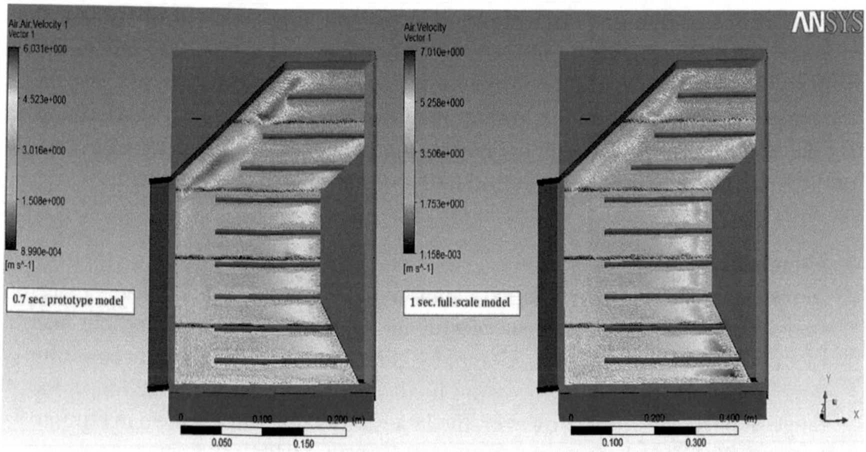

Fig. 8 Velocity vector plot (0.7 s prototype vs. 1 s full-scale model)

Table 3 Turbulent intensity plot (0.7 s prototype vs. 1 s full-scale model)

	Values	
Statistical parameters	Full-scale model	Prototype model
Minimum	0.0046	0.0059
Maximum	3.64	1.72
Area average	0.38	0.3

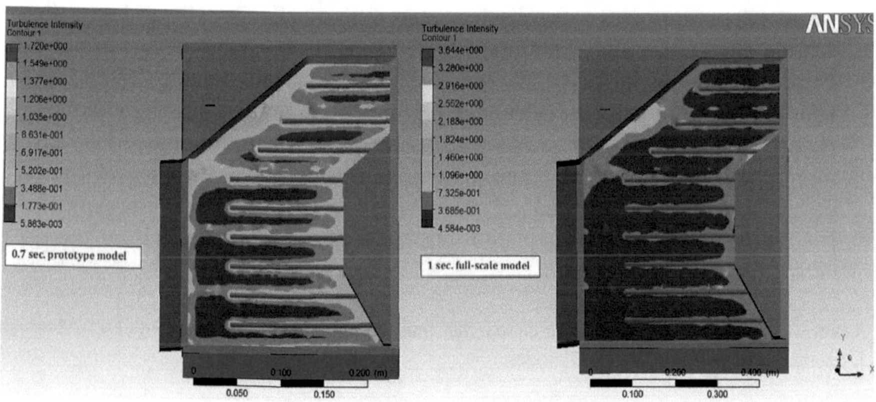

Fig. 9 Turbulent intensity plot (0.7 s prototype vs. 1 s full-scale model)

The minimum, maximum, and area average of turbulent intensity on top of the table deck prior to particle injection for 1 s full-scale and 0.7 s prototype model are listed in Table 3, and turbulent intensity contours are shown in Fig. 9.

The turbulent intensity data proves that the initial assumption of laminar flow was adequate. However it should be noted that the region would be highly turbulent with the injection of particles in the system as the chaotic nature of the particle movement on the deck and the formation of the bed would create a restrictive zone for the airflow passing through the porous media; hence, the intensity values might prove to be nontrivial with the introduction of particles into the system.

Conclusion

Industrial scale dry beneficiation technologies such as the FGX dry density-based separator have been successful in treating plus 6 mm particle size range. A modified Bratney Company air table was used to successfully develop an effective dry processing method to treat finer-size fraction in the range of 6 mm × 1 mm. However, the laboratory table can handle a maximum feed of 200–300 kg/h without overloading the table, and an appropriate scaling method needs to be developed to modify and scale up the table to treat industrial feed rates. To achieve this goal, the art of scale modeling was utilized, scaling laws were derived, and the full-scale model of a laboratory air-based density separator and a prototype model scaled down by the factor of ½ in X, Y, and Z directions were simulated using numerical methods. A user-defined function (UDF) was utilized to vibrate the table in an angular plane. The air velocity and turbulent intensity were used to validate the scaling laws. The variation between the models was found to be minimal and the absolute percent difference of air velocity between the two models was 3.83 %. The air velocity vector and contour plots as well as the turbulent intensity contours in two models displayed similar behavior, all validating the scaling laws. Therefore, to modify the scale of the laboratory air-based density separator to treat industrial feed rates, the air velocity should be kept at a constant value as calculated by the scaling laws, while the volumetric air flow rate needs to be changed based on the deck area ratio. Furthermore, the scaling laws show that the vibration velocity has an integral effect on the separation performance, and its vertical and horizontal components need to be changed by the square roots of the length characteristic ratios.

Acknowledgment The authors acknowledge the invaluable contribution of Dr. Abraham J. Salazar in developing the CFD model.

References

1. Arnold, B.J., Hervol, J.D., Leonard, J.W.: Dry particle concentration. In: Leonard, J.W., Hardinge, B.C. (eds.) Coal Preparation, 5th edn. Society of Mining, Metallurgy and Exploration, Littleton (1991)

2. Rezaee, M., Huggins, F.E., Honaker, R.Q.: Minimizing the environmental impacts of coal waste disposal. In: Craynon, J.R. (ed.) Environmental Consideration in Energy Production, pp. 60–77. Society for Mining, Metallurgy, and Exploration, Englewood (2013). ISBN 978-0-87335-380-9
3. Wotruba, H., Weitkaemper, L., Steinberg, M.: Development of a new dry density separator for fine-grained materials. In: Proceedings of 25th International Mineral Processing Congress (IMPC), Brisbane, 6–10 Sept 2010
4. Honaker, R.Q., Saracoglu, M., Thompson, E., Bratton, R., Luttrell, G.H., Richardson, V.: Upgrading coal using a pneumatic density-based separator. Int. J. Coal Prep. Utiliz. **28**(1), 51–67 (2008)
5. Donnelly, J.: Potential revival of dry cleaning of coal. The Australian Coal Review, pp. 26–30 (1999)
6. Lockhart, N.C.: Review paper: dry beneficiation of coal. Powder Technol. **40**, 17–42 (1984)
7. Patil, D.P., Parekh, B.K.: Beneficiation of fine coal using the air table. Int. J. Coal Prep. Utiliz. **31**(3–4), 203–222 (2011)
8. Napier-Munn1, T.J., Morrison, R.D.: The potential for the dry processing of ores. In: Water in Mining Conference, Brisbane, 13–15 Oct 2003, pp. 247–250 (2003)
9. Gupta, N., Bratton, R., Luttrell, G., Ghosh, T., Honaker, R.Q.: Application of air table technology for cleaning Indian coals. In: Young, C.A., Luttrell, G.H. (eds.) Separation Technologies for Minerals, Coal, and Earth Resources, pp. 199–209. Society for Mining, Metallurgy, and Exploration, Englewood, CO (2012). ISBN 10: 0873353390
10. Honaker, R.Q.: Demonstration of a novel dry coarse coal processing for improved mining economics. Progress report submitted to Div. of Energy, Development & Demonstration Governor's Office of Energy Policy (2007)
11. Bhattacharya, S.: Rotary breakers: prospects of application in India. In: Proceedings: The First Asian Mining Congress, Kolkata, 16–18 Jan 2006, pp. 353–359 (2006)
12. Kelley, M., Snoby, R.: Performance and cost of air jigging in the 21st century. In: Proceedings of 19th International Coal Preparation Conference, Lexington (2002)
13. Weinstein, R., Snoby, R.: Advances in dry jigging improves coal quality. Min. Eng. **59**(1), 29–34 (2007)
14. Li, G., Yang, Y.: Development and application of FGX series compound dry coal cleaning system. China Coal, Technology Monograph, pp. 17–28 (2006)
15. Lu, M., Yang, Y., Li, G.: The application of compound dry separation in China. In: Proceedings of the 20th International Coal Preparation Conference, Lexington, pp. 81–95, 29 Apr–3 May 2003
16. Luo, Z., Chen, Q.: Dry beneficiation technology of coal with an air dense medium fluidized bed. Int. J. Miner. Process. **63**(3), 167–175 (2001)
17. Rubarth, W., Steinberg, M., Wotruba, H., Weitkämper, D.: Evolution in dry separation for environmentally friendly mining AKAFLOW. In: Proceedings of Physical Separation, Falmouth, 16–17 June 2009
18. Dwari, R.K., Rao, K.H.: Fine coal preparation using novel tribo-electrostatic separator. Miner. Eng. **22**(2), 119–127 (2009)
19. Oder, R.R., Hurst, R., Ralston, J.N.: MagMill Processing of Coals at the DTEES Vicksburg Facility. Coal-Gen, Louisville (2008)
20. Kiser, M., Bratton, R., Luttrell, G., Roos, C.: Application of electronic sorting technology for reducing environmental impacts of coal processing. In: Craynon, J.R. (ed.) Environmental Consideration in Energy Production, pp. 22–31. SME, Englewood (2013). ISBN 978-0-87335-380-9
21. Patil, D.P., Ghosh, T., Honaker, R.Q.: Report: Beneficiation of Fine Size Powder River Basin Coal. Center for Applied Separation Technology (CAST) (Project No.: DE-FE0000699) (2013)
22. Emori, R.: Analytical approach to automobile collisions. SAE technical paper 680016 (1968). doi:10.4271/680016

23. Emori, R., Saito, K., Sekimoto, K.: Mokei Jikken no Riron to Ohyou (Scale Models in Engineering), 3rd edn. Gihodo, Tokyo (2000) (in Japanese)
24. Saito, K.: Progress in Scale Modeling: Summary of the First International Symposium on Scale Modeling (ISSM I in 1988) and Selected Papers from Subsequent Symposia (ISSM II in 1997 through ISSM V in 2006), Springer (2008)

Scale-Up of Chemical Looping Combustion

Fang Liu, Saito Kozo, and Kunlei Liu

Abstract The use of chemical looping combustion (CLC) for power generation is a possible pathway to meet demands for developing clean and efficient fossil fuel-based energy systems. CLC is an advanced technology that captures CO_2 inherently and could prove to be an advantageous next electricity generation technology in a carbon-constrained energy scenario. Fluidized bed reactors are used in CLC. The scale-up of laboratory-scale or prototype CLC reactors is necessary to develop an understanding of the potential and efficiencies of CLC because commercial installations would be large, requiring huge amounts of infrastructure, equipment, and investment. In this study, scaling laws were used as a guide to design and then build two different-sized CLC reactors. Testing of the reactors involved a focus on chemical similarities. Comparisons of the performance of both reactors, including the fuel conversion and temperature change, showed good correlations.

Nomenclature

C	Concentration
D	Bed diameter
d_p	Particle diameter
g	Gravity
H	Bed height
M	OC Mass
m	Scale ratio
Nr	Dimensionless reaction number
P	Pressure
Q	Flow rate
T	Temperature
t	Time

F. Liu • K. Liu (✉)
Center for Applied Energy Research, University of Kentucky, Lexington, KY 40511, USA
e-mail: kunlei.liu@uky.edu

S. Kozo
Institute of Research for Technology Development, University of Kentucky,
Lexington, KY 40506, USA

K. Saito et al. (eds.), *Progress in Scale Modeling, Volume II*,
DOI 10.1007/978-3-319-10308-2_19

u_{mf} Minimum fluidization speed
u_o Superficial velocity
V Velocity
X Conversion
β Drag coefficient
ε Volume fraction
ε_H Voidage
κ_v Reaction rate
ρ Density
σ_{ij} Solid-phase stress tensor
τ_{ij} Gas-phase stress tensor
φ Sphericity

Subscripts

g Gas phase
p Particle
s Solid phase

Superscripts

l Big
s Small

Introduction

Studies [1] show that CO_2 atmospheric concentration reached a level of 400 ppm in 2013 or 40 % above preindustrial levels. The contribution of CO_2 from industrial activity to increasing global CO_2 concentrations is widely accepted and points to the need to reduce the emission of this greenhouse gas [2]. Chemical looping combustion (CLC) is one kind of oxy-fuel combustion but has the advantage of in situ oxygen separability; thus CO_2 can be separated from flue gas inherently without the use of energy-intensive external air separation units that are needed for conventional oxy-fuel pulverized coal combustion technology. Accordingly, CLC technology may be a promising method for fossil fuel-based power generation in a carbon-constrained world. A diagram of CLC is shown in Fig. 1.

CLC uses a solid oxygen carrier (OC), usually a metal oxide, to provide oxygen for combusting fuel without the presence of nitrogen from air. The reduced OC is then recycled to an air reactor (AR) to be reoxidized, and then it is reused in the fuel reactor (FR) to provide oxygen in subsequent fuel combustion cycles. Through the use of the OC, the flue gas is separated into two parts. The air reactor gas outlet

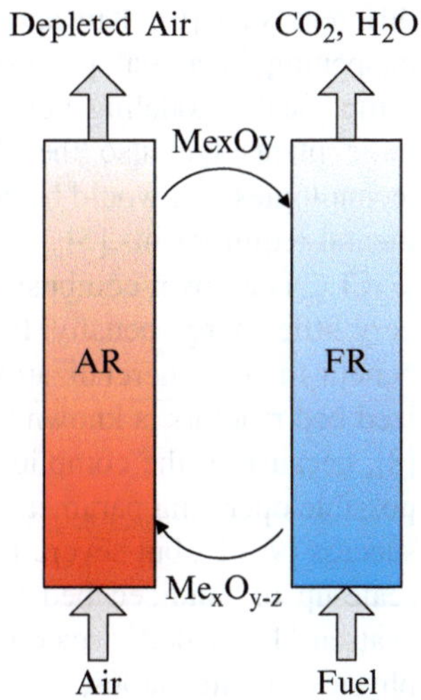

Fig. 1 CLC diagram

contains a high-temperature, oxygen-depleted gas containing mostly N_2. The fuel reactor outlet gas is primarily water vapor and CO_2. Because water vapor can be condensed very easily, an exhaust gas can be made that is highly concentrated in CO_2 and ready for compression and storage.

The reactions in the fuel reactor and the air reactor can be expressed as:

Fuel reactor:

$$\frac{1}{z}\left(2m+\frac{n}{2}\right)\mathrm{Me}_x\mathrm{O}_y + \mathrm{C}_m\mathrm{H}_n \rightarrow \frac{1}{z}\left(2m+\frac{n}{2}\right)\mathrm{Me}_x\mathrm{O}_{y-z} + m\mathrm{CO}_2 + \frac{n}{2}\mathrm{H}_2\mathrm{O}$$

Air reactor:

$$\mathrm{Me}_x\mathrm{O}_{y-z} + \frac{1}{2}z\mathrm{O}_2 \rightarrow \mathrm{Me}_x\mathrm{O}_y$$

The overall reaction is:

$$\mathrm{C}_m\mathrm{H}_n + \left(m+\frac{1}{4}n\right)\mathrm{O}_2 \rightarrow m\mathrm{CO}_2 + \frac{1}{2}n\mathrm{H}_2\mathrm{O}$$

Fluidized bed reactors are used for CLC rather than fixed bed reactors because of their inherent advantages, including better mixing of the gas-solids, that gives a high gas-solid interfacial area and better distribution of temperature gradients and, in fact, uniform temperature distribution due to the fluidized state of the solid's particles [3, 4].

Reactor scale-up is a major task for chemical engineers and is a fundamental step in the realization and optimization of industrial plants [5]. Scale-up is able to transfer information from equipment of one size to similar equipment having a different size [6]. Kuwana [7] gave an example of understanding phenomena happening in a real process by doing a scale modeling study of a laboratory setup. Scale modeling not only implies the capacity of designing and operating large plants but also the skill in developing new and more efficient reaction technologies that would be cost and product quality competitive and meet environmental requirements [5].

CLC is a novel combustion technology that could capture CO_2 inherently with very little energy penalty. Research on scale modeling of CLC would significantly benefit the commercialization of this technology. However, the scale-up of fluidized bed reactors is known to be more difficult than that of other types of reactors [8], because of the complicated fluidization state inside the reactor and the many possible operating parameters. The history of fluidization scale-up has examples of success [9–11], but severe failures have also been reported [4, 8]. Even today, the scale-up of fluidized bed technology is challenging [4]. In other words, just as Matsen [12] stated, "scale-up is still not an exact science, but is rather a mix of physics, mathematics, witchcraft, history and common sense that we call engineering."

The development and commercialization of any new chemical process will cost significant time and effort and require substantial capital expenditures. Using proven techniques based on experience and mathematics and/or design, models can minimize the risk and uncertainty when scaling up fluidized bed technology [14]. Currently, no open literature studies are known that have studied CLC reactor scale-up based on scaling laws. Therefore, the purposes of this study are to gain some experience on CLC fluidized bed reactor scale-up and to find potential correlations that would be beneficial for future CLC reactor scale-up. The focus will be on chemical similarities which in this paper are referred to as combustion similarities.

Scaling Laws

Governing Equations

Mathematical modeling is the most basic approach to scale-up [6]. Anderson and Jackson [15] derived the governing equations for fluidized bed particles. These equations are as follows:

Global continuity equation:

$$\varepsilon_g + \varepsilon_s = 1 \tag{1}$$

Gas-phase continuity equation:

$$\frac{\partial \varepsilon_g}{\partial t} + \frac{\partial}{\partial x_i}\left(\varepsilon_g V_{gi}\right) = 0 \tag{2}$$

Solid-phase continuity equation:

$$\frac{\partial \varepsilon_s}{\partial t} + \frac{\partial}{\partial x_i}\left(\varepsilon_s V_{si}\right) = 0 \tag{3}$$

Gas-phase momentum equation:

$$\rho_g\left[\frac{\partial V_{gi}}{\partial t} + V_{gj}\frac{\partial V_{gi}}{\partial x_j}\right] = -\frac{\partial P}{\partial x_i} - \frac{\partial \tau_{ij}}{\partial x_j} - \beta\left(V_{gi} - V_{si}\right) \tag{4}$$

Solid-phase momentum equation:

$$\rho_s\left[\frac{\partial}{\partial t}\left(\varepsilon_s V_{si}\right) + \frac{\partial}{\partial x_j}\left(\varepsilon_s V_{si} V_{sj}\right)\right] = -\frac{\partial \sigma_{ij}}{\partial x_j} + \rho_s \varepsilon_s g_i + \beta\left(V_{gi} - V_{si}\right) \tag{5}$$

These equations can describe particle movement very well, but they are too complicated to solve even with present computational means [6]. To successfully implement a scaling study, it is therefore necessary to simplify these equations, one approach to which is to achieve hydrodynamic and chemical similarity by using sets of dimensionless numbers which have to be kept constant at both scales [4].

Glicksman [16] derived dimensionless scaling numbers based on the above equations. These numbers are named full set scaling laws. They are shown as follows:

$$\frac{u_o \rho_g}{\mu};\ \frac{u_o^2}{gD};\ \frac{\rho_g}{\rho_p};\ \frac{D}{H};\ \frac{d_p}{D};\ \varphi;\ \text{particle size distribution (PSD)} \tag{6}$$

However, using the full set of Glicksman scaling laws brings many constraints and is very difficult to implement [4]. As a consequence, simplified Glicksman scaling laws [17] have been proposed that utilize the following numbers:

$$\frac{u_o}{u_{mf}};\ \frac{u_o^2}{gD};\ \frac{\rho_g}{\rho_p};\ \frac{D}{H};\ \varphi;\ \text{PSD} \tag{7}$$

The simplified scaling laws eliminate some constraints and give better flexibility in model design, which provides freedom to choose the diameter of the cold flow model without a need for exotic particles and pressures [4]. Hence, in this study, the same oxygen carrier (OC) was used for both reactors. The OC has a composition of 50 wt.% Fe_2O_3, 10 wt. % CeO_2, and 40 wt. % Al_2O_3, with a size range of 150–300 μm. It was produced from one batch of OC. The OC was produced by the method of freeze granulation. OC then was sintered at 1,400 °C for 6 h to gain some mechanical strength to resist attrition.

Improvement in the potential of combustion scaling is possible if important parameters can be kept identical in a test plant and in a full-scale plant to meet combustion similarity. These parameters include bed temperature, total excess-air ratios, primary stoichiometry, fuel, and bed material [6, 18]. An important parameter for combustion similarity is reactor performance as measured by fuel conversion, product distribution, temperature distribution, and selectivity [3, 13, 19]. For this study, reactors were designed and operating parameters chosen based on the simplified set of scaling laws. Reactor performance in a hot model was the focus in this study. Hot model performance also can be used to reflect the fluidization state. Conversion and temperature change from reactor performance parameters were chosen for validating the scaling laws.

Experimental

An important aspect of model experiments is to confirm the similarity between the original phenomena and the scale model counterpart [9, 20]. In this study, a bench scale CLC setup was built and used to conduct experiments; Fig. 2 shows a diagram of the bench scale setup and details of the reactors. A single fluidized bed reactor was used to simulate two reactors by switching a valve that controls the flow of either an oxidizing or reducing gas purged into the reactor from its bottom. Material within the reactor was held in place using a porous gas distributor. A K-type thermocouple was immersed in the bed material to monitor bed temperature.

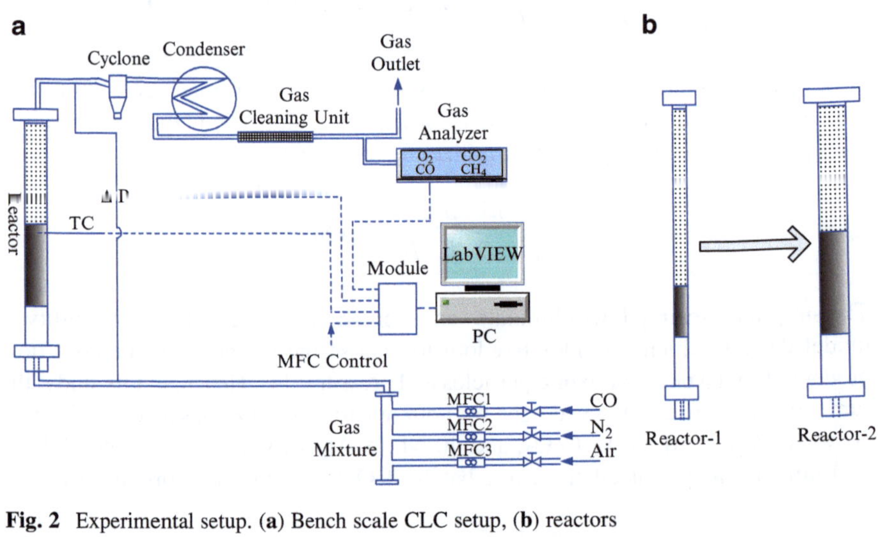

Fig. 2 Experimental setup. (**a**) Bench scale CLC setup, (**b**) reactors

Table 1 Reactor design and operating parameters

	D	H	M	Q	T	P	U_{mf}	k_v	Nr
Units	mm	mm	g	L/min	°C	bar	cm/s	s^{-1}	–
Reactor-1	25.4	3.81	12.5	2.8	970	1.0	2.6	0.00517	0.0011
Reactor-2	50.8	7.62	100	16	970	1.0	4.5	0.00520	0.0015

A Rosemond X-STREAM infrared multichannel gas analyzer was used to measure the composition of the flue gas. A LabVIEW program was written to control the mass flow controllers (MFCs) and record experimental data (i.e., temperature, gas concentrations, and flow rate).

Two reactors, reactor-1 and reactor-2, were fabricated with the scale ratio $m = 2$. Similar to the simplified fluidized bed scaling laws, to build these two reactors, Horio [21] suggested that

$$m = \frac{D^l}{D^s} = \frac{H^l}{H^s}$$

$$\sqrt{m} = \frac{U_o^l}{U_o^s}$$

Following the above equations, the reactors were built to meet hydrodynamic similarity. The two reactors were operated at the same temperature and pressure, and as mentioned previously, the materials in the two reactors were from the same batch. The reaction rate k_v was the same for both reactors. As proposed by Kelkar and Ng [3], the dimensionless reaction number Nr can be used to understand the whole reaction; it will be found that Nr scales with $\sqrt{m}$. Hence, for scaling validation, the time in the two different reactors must be scaled by a factor of $\sqrt{m}$ [4]. Based on the above equations, the reactor's design and operating parameters are determined and are shown in Table 1:

$$Nr = \frac{k_v H \varepsilon_H}{U_o}$$

$$\sqrt{m} = \frac{Nr^l}{Nr^s}$$

$$\sqrt{m} = \frac{t^l}{t^s}$$

Results and Discussion

In this study, the validation parameters monitored using the LabVIEW program by recording data every 0.5 s were the fuel conversion during reduction and temperature changes. The conversion X was calculated by

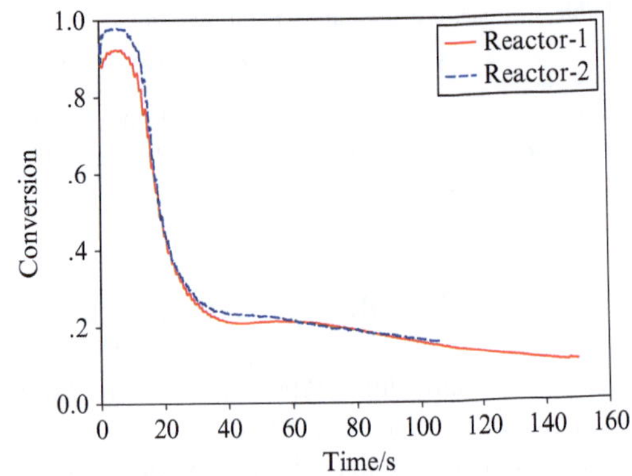

Fig. 3 CO conversion

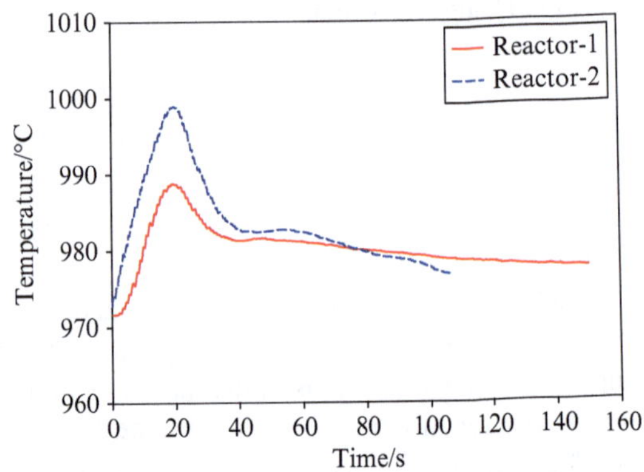

Fig. 4 Temperature change

$$X = \frac{C_{CO_2}}{C_{CO} + C_{CO_2}}$$

These X values versus time, presented in Fig. 3 for both reactors, are very similar. At the beginning of the reaction, the OC induced high conversion because ample amounts of surface oxygen were available to convert CO into CO_2. However, as the reaction proceeded, the amount of oxygen available from the OC became less and less and led to a decrease in conversion. The conversion values decreased because oxygen was being supplied from the bulk of the OC, a process slower than oxygen transfer from the surface. Hence, for these reactors at a time greater than about 25 s, the conversion was controlled by oxygen diffusion from the bulk to the surface of the OC. Overall, the close match of conversion between reactor 1 and reactor 2 validates the scaling laws that were used.

Figure 4 shows that the trends in temperature changes during the reduction reaction within the two reactors were very similar. Although not exactly identical, the temperature differences between the two reactors were very small compared to high temperature used during the testing. Temperature increases were noticed for both reactors because the reactions were exothermic, as shown below. Hence, the data confirms that appropriate scaling laws were used:

$$3Fe_2O_3 + CO = 2Fe_3O_4 + CO_2 \quad \Delta H = -44.4\,kJ/mol$$
$$Fe_2O_3 + CO = 2FeO + CO_2 \quad \Delta H = -1.6\,kJ/mol$$

Summary

Scaling laws of fluidized bed were applied to CLC reactors. Two CLC reactors of different sizes were built, as guided by scaling laws to meet hydrodynamic similarity, and then tested. Combustion similarity was the focus of this study and was monitored by acquiring and analyzing fuel conversion and temperature changes during the CLC reaction process. The comparison of conversion and temperature changes showed excellent combustion similarities in the two reactors, thereby validating the scale modeling method and the scale laws for CLC in fluidized beds. These results suggest the possibility for using scaling laws to scale-up even larger CLC reactors in the future.

Acknowledgments The authors would like to thank Dr. John Stencel for his useful comments.

References

1. Monastersky, R.: Global carbon dioxide levels near worrisome milestone. Nature **497**, 13–14 (2013)
2. Adánez, J., et al.: Ilmenite activation during consecutive redox cycles in chemical-looping combustion. Energy Fuels **24**(2), 1402–1413 (2010)
3. Kelkar, V.V., Ng, K.M.: Development of fluidized catalytic reactors: screening and scale-up. AIChE J. **48**(7), 1498–1518 (2002)
4. Rüdisüli, M., et al.: Scale-up of bubbling fluidized bed reactors—a review. Powder Technol. **217**, 21–38 (2012)
5. Donati, G., Paludetto, R.: Scale up of chemical reactors. Catal. Today **34**, 483–533 (1997)
6. Leckner, B., Szentannai, P., Winter, F.: Scale-up of fluidized-bed combustion—a review. Fuel **90**(10), 2951–2964 (2011)
7. Kuwana, K., Hassan, M.I., Singh, P.K., Saito, K., Nakagawa, J.: Scale-model experiment and numerical simulation of steel teeming processes. SME J. **23**, 1–6 (2008)
8. Werther, J.: Scale-up modeling for fluidized bed reactors. Chem. Eng. Sci. **47**(9–11), 2457–2462 (1992)
9. Emori, R.I., Saito, K., Sekimoto, K.: Scale Models in Engineering (Mokei Jikken no Riron to Ohyou), 3rd edn. Gihodo, Tokyo (2000). Second Print in 2008 in Japanese. ISBN 4-7655-3252-6 C3053
10. Saito, K. (ed.): Progress in Scale Modeling: Summary of the First International Symposium on Scale Modeling (ISSM in 1988) and Selected Papers from subsequent Symposia (ISSM II in 1997 through ISSM V in 2006), Springer (2008)
11. Kunii, D., Levenspiel, O.: Fluidization Engineering, 2nd edn. Butterworth-Heinemann, Boston (1991)

12. Matsen, J.M.: Design and scale up of CFB catalytic reactors. In: Grace, J.R., Avidan, A.A., Knowlton, T.M. (eds.) Circulating Fluidized Bed, pp. 489–503. Chapman & Hall, London (1997)
13. Jazayeri, B.: Successfully scale up catalytic gas-fluidized beds. Chem. Eng. Prog. **91**(4), 26–31 (1995)
14. Knowlton, T.M., Karri, S.B.R., Issangya, A.: Scale-up of fluidized-bed hydrodynamics. Powder Technol **150**(2), 6–6 (2005)
15. Anderson, T.B., Jackson, R.: Fluid mechanical description of fluidized beds. Equations of motion. Ind. Eng. Chem. Fundam. **6**(4), 527–539 (1967)
16. Glicksman, L.R.: Scaling relationships for fluidized beds. Chem. Eng. Sci. **39**(9), 1373–1379 (1984)
17. Glicksman, L.R., Hyre, M., Woloshun, K.: Simplified scaling relationships for fluidized beds. Powder Technol. **77**(2), 177–199 (1993)
18. Leckner, B., Werther, J.: Scale-up of circulating fluidized bed combustion. Energy Fuels **14**(6), 1286–1292 (2000)
19. Kimball, E.E., Geerdink, P., Goetheer, E.L.: Scale-up of fixed-bed chemical looping combustion. In: 2011 Spring Meeting & 7th Global Congress on Process Safety (2011)
20. Emori, R.I., Saito, K.: A study of scaling laws in pool and crib fires. Combust. Sci. Technol. **31** (5–6), 217–231 (1983)
21. Horio, M., Nonaka, A., Sawa, Y., et al.: A new similarity rule for fluidized bed scale-up. AIChE J. **32**(9), 1466–1482 (1986)

Scale Effect on Solid Fuel Regression in CAMUI-Type Hybrid Rocket Motor

Harunori Nagata and Mitsunori Ito

Abstract The objective of this study is to obtain a rule to define a similarity condition under which subscale tests should be conducted to simulate firings of full-scale CAMUI-type hybrid motors. Static firing tests with fuel grains of different scaling have estimated the validity of similarity conditions based on convective heat transfer mechanisms. Fuel grains of all scales consist of four cylindrical polyethylene blocks with two axial ports. Experimental results show that except the fore-end face of the uppermost block and the back-end face of the rearmost block, similarity conditions based on convective heat transfer mechanisms are valid on end faces of fuel blocks. Because there is no end face downstream of the rearmost block, the flow field between fuel blocks with intense turbulence does not exist near the back-end face of the block, resulting in a small convective heat transfer rate. As a result, radiative heat transfer is not negligible on this burning surface and causes an error in the similarity condition. Because the impinging jet onto the fore-end face of the uppermost block is not high-temperature combustion gas but virtually pure oxygen, a similarity about chemical reaction is necessary in addition to those about convective heat transfer to realize a similarity condition.

Nomenclature

A_b	Burning surface area
c	Constant
D	Port diameter
G_p	Mass flow density
H	Impingement distance
h_v	Heat of vaporization of the fuel
l	Constant
m	Constant
$\dot{m}_f$	Fuel flow rate
$\dot{m}_{ox}$	Oxidizer flow rate

H. Nagata (✉) • M. Ito
Division of Mechanical and Space Engineering, Hokkaido University, Sapporo 060-8628, Japan
e-mail: nagata@eng.hokudai.ac.jp

K. Saito et al. (eds.), *Progress in Scale Modeling, Volume II*,
DOI 10.1007/978-3-319-10308-2_20

n	Constant
Nu	Nusselt number
n_r	Order of reaction
P_c	Chamber pressure
$\dot{q}_{\text{con}}$	Convective heat transfer rate per unit area
$\dot{q}_{\text{rad}}$	Radiative heat transfer rate per unit area
$\dot{q}_{\text{total}}$	Heat transfer rate per unit area
Re	Reynolds number
r	Regression rate
St	Stanton number
t	Burning duration
u	Injection velocity
V_{rea}	Volume of reaction zone
v_b	Velocity at the flame location
v_e	Main flow velocity
x	Regression depth
Δh	Enthalpy difference between the gas at the fuel surface and at the flame
δ_{rea}	Reaction zone thickness
κ	Reaction rate constant
μ	Viscosity
ρ	Density of the solid fuel
$\dot{\omega}$	Chemical reaction rate

Subscripts

0	Without blowing from fuel surface
b	With blowing from fuel surface

Introduction

Laboratory of Space System Engineering in Hokkaido University and Uematsu Electric Co., Ltd. have organized a research team to develop a small-scale explosive-free launch system based on a hybrid rocket [1]. A main purpose is to drastically reduce costs of sounding rocket experiments and launches of microsatellites to LEO. A key idea is a new fuel grain design to accelerate the gasification rate of solid fuels and to increase the thrust. CAMUI comes from the abbreviation of the new combustion method of "cascaded multistage impinging jet." Figure 1 shows the basic idea of CAMUI-type fuel grain [2]. By separating conventional cylinder-shape solid fuel grain with a central port into multiple cylinder blocks, end faces of all cylinder blocks burn concurrently. The combustion gas repeatedly collides with surfaces of fuel blocks to accelerate the heat transfer to the fuel.

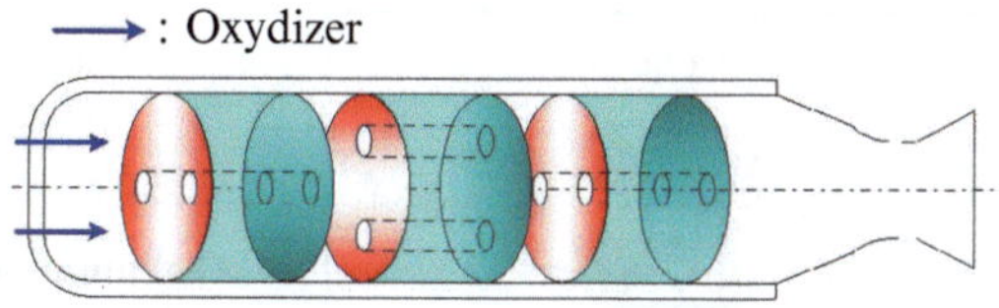

Fig. 1 Basic concept of CAMUI fuel grain

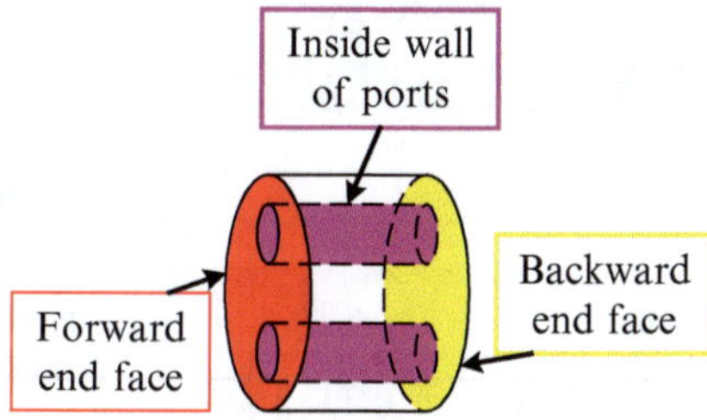

Fig. 2 Burning surfaces on a block of CAMUI fuel grain

The solid fuel gasification rate is a major internal ballistics parameter of a hybrid rocket combustor, giving the fuel flow rate; $\dot{m}_f = \rho A_b r$, where $\dot{m}_f$, ρ, A_b, and r are fuel flow rate [kg/s], density of the solid fuel [kg/m^3], burning surface area [m^2], and the regression rate of the burning surface [m/s], respectively. In a conventional-type fuel grain with a single port, the burning surface is the inside wall of the port and the regression rate is virtually uniform except a small part in the upstream [3]. On the other hand, a fuel block in CAMUI-type grain has three burning surfaces as Fig. 2 shows, i.e., a forward end face, inner walls of ports, and a backward end face. Because regressions of these surfaces progress simultaneously with different regression rates among them, temporal variation of grain geometry is complicated. As a result, predicting the progress of the grain geometry and temporal variation of fuel flow rate is not easy. A probable approach is to assume that the three burning surfaces, fore-end and back-end faces and inside walls of two ports, regress uniformly and independently with each other, keeping edges between two faces a right angle. Previous experimental researches show that the amount of fuel consumption calculated from this simple model, using mean regression depth of each burning surface, includes error up to 10 %. Accordingly, a simplified empirical method based on this model is not sufficient to accurately predict combustion characteristics of a motor. Static firing tests are necessary to correct for variations of the predicted fuel regression progress. Figure 3 shows the detail of the design procedure of a fuel grain: Design an optimum geometry of fuel blocks based on the simplified regression model and empirical regression formula for each burning surface. Fine-tune the grain geometry on the basis of results of static firing tests by a subscale motor. Scale the fine-tuned geometry up to a full-scale size and make a design verification test by a full-scale motor. Because many static firing tests are necessary to fine-tune the grain geometry, employing a subscale motor is preferable to save time and expense. However, the different motor scale likely changes internal ballistics, and the optimum grain geometry for a subscale motor may not be available for a full-scale motor. Indeed, some researchers have investigated scaling effects in hybrid motors for conventional single-port grain geometries and

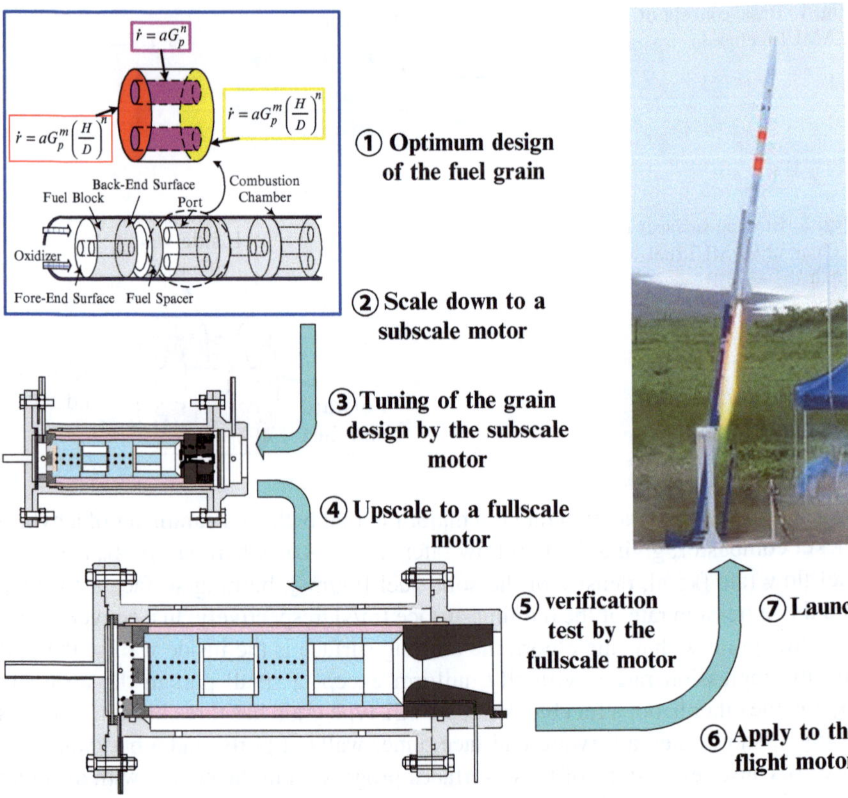

Fig. 3 Design procedure of CAMUI-type fuel grain

have concluded that scaling of hybrid rocket motors should be done under appropriate similarity conditions [4–7]. The main objective of this study is to obtain a rule to define a similarity condition under which subscale tests should be conducted to simulate firings of full-scale CAMUI-type hybrid motors. Static firing tests with fuel grains of different scaling shown in Fig. 4 have been conducted to estimate the validity of similarity conditions based on convective heat transfer mechanisms.

Scale Effect

Regression rate of a burning surface is proportional to the heat transfer rate per unit area:

$$r = \frac{\dot{q}_{\text{total}}}{\rho h_v} \tag{1}$$

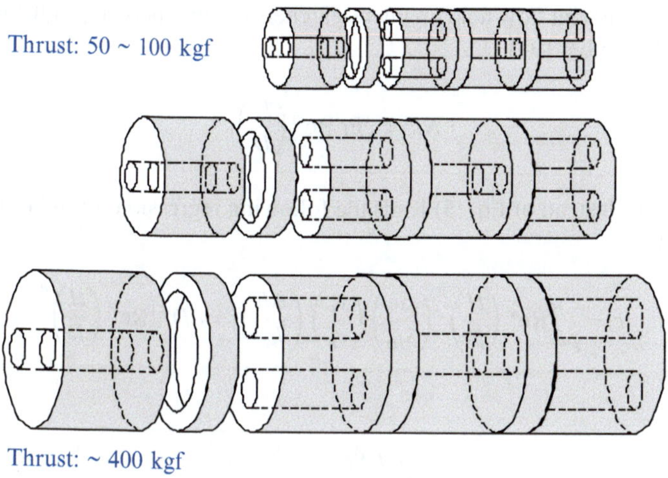

Fig. 4 CAMUI-type fuel grains of different scales

Hybrid motors, including our CAMUI type, are characterized by a turbulent internal flow with dominant boundary layer effects. Therefore, heat transfer in hybrid combustors mainly arises from turbulent convection with minor contribution of radiation. Assuming that chemical reaction rate is virtually infinite comparing with species transfer rate and the flame has a zero thickness, a theoretical derivation gives the following relational expression [8]:

$$r = \frac{c}{\rho}\left(\frac{\mu^{1-m}}{D^{1-m}}\right) G_p{}^m \left(\frac{St}{St_0}\right)\left(\frac{v_e}{v_b}\right)\left(\frac{\Delta h}{h_v}\right) \tag{2}$$

In a conventional hybrid rocket, the following equation gives the Reynolds number:

$$\mathrm{Re} = \frac{G_p D}{\mu} \tag{3}$$

Substituting Eq. (3) into Eq. (2),

$$\dot{r} = \frac{c\mu}{\rho D}\mathrm{Re}^m \left(\frac{St}{St_0}\right)\left(\frac{v_e}{v_b}\right)\left(\frac{\Delta h}{h_v}\right) \tag{4}$$

The first half of the right-hand side, from c/ρ to Re^m, represents the convective heat transfer characteristics of the flow, and numerical values of c and m equals to those in the following equation, which gives Nusselt number in a circular duct:

$$Nu = c\mathrm{Pr}^l\mathrm{Re}^m \tag{5}$$

In a jet impingement flow field, many researchers use the following equation to give the Nusselt number [9–11]:

$$Nu = c\mathrm{Pr}^l\mathrm{Re}^m\left(\frac{H}{D}\right)^n \tag{6}$$

Using Eq. (6) instead of Eq. (5) leads the following regression formula for forward end faces:

$$r = \frac{c\mu}{\rho D}\mathrm{Re}^m\left(\frac{H}{D}\right)^n\left(\frac{St}{St_0}\right)\left(\frac{v_e}{v_b}\right)\left(\frac{\Delta h}{h_v}\right) = aD^{-1}\mathrm{Re}^m\left(\frac{H}{D}\right)^n \tag{7}$$

where

$$a = \mu\frac{c}{\rho}\left(\frac{St}{St_0}\right)\left(\frac{v_e}{v_b}\right)\left(\frac{\Delta h}{h_v}\right) \tag{8}$$

In a series of static firing tests, we employed common H/D and Reynolds number among different scales. Assuming that the difference of motor scaling does not affect the numerical value of a,

$$r \propto D^{-1} \tag{9}$$

Accordingly, regression rates of forward end faces are inversely proportional to the motor scaling. Because we could not find an appropriate formula giving Nusselt number for backward end faces, we assumed Eq. (6) for backward end faces too. As a result, regression rates of backward end faces are also inversely proportional to the motor scaling.

Similarity Conditions

To equalize Reynolds number among firing tests with different motor scaling, oxidizer flow rate is proportional to motor scaling:

$$\dot{m}_{ox} \propto D \tag{10}$$

By scaling the burning duration according to

$$t \propto D^2 \tag{11}$$

the regression depth x [m] becomes proportional to the motor scaling:

$$x = r\left(\propto \frac{1}{D}\right) * t\left(\propto D^2\right) \propto D \tag{12}$$

Because the initial geometry is similar among motors of different scaling and the regression depth is proportional to the motor scaling, geometries of fuel grains after static firing tests by different motor scaling should be similar among them. Accordingly, the degree of similarity of fuel grains after static firing tests shows the validity of similarity conditions Eq. (7) shows.

Static Firing Tests

Figure 5 shows the appearance of fuel grains for motors of three different scales. Fuel blocks are made of polyethylene. Homothetic ratio among scales S, M, and L are 1:1.5:2, with outer diameters of 60, 90, and 120 mm, respectively. Fuel grains of all scales consist of four cylindrical blocks. Figure 6 shows the outline of the facility for static firing tests with various scaling. Three combustion chambers of S, M, and L scales are in similarity shape from injector to the nozzle inlet. Oxygen flows into combustion chambers in a liquid phase. A nichrome wire on the fore-end face of the uppermost block ignites the fuel by electrical heating.

Figure 7 shows comparisons of regression distributions on end faces after static firing tests with experimental conditions that Table 1 summarizes. Reynolds numbers of three scales are nearly identical to 75,000, within a range of 74,000–77,000. Square roots of burning durations are 1.95, 2.98, and 3.91 for S, M, and L scales, and the ratio among them (1:1.53:2.01) is nearly identical to the ratio of scaling. The ratio of LOX flow rates among three scales is 1:1.45:2.04, also nearly equal to the ratio of scaling. Nozzle throat area is proportional to motor scaling, resulting in nearly identical chamber pressures among three scales. Fuel blocks are numbered from upstream to downstream. Upper and lower dots in each figure show regression distributions along broken lines; Fig. 8 shows on forward and backward end faces, respectively. Horizontal and vertical axes are distance from grain axis normalized by the radius of each grain and regression depth normalized by the initial axial length of each block, respectively. Except the for-end face of the 1st block (uppermost) and the back-end face of the 4th block (rearmost), regression distributions are in

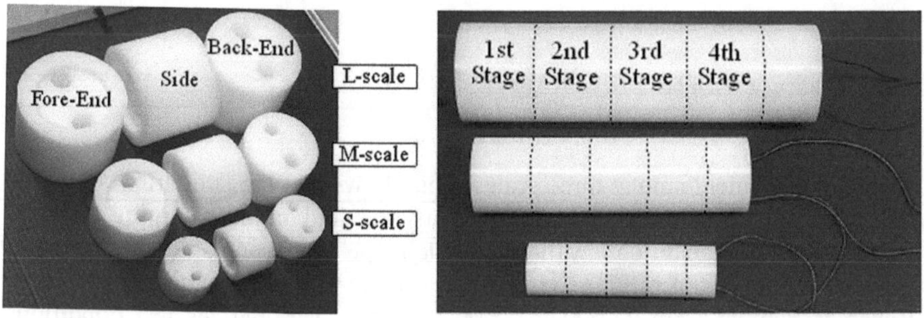

Fig. 5 Appearance of fuel grains

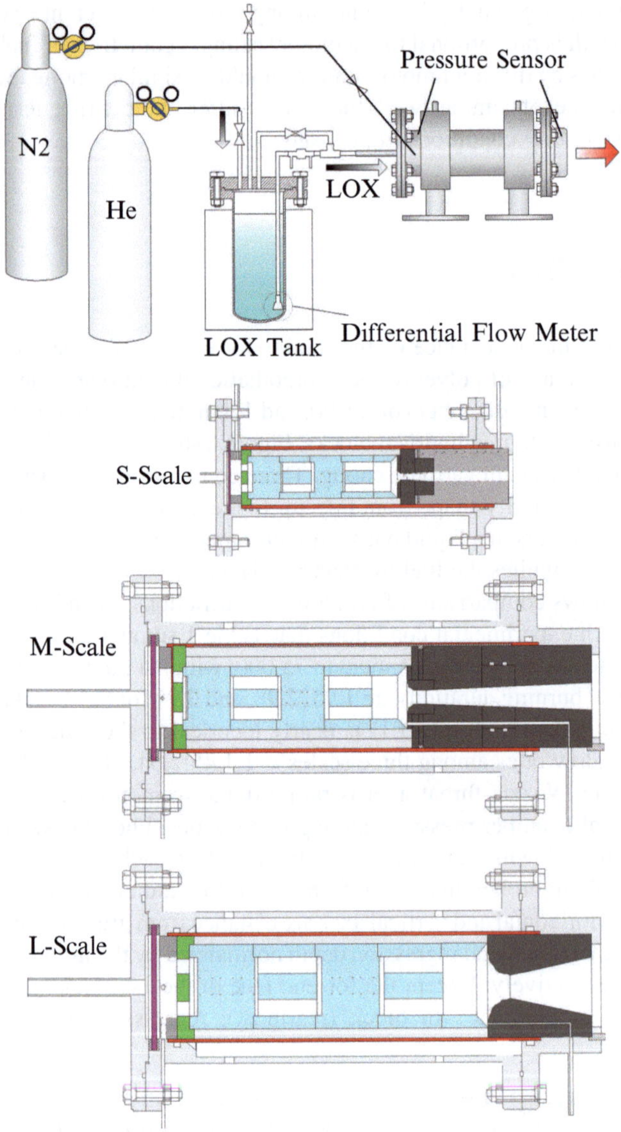

Fig. 6 Outline of test motors

excellent agreement among three scales. This shows that the similarity conditions based on convective heat transfer mechanisms are valid on these burning surfaces.

The similarity condition is valid in a wide range of Reynolds number. Figure 9 shows comparisons of mean regression rates between S and L scales with various Reynolds number from 33,000 to 75,000. Table 2 summarizes test conditions. Horizontal and vertical axes in Fig. 9 show the regression rate of S scale and the

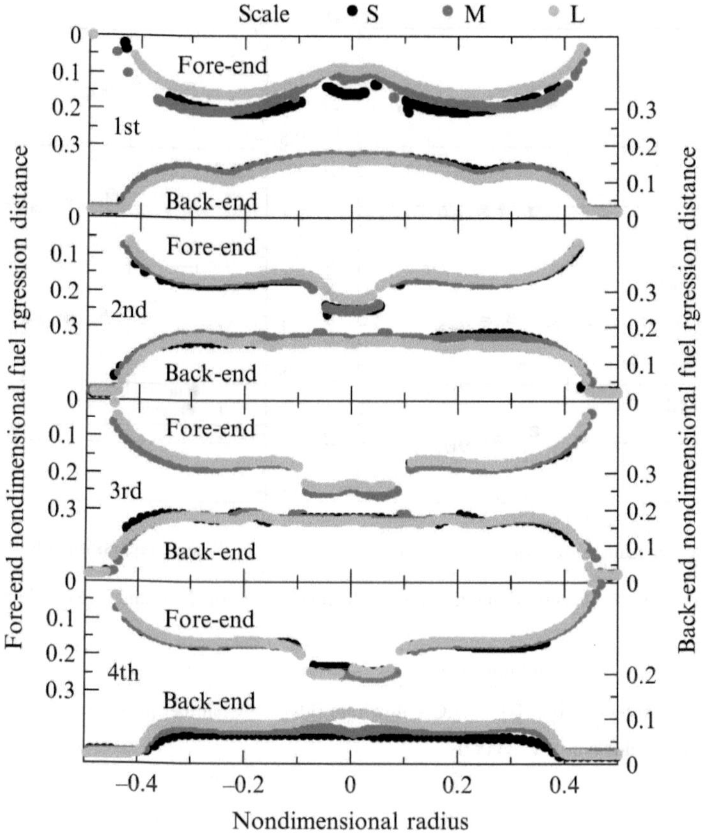

Fig. 7 Regression distributions

Table 1 Experimental conditions (series A)

Scale	Burning duration [s]	LOX flow rate [g/s]	Chamber pressure [MPa]	Re
S (1)	3.8	105	1.0	77,000
M (1.5)	8.9	152	1.0	74,000
L (2)	15.3	214	1.1	74,000

Fig. 8 Measurement region

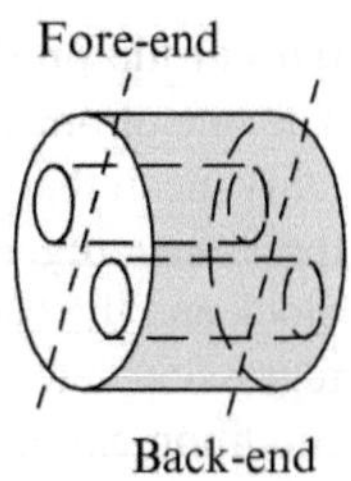

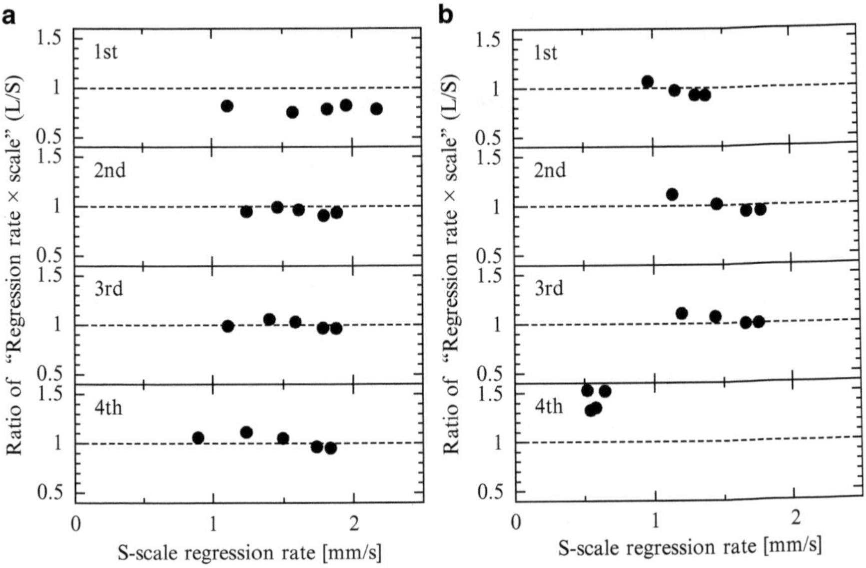

Fig. 9 Comparison of regression rates between S and L scales. (**a**) Fore-end face (**b**) Back-end face

Table 2 Experimental conditions (series B)

Scale	Burning duration [s]	LOX flow rate [g/s]	Chamber pressure [MPa]	Re
Average Reynolds number: 33,000				
S (1)	3.9	29	0.98	32,000
L (2)	15.6	61	1.1	34,000
Average Reynolds number: 46,000				
S (1)	4.0	52	1.0	44,000
L (2)	16.5	110	1.1	48,000
Average Reynolds number: 62,000				
S (1)	4.2	81	1.1	62,000
L (2)	16.9	169	1.0	62,000
Average Reynolds number: 69,000				
S (1)	3.8	94	0.95	70,000
L (2)	15.3	189	0.94	68,000

ratio of the product of regression rate and motor scaling between S and L scales. The line of unity means that the similarity condition is valid. Dots under the line of unity mean that the regression rate of L scale is smaller than the value predicted by the S-scale regression rate, and vice versa. Again, except the fore-end face of the 1st block and the back-end face of the 4th block, dots lie near the line of unity. L-scale regression rates of the 1st block fore-end face are less than the value similarity condition expects. On the contrary, L-scale regression rates of the 4th block back-end face exceed the expected value.

Radiation Effect

The similarity condition Eq. (7) describes assumes that the heat transfer mainly arises from turbulent convection with minor contribution of radiation. Therefore, if radiation effect is not negligible, it can cause an error in the similarity condition. The heat transfer from the combustion gas to the solid fuel is the sum of convective and radiative heat transfers:

$$\dot{q}_{\text{total}} = \dot{q}_{\text{con}} + \dot{q}_{\text{rad}} \tag{13}$$

In a similarity condition with a scale ratio between S and L scales of 2, for instance, whereas $\dot{q}_{\text{con}}$ of L scale is a half of that of S scale, $\dot{q}_{\text{rad}}$ of L scale is almost the same as that of S scale if the flame temperature does not change with motor scaling. As a result, $\dot{q}_{\text{total}}$ of L scale exceeds a half of S scale, and dots in Fig. 9 lie over the line in this case. Consequently, dots in Fig. 9 shall move upward from the line of unity more as the radiative heat transfer becomes considerable. Figure 10 shows comparisons of mean regression rates between S and L scales in similarity conditions with lower (0.5 MPa) and higher (2.0 MPa) chamber pressures. Table 3 summarizes test conditions of this series. By increasing (or decreasing) chamber pressure without changing Reynolds number, we can increase (or decrease) the radiative heat transfer rate without changing the convective heat transfer rate. A tendency that dots of higher pressures moving upward from the line is clear in Fig. 10. Because there is no fuel block downstream of the 4th block, the flow field between fuel blocks with intense turbulence does not exist near the back-end face of the 4th block, resulting in a small convective heat transfer rate. The small convective heat transfer appears in small regression rates of S scale (horizontal axis) comparing with other burning surfaces. As a result, radiative heat transfer is not negligible on this burning surface, and dots in Fig. 9 lie over the line of unity.

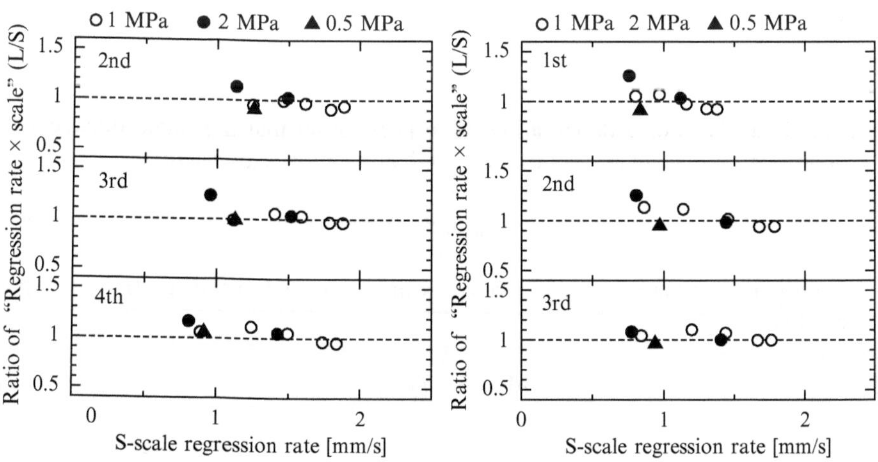

Fig. 10 Effect of chamber pressure on the similarity condition

Table 3 Experimental conditions (series C)

Scale	Burning duration [s]	LOX flow rate [g/s]	Chamber pressure [MPa]	Re
Chamber pressure: 2 MPa				
S (1)	4.3	80	2.0	58,000
L (2)	16.8	173	2.0	62,000
S (1)	3.9	31	1.7	27,000
L (2)	16.2	63	1.8	33,000
Chamber pressure: 0.5 MPa				
S (1)	4.0	30	0.54	35,000
L (2)	16.0	62	0.50	32,000

Effect of Chemical Reaction

Because on the fore-end face of the first block the regression rate of L scale is smaller than the value predicted by the S-scale regression rate as Fig. 9 shows, this disagreement is not due to the radiation effect. A distinctive condition about this burning surface is that the impinging jet is not high-temperature combustion gas but virtually pure oxygen. Therefore, it is reasonable to think of the effect of chemical reaction as a cause of this disagreement. To investigate the effect of chemical reaction, we added a similarity about chemical reaction to the similarity condition and conducted static firing tests. Ben-Arosh et al. [5] have studied a similarity condition about chemical reaction on a burning surface of conventional tubular solid fuel grain. We adopted this methodology to our CAMUI-type fuel grain: The volume of the reaction zone over the fore-end face of the 1st block, V_{rea}, is proportional to the square of characteristic length and reaction zone thickness:

$$V_{\text{rea}} \propto D^2 \delta_{\text{rea}} \tag{14}$$

The rate of chemical reaction described by a global nth order single step is expressed by

$$\dot{\omega} = \kappa {P_c}^{n_r} \tag{15}$$

A simple analysis of a deflagration wave [12] shows that the flame thickness is proportional to the square root of the chemical reaction rate. Therefore,

$$\delta_{\text{rea}} \propto {P_c}^{-n_r/2} \tag{16}$$

Accordingly, the mass of the reactant mixture converted in the reaction zone per second is

$$\dot{\omega} V_{\text{rea}} \propto \kappa D^2 {P_c}^{n_r/2} \tag{17}$$

Using the equation of state for perfect gas along with the constancy of temperature, the incoming mass flux into the reaction zone is

$$\dot{m} \propto P_c u D^2 \tag{18}$$

To establish a similarity condition concerning chemical reaction between two different scales, the ratio between $\dot{\omega} V_{\mathrm{rea}}$ and $\dot{m}$ should be constant among various scales:

$$\frac{\dot{\omega} V_{\mathrm{rea}}}{\dot{m}} \propto \frac{\kappa D^2 {P_c}^{n_r/2}}{P_c u D^2} \propto \frac{{P_c}^{n_r/2-1}}{u} = \mathrm{const} \tag{19}$$

For $n_r = 2$ Eq. (19) does not depend of pressure, and fortunately, typical bimolecular reactions including the case for the overall hydrocarbon-oxygen reactions are generally of second order, representing n_r of 2. As a result, Eq. (19) reduces to the following requirement:

$$u = \mathrm{const} \tag{20}$$

Accordingly, the requirement for the similarity of chemical reaction between different motor scaling is the same oxidizer flow velocity at corresponding points of motors. We can satisfy this requirement easily by using the same oxidizer injection velocity. Figure 11 shows comparisons of regression distributions between S and L scales with various velocity ratios, which is defined by L-scale injection velocity divided by S-scale injection velocity. Table 4 summarizes the experimental conditions. To equalize the Reynolds number, chamber pressure of each scale is inversely proportional to the motor scaling for each pair. Therefore, the similarity condition about convective heat transfer Eq. (7) describes is valid for each pair. Despite this, the regression distribution is not similar with each other except the pair with the velocity ratio of near unity. The regression depth of L scale with the velocity ratio exceeding unity is less than expected, and Fig. 6 belongs to this case. Figure 11 clearly shows that we can obtain a similarity condition on the fore-end face of the 1st block by introducing the similarity about chemical reaction in addition to those about convective heat transfer.

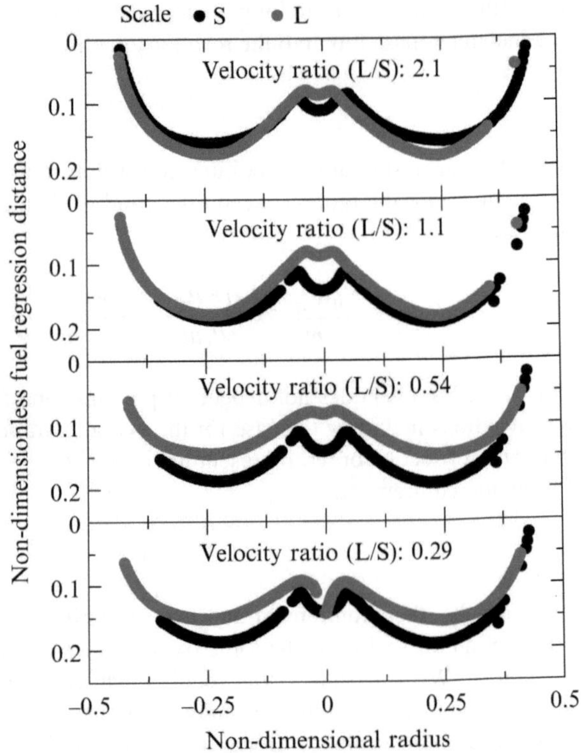

Fig. 11 Regression distributions with various velocity ratio

Table 4 Experimental conditions (series D)

Scale	Burning duration [s]	LOX flow rate [g/s]	Chamber pressure [MPa]	Re
Velocity ratio (L/S): 2.1				
S (1)	4.3	80	2.0	58,000
L (2)	16.8	158	0.50	65,000
Velocity ratio (L/S): 1.1				
S (1)	4.2	81	1.1	62,000
L (2)	16.8	158	0.50	65,000
Velocity ratio (L/S): 0.54				
S (1)	4.2	81	1.1	62,000
L (2)	16.9	169	1.0	62,000
Velocity ratio (L/S): 0.29				
S (1)	4.2	81	1.1	62,000
L (2)	16.8	173	2.0	62,000

Conclusion

Static firing tests with fuel grains of different scaling have estimated the validity of similarity conditions based on convective heat transfer mechanisms. Results show that except the fore-end face of the 1st block (uppermost) and the back-end face of the 4th block (rearmost), similarity conditions based on convective heat transfer mechanisms are valid on end faces of fuel blocks. Because there is no fuel block downstream of the 4th block, the flow field between fuel blocks with intense turbulence does not exist near the back-end face of the 4th block, resulting in a small convective heat transfer rate. As a result, radiative heat transfer is not negligible on this burning surface and causes an error in the similarity condition. Because the impinging jet onto the fore-end face of the 1st block is not high-temperature combustion gas but virtually pure oxygen, a similarity about chemical reaction is necessary in addition to those about convective heat transfer to realize a similarity condition.

Acknowledgments This research was partially supported by the Ministry of Education, Science, Sports and Culture, Grant-in-Aid for Scientific Research (B), 21360410, 2009.

References

1. Nagata, H., Ito, M., Maeda, T., Watanabe, M., Uematsu, T., Totani, T., Kudo, I.: Development of CAMUI hybrid rocket to create a market for small rocket experiments. Acta Astron. **59**(1–5), 253–258 (2006)
2. Nagata, H., Okada, K., San'da, T., Akiba, R., Satori, S., Kudo, I.: New fuel configurations for advanced hybrid rockets. In: 49th International Astronautical Congress, IAF-98-S.3.09 (1998)
3. Chiaverini, M.J., Kuo, K.K. (eds.): Fundamentals of Hybrid Rocket Combustion and Propulsion, AIAA, 1st edn. (2007)
4. Estey, P., et al.: AIAA, SAE, and ASME, Joint Propulsion Conference, 27th, AIAA-91-2517 (1991)
5. Ben-Arosh, R., Gany, A.: Similarity and scale effects in solid-fuel ramjet combustors. J. Propul. Power **8**(3), 615–623 (1992)
6. Gany, A.: AIAA, ASME, SAE, and ASEE, Joint Propulsion Conference and Exhibit, 32nd, AIAA Paper 96–2846 (1996)
7. Swami, R.D., Gany, A.: Analysis and testing of similarity and scale effects in hybrid rocket motors. Acta Astron. **52**, 619–628 (2003)
8. Humble, R., et al.: Space Propulsion Analysis and Design. McGraw-Hill, New York (1995)
9. Viskanta, R.: Heat transfer to impinging isothermal gas and flame jets. Exp. Thermal Fluid Sci. **6**, 111–134 (1993)
10. Baukal, C.E., Gebhart, B.: A review of empirical correlations for flame impingement heat transfer. Int. J. Heat Fluid Flow **17**, 386–396 (1996)
11. San, J.-Y., Lai, M.-D.: Optimum jet-to-jet spacing of heat transfer for staggered arrays of impinging air jets. Int. J. Heat Mass Transfer **44**, 3997–4007 (2001)
12. Williams, F.A.: Combustion Theory, 2nd edn. Benjamin/Cummings, Menlo Park (1985)

Scale Effect Analysis for Locomotion Systems in Different Gravity Fields

Kazuya Yoshida and Havard Lund

Abstract This paper presents surface exploration strategies for possible future asteroid missions. As a pioneer of the asteroid sample-return missions, Hayabusa was developed by the Institute of Space and Astronautical Science, Japan, and launched in May 2003. Various technologies developed for Hayabusa will be useful for current and future exploration of minor bodies in our solar system. Because the gravitational force of an asteroid is very small, it is difficult for a spacecraft to stay on the asteroid surface stably; therefore, a touchdown sampling sequence was developed. A micro-rover for surface locomotion was also developed and deployed over the asteroid surface. This study investigates possible designs for a future locomotion system over the microgravity surface and compares different types of candidate grabbing forces as a function of the representative size.

Introduction

Throughout the solar system, a vast amount of minor bodies exists. In recent years, there has been an increased interest in studying these bodies for several reasons. First, minor bodies such as asteroids are believed to be remains from the creation of the solar system, and thus, by studying them, researchers can learn more about the conditions surrounding the birth of the solar system. Second, the collision of asteroids with the Earth is linked to the extinction of certain species on the Earth. A widely believed example of this is the extinction of dinosaurs. To prevent such a disaster in the future, more information must be gathered about these objects before any mitigating actions can be put into effect. Third, many of the Earth's resources cannot be replenished and at some point in the future, they will be exhausted; therefore, a new resource is required and asteroids appear to be a vast source of raw materials existing in Earth's neighborhood.

K. Yoshida (✉) • H. Lund
Department of Aerospace Engineering, Tohoku University, Aoba 6-6-01, Sendai 980-8579, Japan
e-mail: yoshida@astro.mech.tohoku.ac.jp

K. Saito et al. (eds.), *Progress in Scale Modeling, Volume II*,
DOI 10.1007/978-3-319-10308-2_21

Several missions have been launched to explore the minor bodies of the solar system. For example, the European Giotto mission conducted fly-by observation of comet 1P/Halley in 1986 and unveiled the shape of the comet's nucleus. NASA's NEAR-Shoemaker mission conducted detailed orbital observation of asteroid 433 Eros in 1999–2000. NASA's Stardust, which was launched in 1999, conducted a fly-by with comet 81P/Wild in 2004, captured particles from the comet's tail, and brought them back to the Earth in 2006. With the aim of studying the interior composition of comet 9P/Tempel, NASA's Deep Impact probe successfully impacted the comet's nucleus and then analyzed the large amount of dust that was ejected from the surface of its nucleus.

Most recent challenging missions include Hayabusa and Rosetta. Hayabusa is a Japanese probe launched in 2003 to visit asteroid 25143 Itokawa and collect material samples from its surface and bring them back to the Earth. The probe successfully returned to the Earth in June 2010, and more than 1,000 soil particles from asteroid Itokawa were confirmed inside its reentry capsule. Rosetta is a European spacecraft mission launched in 2004 to study the comet 67P/Churyumov-Gerasimenko. It consists of two main elements: the Rosetta space probe and the Philae lander. Philae is scheduled to land on the surface of the nucleus in 2014. Two harpoons will be fired into the comet to prevent the lander from bouncing off while additional drills are used to further secure the lander to the comet. Philae will then conduct in situ analysis of the nucleus.

Hayabusa and Minerva

Hayabusa (Fig. 1) was successfully launched in 2003 and encountered Itokawa in 2005. Detailed scientific observation was conducted, including imaging and mapping of the surface and identification of the gravitational field [1]. On November 19 and 25 (UT), Hayabusa conducted sample acquisition from the asteroid surface.

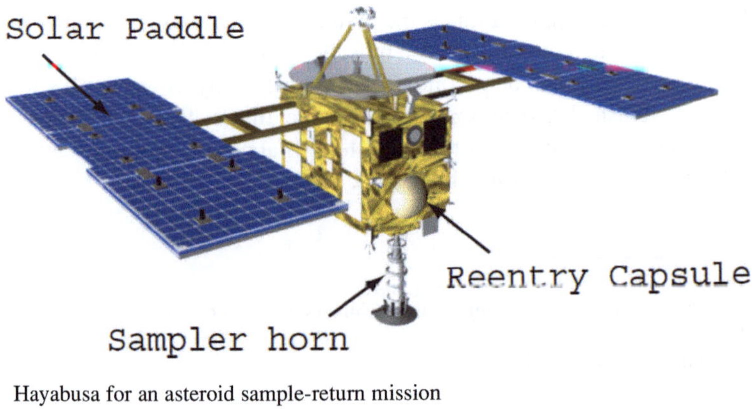

Fig. 1 Hayabusa for an asteroid sample-return mission

Owing to the extremely tiny size (approximately 550 m in the longest axis) and small mass of the asteroid, the gravitational acceleration on Itokawa's surface is of the order of 10–5 G. Such a small gravitational field makes it very difficult for the probe to remain on the asteroid's surface. It can easily jump off from the surface. For this reason, Hayabusa adopted a *touch-and-go* sequence with high-speed projectiles to crash onto the surface and secure samples in a very short surface contact time (1–2 s) [2–4].

Hayabusa carried a tiny mini-spacecraft (weighing only 590 g and approximately 10 cm tall and 12 cm in diameter) named Minerva. It was designed to take advantage of Itokawa's very low gravity by using an internal flywheel assembly to hop across the surface of the asteroid, obtaining images from its cameras and sending them back to Hayabusa whenever the two spacecraft were in sight of one another [5].

Minerva was deployed on November 12, 2005. The lander release command was sent from the Earth, but before the command could arrive, Hayabusa's altimeter measured its distance from Itokawa to be lower than intended then started an automatic altitude keeping sequence. As a result, Minerva was released while the probe was ascending with upward velocity, so that it escaped Itokawa's gravitational pull. It did not reach the surface of Itokawa. If it had been successful, Minerva could have used the reaction wheel in its body to travel on the microgravity surface by hopping and tumbling.

Design of Future Asteroid Exploration Rovers

Surface Mobility in a Different Gravitational Field

The design of Minerva raised an interesting problem of locomotion mechanisms in a small gravitational field. Figure 2 summarizes typical designs for surface locomotion systems. As evidenced by terrestrial motor vehicles and Moon/Mars rovers, a wheel traction system is the best option in a 1G terrestrial gravitational field or fields down to 1/10G. It should be noted that the smaller the gravitational reaction forces against the wheels, the smaller are the resulting traction forces.

In the smaller gravitational field, a Nanorover type of design can be advantageous. A Nanorover is a 1-kg small robot equipped with four wheels at the end of swingable struts. It was developed by NASA/JPL and intensely discussed as an optional payload candidate on Hayabusa, but the development of the flight model was canceled owing to budgetary constraints. As its design principle, a Nanorover swings its struts down on the surface while rotating its wheels. Such downswing actions will be useful to produce the forces necessary to push the wheels onto the surface while yielding locomotive forces tangent to the surface. However, the downswing forces inevitably push the robot body up into space. Hence, the entire robot will hop up, and if the hopping velocity is smaller than the escape velocity

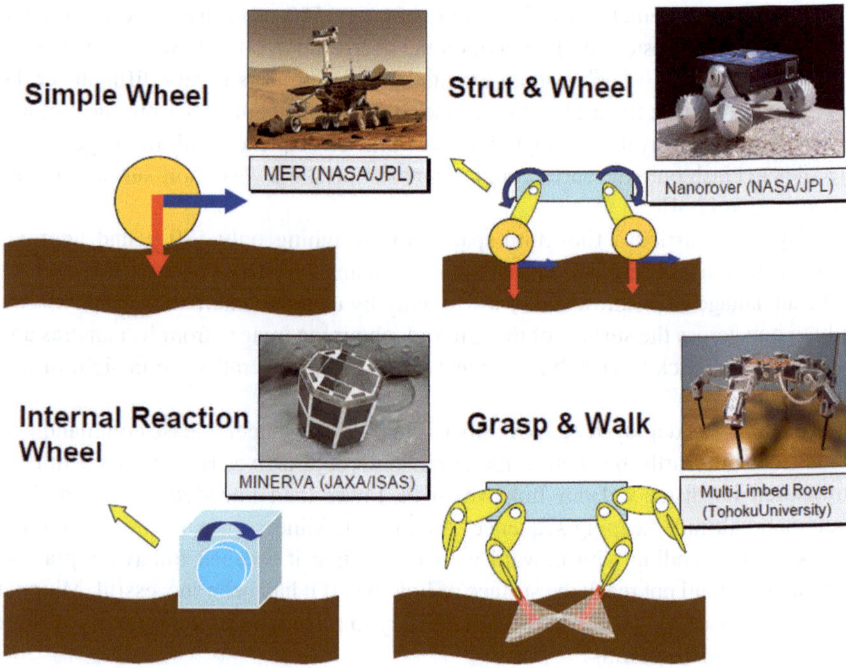

Fig. 2 Possible designs for a surface locomotion system in different gravitational fields

from the planetary body, it will return to the surface after a while, continuing to bounce and tumble until the kinetic energy is dissipated. Minerva did not have struts or wheels in its exterior but a single axis wheel inside. Owing to the reaction of the wheel, the entire body would start to tumble and hop up if edges or spikes around its cylindrical body kick the planetary surface. Both Nanorover and Minerva designs rely on opportunistic tumbling motion. Their traveling direction may be roughly indicated, but detailed point-to-point navigation will be difficult.

As an alternative design, articulated robots have a potential advantage in microgravity locomotion [3]. The idea is to hold the surface by articulated limbs or grippers while walking on the surface. This idea was inspired by rock climbing. In principle, rock climbing would become much easier in a microgravity field. The complexity of the articulated mechanism design would be an issue when creating such a robot. Even without a dexterous gripper, but instead utilizing 6 3-DOF legs like an insect, such a robot can cling to the surface using 3 legs while the other 3 legs swing forward. By repeating a 3-holding and 3-swinging gate, the robot will be able to keep to a static walk [6].

A key element of the holding and walking strategy is the friction force at the tip of each limb. The smaller the friction, the more difficult it will be to maintain static holding. The effective use of electrostatic forces [7] or van der Waals forces [8] to generate additional sticking forces or to control the magnitude of friction is an interesting option for future research.

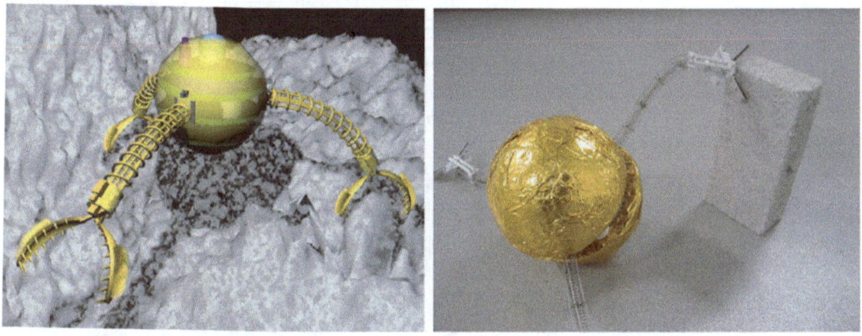

Fig. 3 A conceptual drawing (*left*) and a laboratory prototype (*right*) of "Cliff Hanger, Rock Climber Rover"

The Cliff Hanger, Rock Climber Rover

A conceptual design of a newly proposed robot is depicted in Fig. 3 alongside a photograph of a laboratory prototype.

The robot has multiple limbs with a dedicated sticker at the end and walks over the surface using these limbs. The design of the limb can be like an articulated manipulator arm. Considering that it does not need to support gravitational load, however, it should be light weight, slim, and compact. The limb can be like the multi-DOF forceps that are used in laparoscopic surgery.

Like laparoscopic forceps, the end tip of the limb has jaws to pick or pinch an object. The idea is to use the jaws to hold the surface of the asteroid. As discussed in the following section, claws sized in proportion to the scale of surface roughness will help ensure the holding capability of the jaws.

In addition to their use for locomotion, the limbs will be used to pick up rock fragments and to scoop soft regolith if it exists. The rover may also need to "row" in a pond of regolith between boulders. As for the preparation of in situ analysis, the limbs will be useful to brush the surface for specimens.

The mission concept of the boulder exploration is depicted in Fig. 4. Exposed surface boulders contain direct information about the asteroid's interior down to the depth of their size. The proposed rover can provide crawling capability over random boulders, cliffs, grooves, and regolith ponds.

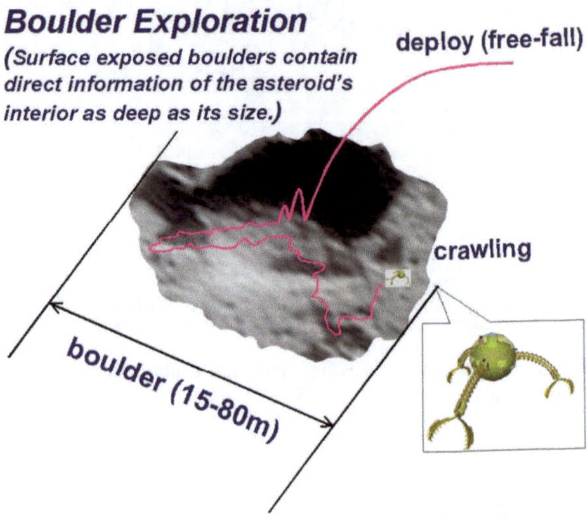

Fig. 4 A mission concept for boulder exploration

Grabbing Forces

Candidates for the Grabbing Sticker

The grabbing force is key to achieve surface locomotion. In this section, four fundamental forces are compared and the feasibility of the claw-like mechanical sticker is discussed.

As candidates for sticking forces that may work on a micro-G surface, we compared the van der Waals force, electrostatic force, and universal gravity to the holding force of mechanical claws.

The van der Waals force is known as the intermolecular force. In the case of two molecules, the attracting magnitude is inversely proportional to the sixth power of the distance between them. However, as the summation of those forces shows, the force between two parallel surfaces is inversely proportional to the third power of the distance between the surfaces:

$$F_v = \frac{A}{6\pi L^3} \text{(per unit area)} \tag{1}$$

where L is a representative length of the distance between parallel surfaces and A is known as the Hamaker constant.

Electrostatic force works if there is electrical charge or potential field. The magnitude of the force between two parallel surfaces is

$$F_e = \frac{\varepsilon_0 V^2}{2L^2} \text{ (per unit area)} \tag{2}$$

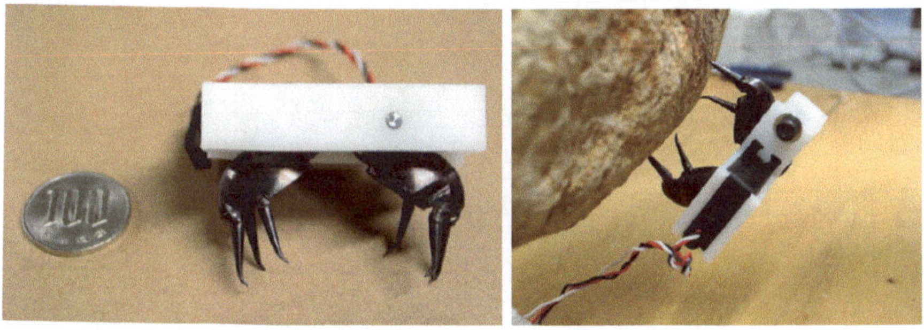

Fig. 5 Photo of a claw prototype

where V is the voltage (electrical potential) and ε_0 is known as the dielectric coefficient or vacuum permittivity.

The universal gravity force between two bodies is commonly given as

$$F_g = G\frac{Mm}{r^2} \tag{3}$$

where r is not the distance of the gap but the distance of the centroid of two bodies and G is known as the gravity constant.

As for the clamping force of claws, the following model is considered

$$F_c = W_{\max}\mu \sin^2\theta \tag{4}$$

where μ is the friction coefficient and θ is the inclination of the surface. $W_{\max}$ is the force when the claw has a maximum bending displacement, $x_{\max}$. If the claw is modeled as a uniform cantilever, the relationship between the bending displacement and force is expressed as

$$W_{\max} = \frac{3EIx_{\max}}{l^3} \tag{5}$$

where E and I are Young's modulus and geometrical moment of inertia, respectively. l is the length of the cantilever.

Figure 5 shows a prototype of a mechanical claw that successfully grabs the surface of a stone [3].

Scale Effect Analysis

For a fair comparison of the above four possible grabbing forces, a contact surface model with normalization by a representative length is introduced. The magnitude of the forces is evaluated according to the representative length.

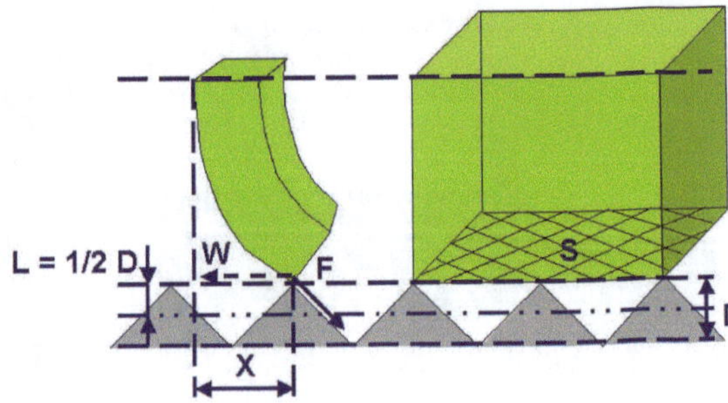

Fig. 6 Contact model on a rough surface

As shown in Fig. 6, the roughness of the asteroid surface is modeled by uniform ridges with the height (or depth) of D and the width of $2D$. For the evaluation of the van der Waals, electrostatic, and gravitational forces, the robot is assumed to be sitting on the ridges. The covering area S is greater than a single ridge. In such a case, the mean distance between the asteroid and the robot is $D/2$.

If the representative length is set at $L = D/2$, and the size of the robot is assumed to a cube with $100L$ sides, then $S = L^2 \times 10^4$.

As for the claw force, the maximum deformation of the claw is assumed to be $x_{\max} = D = 2L$. If the deformation is greater than this, the claw will lose contact with a current ridge and slip to a neighboring ridge. Here, the claw's length and its geometrical moment of inertia are assumed to be $l = 10L$ and $I = D^4/12 = 4L^4/3$. Then, Eq. (4) becomes

$$F_c = 8\mu E L^2 \sin^2\theta \times 10^{-3} \tag{6}$$

Here, the constants are listed as

$$A = 10^{-19} \text{ [J]}$$
$$\varepsilon_0 = 8.85 \times 10^{-12} \text{ [F/m]}$$
$$V = 10^3 \text{ [volt]}$$
$$G = 6.67 \times 10^{-11} \text{ [m}^3\text{/kgs]}$$
$$R = 1.0 \times 10^3 \text{ [m]}$$
$$M = 5.0 \times 10^{12} \text{ [kg]}$$
$$\rho = 10^3 \text{ [kg/m}^3\text{]} = 1 \text{ [g/cm}^3\text{]}$$
$$\mu = 0.5$$
$$\theta = 45 \text{ [deg.]}$$
$$E = 6.9 \times 10^{10} \text{ [N/m}^2\text{]}$$

Figure 7 depicts the result of comparison of four forces for the scale L from 10^{-9} to 10^3 [m]. The figure clearly indicates the scale effect of the forces.

Fig. 7 Scale effect of grabbing forces

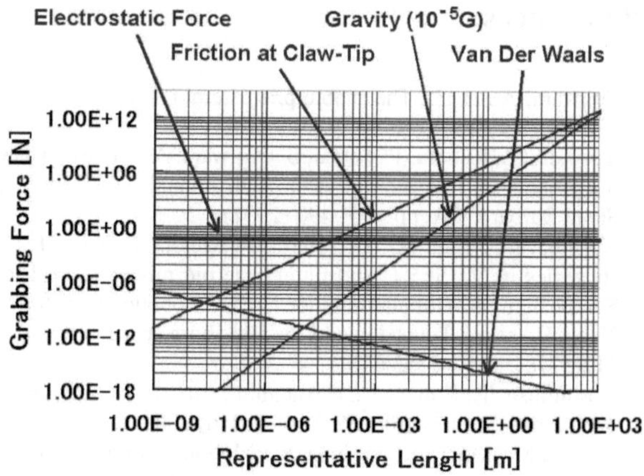

The van der Waals force is almost always smaller than the others across all scales. The electrostatic force is dominant in the scale of $L < 10^{-5}$ [m]. The claw force is then dominant in the scale of $10^{-5} < L < 10^3$ [m], which is a very wide range covering from 10 μm to 1 km. Finally the gravitational force dominates in the scale of $L > 10^3$ [m], i.e., when the robot is of an equivalent size or larger than the asteroid.

As a feasible size in a practical mission, this paper assumes a robot with a size on the order of a 0.1 [m] (10 cm) cube or sphere, having a mass of 1–10 [kg], with claws that are 0.01 [m] (1 cm) long and 0.001 [m] (1 mm) thick, which can grip the surface having a roughness of ±0.001 [m] (1 mm). In this scale, the friction force at the claw tip is the most feasible solution.

Conclusions

This study investigated possible designs of a surface locomotion system for different gravitational fields. With large gravity fields such as from 1G to 1/10G, wheel-like locomotion systems are the most advantageous. However, on the surface of an asteroid with 10^{-5} G, for example, our scale model analysis shows that the gravitational pull is too small for wheeled systems. In this scenario, the most feasible design is a rock-climber-type articulated robot using friction forces and claw-type grippers to grab the surface. The results also suggest that a gripper system using electrostatic forces will work in a robotic system of the order of micrometers.

References

1. Fujiwara, A., et al.: The rubble-pile asteroid Itokawa as observed by Hayabusa. Science **312** (5778), 1330–1334 (2006)
2. Yoshida, K., Kubota, T., Sawai, S., Fujiwara, A., Uo, M.: MUSES-C touch-down simulation on the ground. In: Proceedings of the 11th Annual AAS/AIAA Space Flight Mechanics Meeting, Santa Barbara, CA, pp. 481–490 (2001)
3. Yoshida, K., Nishimaki, Y., Maruki, T., Kubota, T., Yano, H.: Sampling and surface exploration strategies in MUSES-C and future asteroid missions. In: Proceeding of the 7th International Symposium on Artificial Intelligence and Robotics & Automation in Space (2003)
4. Yano, H., et al.: Touchdown of the Hayabusa spacecraft at the Muses Sea on Itokawa. Science **312**(5778), 1350–1353 (2006)
5. Yoshimitsu, T., Kubota, T., Nakatani, I., Adachi, T., Saito, H.: Micro-hopping robot for asteroid exploration. Acta Astron. **52**(2), 441–446 (2003)
6. Chacin, M., Mora, A., Yoshida, K.: Motion control of multi-limbed robots for asteroid exploration missions. In: Proceedings of IEEE International Conference on Robotics and Automation (ICRA 09), pp. 3037–3042 (2009)
7. Wagner, R.: Grippers for space locomotion. In: International Conference on Robotics and Automation, a Presentation to the Orbital Robotics Workshop, Pasadena, CA (2008)
8. Sitti, M, Fearing, R.S.: Synthetic Gecko foot-hair micro/nano-structures for future wall-climbing robots. In: Proceedings of IEEE International Conference on Robotics and Automation (ICRA 03), pp. 1164–1170 (2003)

Scale Modeling of Flame Spread Over PE-Coated Electric Wires

Yuji Nakamura, Keisuke Azumaya, Junya Iwakami, and Kaoru Wakatsuki

Abstract A universal correlation between the flame spread rate and the flame length formed along the electric wire was studied experimentally based on scale modeling concept. In the first place, we studied the burning behavior of research-graded wire (i.e., controlled wire, polyethylene-coated metal thin rod) in order to examine the precise effect of the total pressure (30–100 kPa), the core material (nickel chrome, iron, copper), and the scale (e.g., diameter, coating thickness, etc.) on the spread rate. It turned out that the flame shape was not the only primary factor involved in determining the spread rate, implying that the heat transfer process in solid phase is essential to consider. The simplest 1-D heat transfer model along the core was introduced, and two kinds of non-dimensional groups (i.e., Peclet number (Pe) and one to describe the radial direction of heat transfer process: Λ) were found necessary to preserve the similarity. By introducing two length scales to represent the processes in gas and solid phases, all measured data were found to have collapsed into the single line in Pe-Λ plane, suggesting that flame spread behavior would be predictable based on their correlation. This correlation curve is justified with the spread data obtained using practical electric wire and cables (with/without sheath), confirming that scale modeling of flame spreading over the electric wire was successful.

Y. Nakamura (✉)
Department of Mechanical Engineering, Toyohashi University of Technology, 1-1 Hibarigaoka, Tempaku, Toyohashi 441-8580, Japan

Division of Mechanical and Space Engineering, Hokkaido University, N13 W8, Kita-ku, Sapporo 060-8628, Japan
e-mail: yuji@me.tut.ac.jp

K. Azumaya • J. Iwakami
Division of Mechanical and Space Engineering, Hokkaido University, N13 W8, Kita-ku, Sapporo 060-8628, Japan

K. Wakatsuki
National Research Institute of Fire and Disaster, 4-35-3 Jindaiji-higashi-machi, Chofu, Tokyo 182-8508, Japan

K. Saito et al. (eds.), *Progress in Scale Modeling, Volume II*,
DOI 10.1007/978-3-319-10308-2_22

Nomenclature

A	Cross-sectional area [mm^2]
a	Thermal diffusivity [mm^2/s]
C	Heat capacity [J/(kgK)]
d	Diameter [mm]
K	System constant [1/s]
L	Length scale along the wire axis [mm]
p	Pressure [kPa]
Q	Heat flow [W]
R	Length scale perpendicular to the axis [mm]
r	Radial direction [mm]
Δr	Radial direction difference [mm]
Pe	PECLET number [−]
T	Temperature [K]
ΔT	Temperature difference [K]
t	Time [s]
Δt	Small time difference [s]
V_{sp}	Flame spread rate [mm/s]
x	Axial direction of the wire [mm]
Δx	Small distance in axial direction [mm]
λ	Thermal conductivity [J/(mKs)]
ρ	Density [g/cm^3]
Λ	Non-dimensional number appeared in Eq. (3)

Subscripts

c	Wire core
flame	Flame
loss	Lost component
s	Solid (polyethylene: PE)
*	Non-dimensional expression

Introduction

Background of "Wire Combustion"

Wire combustion is the simplest unit of electric fire so that its comprehensive understanding will contribute directly to improve the fire safety design of electric facilities. Even though it is "simplest," the burning event includes complex physical and chemical processes, such as time-dependent, multidimensional heat and mass transfer with chemical reactions in gas/liquid/solid phases. In order to fulfill the aforementioned practical demands, there are two steps to be taken; the first is to gain the required fundamental knowledge about wire combustion (fundamental research); the second is to address the question of how to apply this fundamental knowledge to the actual problem (applied research). In general, deep understanding of the effect/role of each process on the burning character using practical cable is difficult since practical wire/cable includes many kinds of additives so that the burning character varies and cannot be well controlled. For this purpose, "controlled wire," which is specially designed for the specific test purpose, is often used. So far, several attempts have been made with "controlled wire" from various aspects and important fundamentals have been made available in the literature [1–19].

What Should We Consider to Model the Wire Burning?

Figure 1 shows a representative still picture of burning polyethylene (PE) coated over a thin metal rod [8]. Flame spreads from left to right nearly steadily, and decomposition/oxidative reactions occur at the molten surface, causing decomposition gas to issue into the atmosphere to sustain the flame. As mentioned above, very complex physics is involved in the burning event. The molten PE experiences several forces, namely, its surface tension force, surface interactive force between molten PE and the metal rod, and the gravitational force, and, accordingly, by these the deformed shape of molten PE is determined. Flame covers the molten PE and the upstream edge preheats the unburned PE beneath it (i.e., pyrolysis front). The downstream flame edge, on the other hand, contacts the metal rod, and then the extensive heat would be transferred toward the molten PE, with the result that the formation of bubbles can be identified there. Bubbles move in the molten phase during the burning event, and sometimes come to the surface, then burst to release the gas jet into the atmosphere. In this way, the shape of the molten PE is not really steady but fluctuates in time. More importantly, it is found that the molten PE grows in size gradually in time and eventually falls off periodically due to imbalance of gasification rate and melting rate.

Although many processes simultaneously occur in the combustion event, major/primary factor to determine the burning rate, i.e., the flame spread rate, should be

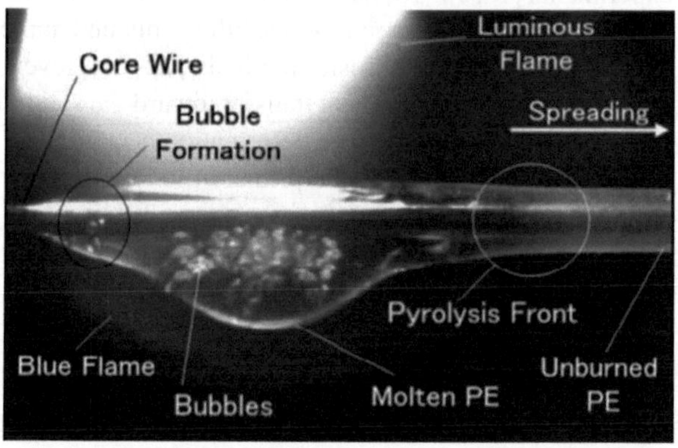

Fig. 1 Direct picture of burning of the controlled wire in atmospheric air [8]

the slowest time scale phenomenon in the system. As is well known, the chemical reaction is fast enough so that the decomposition/oxidative reactions of molten PE as well as the gaseous oxidative reaction to govern the heat release in the system are not the system-controlling processes. Similarly, bubbles move quickly so that internal convection in the molten phase is not the one. Combustion takes place under gravimetric environment so that buoyancy-driven flow in gas phase is fast enough. The flame and molten PE appearing in the system are less sooty and relatively small; radiation heat transfer might be negligible. Eventually it is suspected that heat conduction in both gas and solid phases might be the most likely candidate for system-controlling parameter. In this sense, our first target should be to investigate the effect of heat transfer processes on the burning rate.

Brief Review of the Previous Work

It is known that heat transferred not only from the flame but also through the metal wire does modify the burning character. Bakhman et al. have revealed that a highly conductive metal wire can give faster flame spread [1, 2]. On the contrary, Crescitelli et al. have pointed to the presence of metal beneath the combustible solid acting as a heat sink so that the flame spread is prohibited [3]. To simplify the system from 3-D to 2-D, a series of burning tests of controlled wire under microgravity have been made by several groups [4–7, 17, 19]. Although utilizing microgravity is quite useful for simplification purposes, there are severe limitations (test volume, chance) we must accept. To avoid such difficulties, authors have proposed utilizing low pressure to obtain a combustion event physically similar to microgravity [8–11, 14–16] via Grashof number similarity [20]. This allows us to access the flame behavior more precisely with high reproducibility and precision so that deep understanding of the physics of wire combustion can be obtained. Based on the measured data, it has been discovered that 1-D model of heat transfer along the wire can work well to reveal the characteristics of the burning process. By using such an assumption, the existence of two modes ("flame-driven" and "wire-driven") has been clearly identified [13], and the transition from ignition to flame spread has been also modeled satisfactory [18]. Moreover, non-steadiness of the system which is clearly observed in the experiments [10, 14–16] has been formulated via simplified model [21]. Deformation of molten polymer has been successfully simulated numerically with the combination of VOF and enthalpy-porosity methods [22–24], revealing that the presence of the deformation can promote heat transfer toward the solid PE to enhance further melting [23]. Because the free-surface deformation is negligible, it's unnecessary to consider its effect when developing the primary model of the wire burning process.

Target and Objective of the Present Study

Based on the past rich attempts as described above, it's expected that the 1-D treatment might work to represent the wire combustion characteristic, e.g., flame spreading rate. Nevertheless, there has never been research conducted that took into consideration how the small lab-sized wire compares to the much thicker cable that's currently in use. This study uses scale modeling methodology in order to more accurately measure the combustion results in the lab and predict the actual burning behavior of the said cable.

Experiment

Test Facility

Two series of experiments using controlled wire and practical cable/wire were performed in this study. Experimental apparatus used is essentially same as in previous studies [8–11, 14–16]. Experimental apparatus is schematically illustrated in Fig. 2. Combustion chamber is 365 mm (L) × 260 mm (W) × 180 mm (H), equipped with two gas lines: one comes from a gas supplying system and the other goes to vacuum pump. Considered total pressure is varied in the low atmospheric range (30, 40, 60, 80, 100 kPa), while gas composition is fixed at 79 vol.% N_2 and 21 vol.% O_2 throughout the study. For tests of practical wire, the apparatus is essentially the same, but experiments are performed in the open atmosphere (namely, under 101 kPa conditions) by opening the upper lid.

Coil heater is set on the left end of the sample to initiate the ignition and subsequent flame spread along the wire. A whole spread event is recorded by a digital video camera (Sony DCR-TRV900, 30 frames per second, denoted "DV camera" hereafter). From the obtained video image, we can calculate the spread rate under the applied condition. Although non-steady spread behavior is often found in subatmospheric range [10, 14–16] as well as in microgravity [19], average spread rate is our focus in this study. All experiments were performed in a darkened room to avoid any possible noise in the observed images.

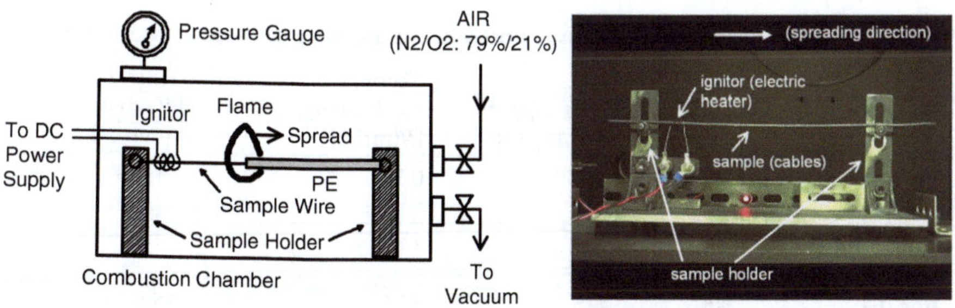

Fig. 2 Schematic diagram of experimental apparatus used in this study for controlled wire [8–11, 14–16] (*left*). Direct picture of burning test of practical wire (*right*)

Tested Sample

Polyethylene (PE)-coated metal wires are used as controlled wire, whereas commercially available eco-material cable/wires are used as the practical samples in this study. For the controlled wire, three types of core materials are used here: namely, nickel chrome (NiCr), iron (Fe), and copper (Cu). Core diameter as well as the coating thickness of PE has been modified to form ten types of samples to be tested. Details of the tested samples are summarized in Table 1, and thermophysical/transport properties (e.g., density, heat capacity, thermal conductivity, etc.) of the materials used in the sample are listed in Table 2.

As the practical sample, a low-toxic material with higher thermal resistance as well as low caloric material regulated by the Japanese Electric Wire and Cable Makers' Association (JCMA) (so-called eco-material cable (EM cable) [25]) is used in this study. In addition to its low-toxicity character (e.g., low smoke, halogen free), EM cable has the feature of higher heat resistance; therefore, the allowable current becomes larger as compared to that of conventional cables (e.g., insulated by polyvinyl chloride, PVC, or vinyl cable, VV). Besides, its recycle performance is

Table 1 Summary of tested samples (controlled wire) [16]

Name	Core material	Core dia. [mm]	Cable O.D. [mm]	PE thickness [mm]
NiCr ϕ1.0	Nickel–chrome alloy	0.7	1.0	0.15
NiCr ϕ1.1	Nickel–chrome alloy	0.5	1.1	0.3
NiCr ϕ0.8	Nickel–chrome alloy	0.5	0.8	0.15
NiCr ϕ0.7	Nickel–chrome alloy	0.3	0.7	0.2
Fe ϕ1.0	Iron	0.7	1.0	0.15
Fe ϕ0.8	Iron	0.5	0.8	0.15
Fe ϕ0.7	Iron	0.3	0.7	0.2
Cu ϕ1.0	Copper	0.7	1.0	0.15
Cu ϕ0.8	Copper	0.5	0.8	0.15
Cu ϕ0.7	Copper	0.3	0.7	0.2

Table 2 Lists of physical/thermal/transport properties of the materials used in the samples

	Density, ρ [g/cm^3]	Heat capacity, C [J/(kgK)]	Thermal conductivity, λ [J/(mKs)]	Thermal diffusivity, a [mm^2/s]
Polyethylene (PE)	0.92	2,300	0.33	1.56
Nickel–chrome alloy (NiCr)	7.85	470	15	4.07
Iron (Fe)	7.85	475.1	58.1	15.6
Copper (Cu)	8.89	385.1	376.7	110.0

All values are evaluated at standard state (300 K)

Table 3 Specification of eco-material (EM) cable/wire [26]

	Single-core wire (1C)	4-bundled cables (4C)	10-bundled cables (10C)
Product name (manufacturer)	HS&T Cable; EM-CEE/F (Sumiden Hitachi Cable Ltd.)		
Cross-sectional area of conductor (copper) (outer dia.) [thickness of insulator]	1.25 mm^2 (1.35 mm) [0.8 mm]	1.25 × 4 mm^2 (1.35 mm) [0.8 mm]	1.25 × 10 mm^2 (1.35 mm) [0.8 mm]
Outer dia.	3 mm	10.5 mm	15.5 mm
Volume % of metal part	17.7 %	5.8 %	6.6 %
Resistance (at room temp.)	16.8 Ω/km	16.8 Ω/km	16.8 Ω/km

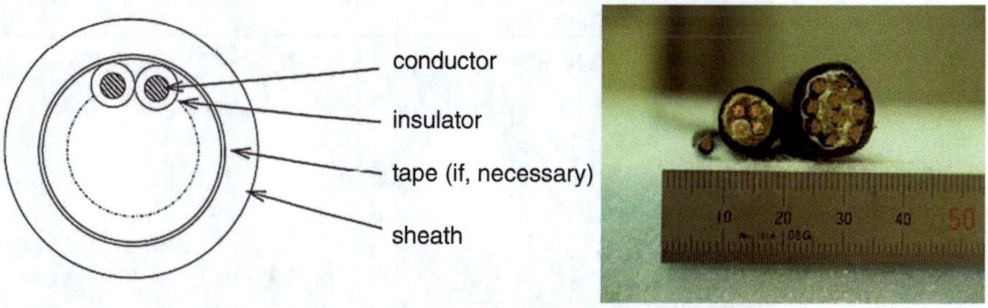

Fig. 3 Schematic illustration of the cross-sectional view of eco-material cable [26] (*left*) and the direct photo of the cable/wire (*right*); from *left* to *right*, 1C, 4C, and 10C are shown. Note that 1C is the combination of conductor and insulator without sheath, whereas 4C and 10C have sheath surrounded by the bundled 1C wires

remarkably good considering that since the insulator mainly consists of polyethylene, PE (by the way sheath material is not pure PE, but ethylene vinyl acetate (EVA) copolymer). Details of features and structure of the cable are summarized in Table 3 and Fig. 3, respectively.

Results and Discussion

Flame Shape Formed over the Various Types of Wire in Subatmospheric Pressure

Figure 4 shows the direct images of the spreading flame with various controlled wires in subatmospheric pressure. As indicated, various flame heights as well as widths are indicated. Although it may be hard to detect in the images, the size of molten PE appears differently according to the samples. In all samples, however, flame shapes tend to be shorter and wider as the pressure decreases. This fact indicates that the fuel mass flux ejected from the molten PE decreases as the

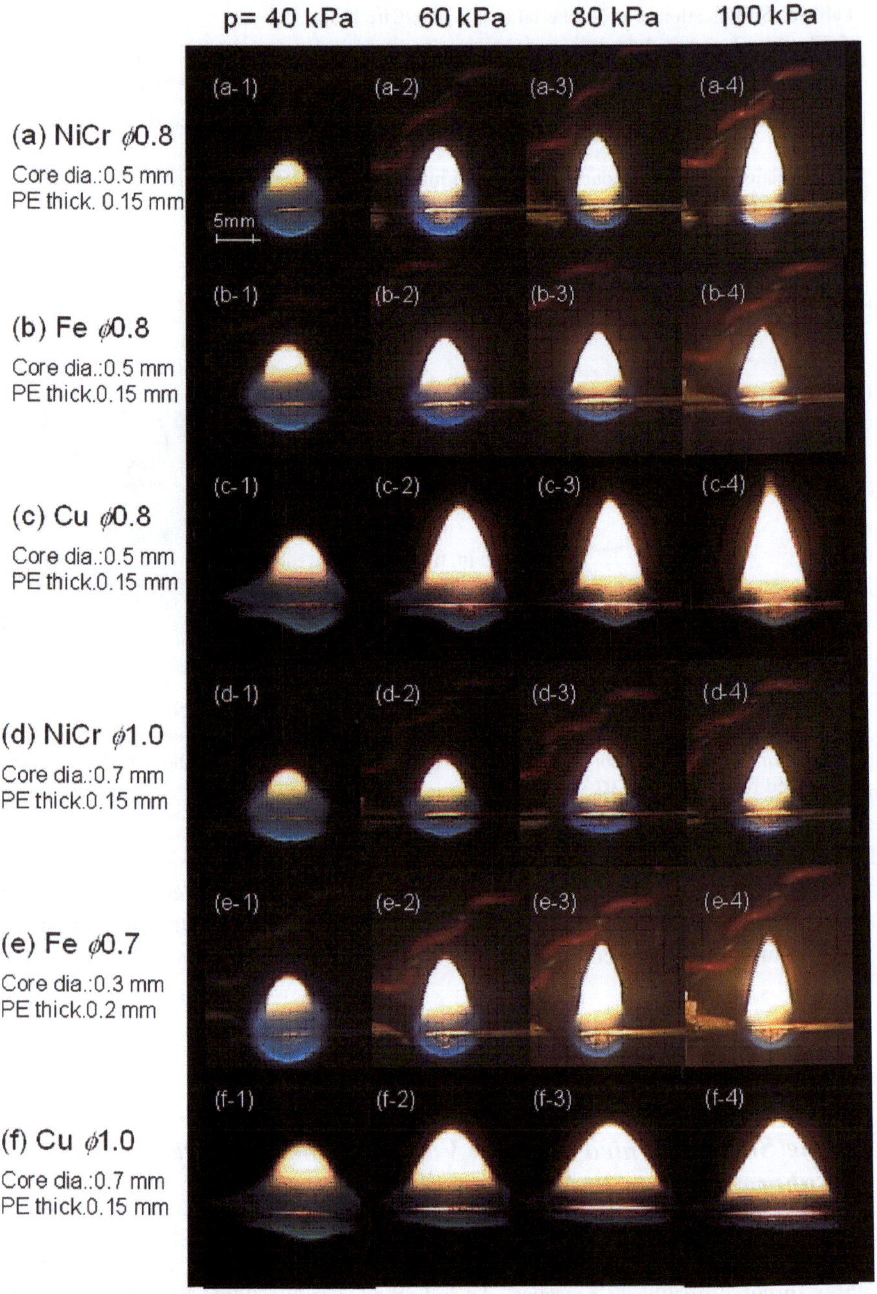

Fig. 4 Direct flame images with selected controlled wires in subatmospheric pressure (flame spreads from *left* to *right*) (Pictures were taken at $t = 15$ s after ignition [16])

pressure decreases. Unlike NiCr and Fe wires, Cu wires give an elongated flame width. This trend is quite consistent even at low pressure [see (f) showing that wider blue flame covers the wire].

As noted in the previous studies, since the thermal conductivity of NiCr is 1/5 that of Fe wire, the flame width is smaller in the former case. However, once we use the thicker core of NiCr to enlarge the amount of heat transferred along the wire, this trend changes obviously. Interestingly, by using larger diameter of NiCr wire (d =1.0 mm) and moderate diameter of Fe (d =0.8 mm), flame shape becomes nearly identical in the entire range of pressure considered in this study [see flames with (b) Fe ϕ0.8 and (d) NiCr ϕ1.0 shown in Fig. 4]. This might be because the amount of burning material is similar. In this way, it is important to stress that we can achieve a static similarity (i.e., flame shape is identical) by controlling the physical scales of the controlled wire.

Flame Spread Rate

Figure 5 shows the flame spread rate (with error bar) under the conditions studied here. As shown in the figure, some samples show little dependency of the spread rate on pressure, but the others do have negative correlation between the spread rate and the pressure; as pressure decreases, the spread rate becomes large as noted previously [8].

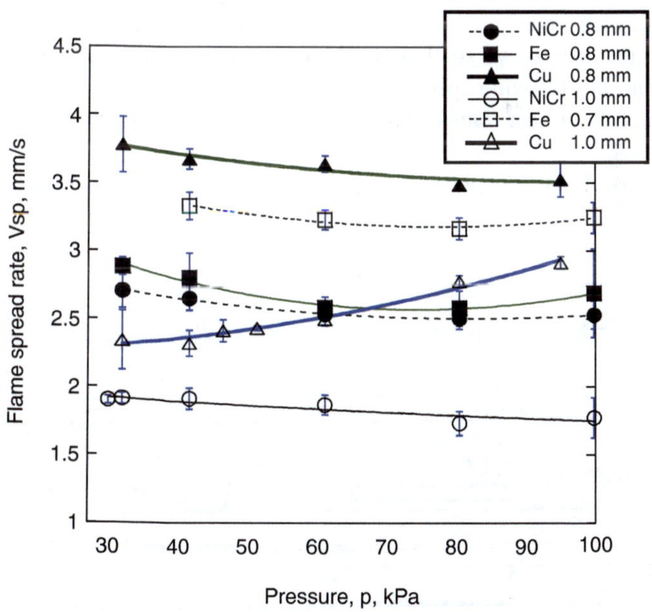

Fig. 5 A dependence of pressure on flame spread rate with various types of samples [16]

It is understood in the comparisons between NiCr ϕ1.0 and Fe ϕ0.8 cases (note that the flame shapes are nearly identical under these conditions; see Fig. 4) that the spread rate is different and the flame shape cannot solely determine the spread rate. This is an important experimental evidence that the dynamic similarity obeys the different rule; hence, the similarity law to govern the heat transfer in both gas and solid phases needs to be taken into account to have perfect similarity.

1-D Heat Conduction Model

In Fig. 6, the major route of heat flow during the spreading event is schematically shown. For simplicity, the contribution of heat from the flame [Q(flame) in Fig. 6] is neglected at first (this will be considered later). In previous studies [13], we assumed that heat conduction through the metal core could play an important role in the spread behavior and the coated PE is thin enough to have the same temperature at the localized core. In this case, it is possible to consider that the wire combustion should be simplified into "purely 1-D heat conduction problem" along the wire. However, if the spread rate is fast enough (i.e., the residence time of the flame over the unburned solid is long enough) or the combustible insulator is thick enough, the temperature of PE might not be the same as that of the core so that the "thermally thin" assumption would no longer be valid. In this study, let us assume the sample is "thermally thick" and that conduction toward the radial direction (from the core to PE) is taken into account. Namely, the heat transferred along the wire should distribute into both the core through axial direction and PE through radial direction. Now, we shall study the one-dimensional heat transfer model of moving materials (moving speed is V_{sp}, flame spread rate).

The equation of heat transfer along the moving wire can be written as follows:

$$\rho_c C_c \Delta T \times A_c \Delta x = \dot{Q}_x A_c \times \Delta t - \dot{Q}_{x+\Delta x} A_c \times \Delta t - \dot{Q}_{\text{loss}} \times 2\pi r_c \Delta x \times \Delta t \quad (1)$$

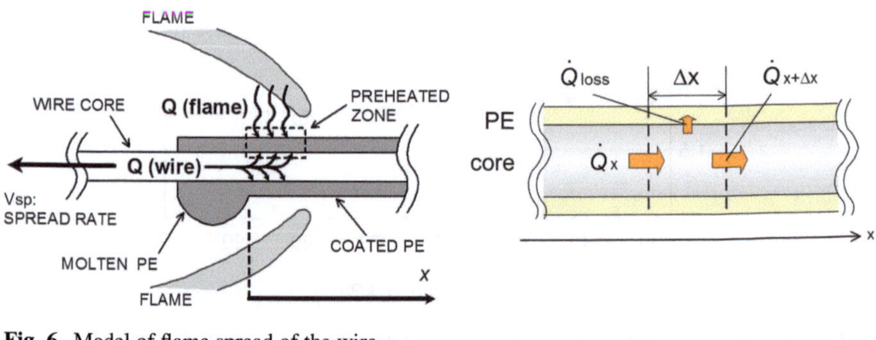

Fig. 6 Model of flame spread of the wire

where x, r_c, and A_c stand for axial coordinate, radius of the core, and cross-sectional area of the core ($=\pi r_c^2$), respectively. ρ_c, C_c, ΔT, Δt, λ_c, $\dot{Q}_{\mathrm{loss}}$, and $\dot{Q}_x$ indicate density of the core, heat capacity of the core, change in temperature along the wire (core), time difference, thermal conductivity of the core, heat lost from the core to PE, and heat transferred along the wire (core), respectively.

By adopting Fourier law as thermal conduction in axial and radial directions and the definition of flame spreading rate as the first derivative of x:

$$\frac{\Delta x}{\Delta t} = |V_{sp}|,\ \dot{Q}_x = -\left(\lambda \frac{\Delta T}{\Delta x}\right),\ \dot{Q}_{\mathrm{loss}} \approx -\lambda_s \left(\frac{\Delta T}{\Delta r}\right)_{r=r_c},$$

where λ_s stands for the thermal conductivity of PE. Considering the limit of $\Delta \to 0$, then Eq. (1) becomes the following differential equation:

$$\rho_c C_c |V_{sp}| \frac{\mathrm{d}T}{\mathrm{d}x} A_c = \lambda_c \frac{\mathrm{d}^2 T}{\mathrm{d}x^2} A_c + \lambda_s \left(\frac{\mathrm{d}T}{\mathrm{d}r}\right)_{r=r_c} \cdot 2\pi r_c. \tag{2}$$

Let us perform a non-dimensional procedure to the above equation with two representative length scales, L for axial direction (x) and R for radial direction (r), and denote non-dimensional quantities with *; Eq. (2) can be transformed as follows:

$$\frac{\mathrm{d}^2 T^*}{\mathrm{d}x^{*2}} = \frac{L|V_{sp}|}{\frac{\lambda_c}{\rho_c C_c}} \cdot \frac{\mathrm{d}T^*}{\mathrm{d}x^*} - \frac{L^2}{R}\frac{2\pi r_c}{A_c}\left(\frac{\lambda_s}{\lambda c}\frac{\mathrm{d}T^*}{\mathrm{d}r^*}\right)_{r=r_c}. \tag{3}$$

If we introduce the following non-dimensional number, Pe and Λ,

$$Pe \equiv \frac{L|V_{sp}|}{\frac{\lambda_c}{\rho_c C_c}} = \frac{L|V_{sp}|}{a_c}\left(\because a_c = \frac{\lambda_c}{\rho_c C_c}\right),\ \Lambda \equiv \frac{L^2}{R}\frac{2\pi r_c}{A_c}\frac{\lambda_s}{\lambda_c},$$

then Eq. (3) can be expressed as follows:

$$\frac{\mathrm{d}^2 T^*}{\mathrm{d}x^{*2}} = Pe \cdot \frac{\mathrm{d}T^*}{\mathrm{d}x^*} - \Lambda \cdot \left(\frac{\mathrm{d}T^*}{\mathrm{d}r^*}\right)_{r=r_c}. \tag{4}$$

Two representative length scales (L, R) should be referred as the (observed) flame width and the radius of the (entire) wire, respectively. The reason to consider L from gaseous process, even though the present equation is for solid phase, is to take into account the contribution of Q(flame) which was neglected in the model development. As noted previously, regarded as the preheating of the unburned PE, the contributions from the flame and through the core are both important. In this way, thermal status of the solid can be affected by both gas- and solid-phase heat transfers. In Fig. 7, the relation between Pe and Λ is summarized for all controlled wires considered in this study.

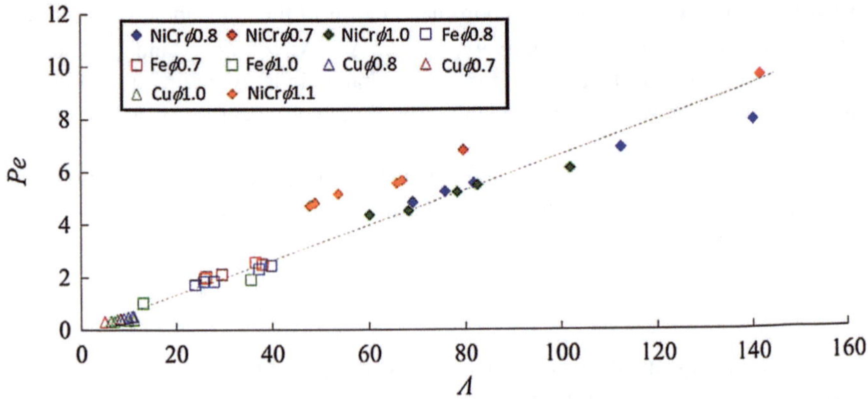

Fig. 7 Relation between *Pe* and Λ in Eq. (4) for various controlled wires

As seen in the figure, all data collapsed into a single line, suggesting that there is very solid linear relationship between *Pe* and Λ. Namely,

$$\frac{\Lambda}{Pe} = \frac{L^2}{R} \frac{2\pi r_c}{A_c} \frac{\lambda_s}{\lambda_c} \cdot \frac{a_c}{L|V_{sp}|} = K \cdot \frac{L}{|V_{sp}|} = \text{const}, \left(K \equiv \frac{2\pi r_c a_c \lambda_s}{RA_c \lambda_c} \right). \quad (5)$$

Since K is a wire-dependent constant (namely, wire property), the flame width (L) and the spread rate (V_{sp}) are always linearly correlated. Moreover, L/V_{sp} has the unit of time (s) and represents the residence time of the flame along the wire. In other words, the experimental fact provides us the important information about how residence time is determined by the wire properties (=K). Now, we consider whether this trend is applicable only to the controlled wires or whether the same scenario is valid for the practical cables/wires as well.

Burning Behavior of the Practical Cable/Wire

In Figs. 8, 9, and 10, the time sequences of the burning behaviors of practical cable/wire (1C, 4C, and 10) are shown. Note that the applied time of initial heating by coil heater is varied depending on the condition till to achieve the successful ignition.

For 1C wire (Fig. 8), the applied time for successful ignition is 10 s. Although the flame height and width is slightly larger at early stage of ignition, it calms down and becomes a steady value after 55 s from the ignition. Flame also moves nearly steadily once steady burning is attained. In this way, the flame width and the spread rate can be recognized as the system eigenvalues.

For 4C cable (Fig. 9), the applied time for successful ignition is four times longer (40 s) than that for 1C wire; it is somehow expected since the 4C cable consists of four of 1C wire. Since the cable holding mechanism does not work well, the cable

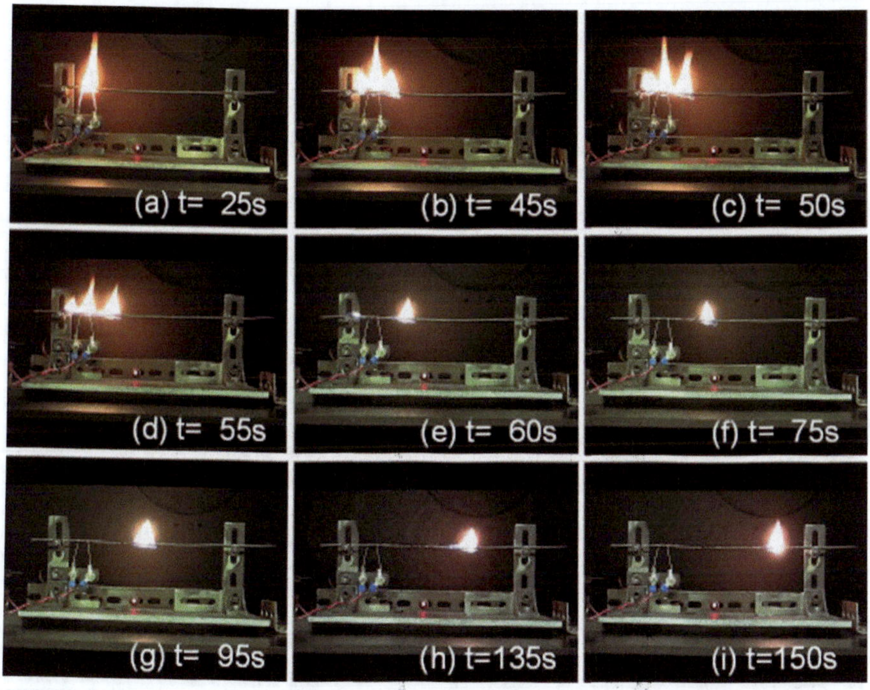

Fig. 8 Time sequence of ignition to flame spreading of 1C wire. Total duration of ignition stage is 10 s

bends apparently. Nevertheless, overall flame behavior eventually becomes nearly steady. It is clearly found in the presence of the burnout point at the left edge of the flame. Moreover, the flame width holds nearly constant value. Hence, again, the flame width and the spread rate can be recognized as the system eigenvalue, similar to 1C wire.

For 10C cable (Fig. 10), the applied time for successful ignition is more than 10 times longer (130 s) than that for 1C wire. Although the burning behavior is not laminar and much fluctuation is found, again, the clear burnout point is found during the burning event so that the flame width can be defined. Although the spread rate fluctuates, averaged spread rate can be defined. In this way, again, the flame width and the spread rate can be recognized as the system eigenvalue, similar to 1C wire and 4C cable.

Validity of Pe-Λ *Correlation: Can We Predict the Burning Behavior of the Electric Wire?*

From Figs. 8, 9, and 10, it is understood that we could define L (flame width) and V_{sp} (flame spread rate) for each practical cable/wire as defined in controlled wires. This

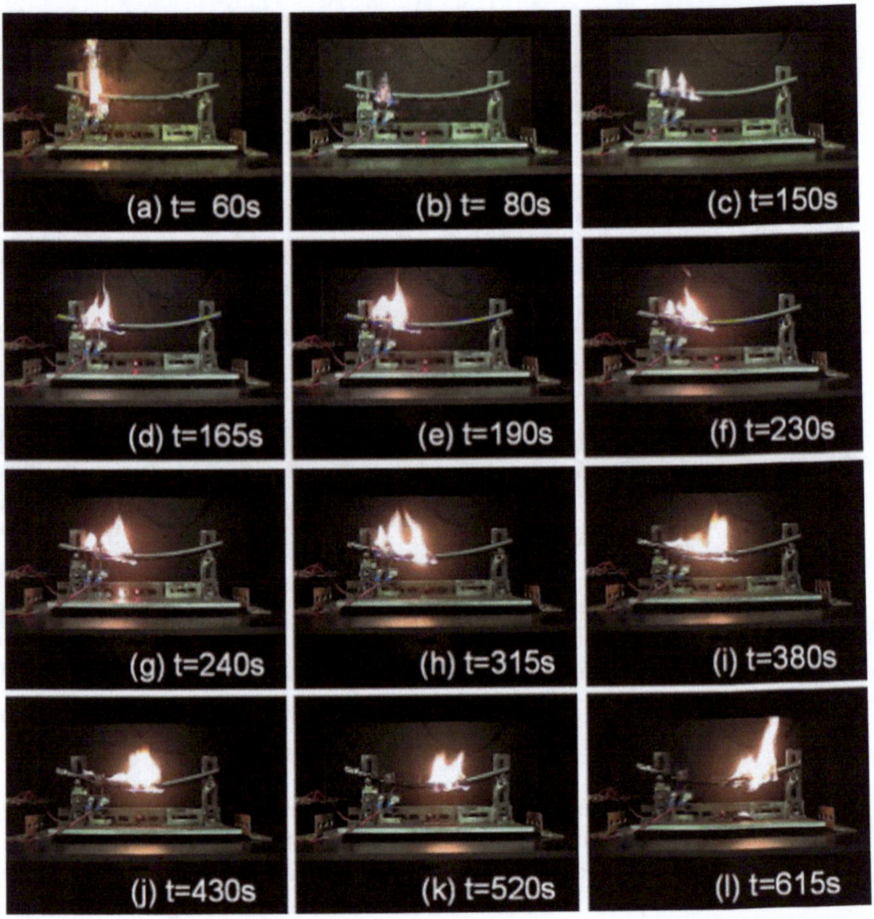

Fig. 9 Time sequence of ignition to flame spreading of 4C cable (four-bundled 1Cs with sheath). Total duration of ignition stage is 40 s

means that we could know *Pe* and Λ for each practical cable/wire as well so that we can plot in *Pe*-Λ plane as is done in Fig. 7. The result is summarized in Fig. 11.

It is found that the larger bundled cable (i.e., 10C cable) tends to give the slightly "faster" spread rate than the linear line as shown in Fig. 11. This might be caused by a combination of several potential reasons as follows. As the core becomes thicker, radial thermal distribution might occur in the core and radiation from the flame to the unburned fuel in order to preheat there. In this sense, 1-D model might not be suitable. In addition, thicker cable has a lower percentage of metal inclusion (see Table 3), which works as the heat sink to weaken the burning. Although the possibilities need to be validated, due to various ambiguities in this work—e.g., model is too simplified, cable is not perfectly horizontal (due to its weight), only visual flame image is used for discussion—no further argument is made here;

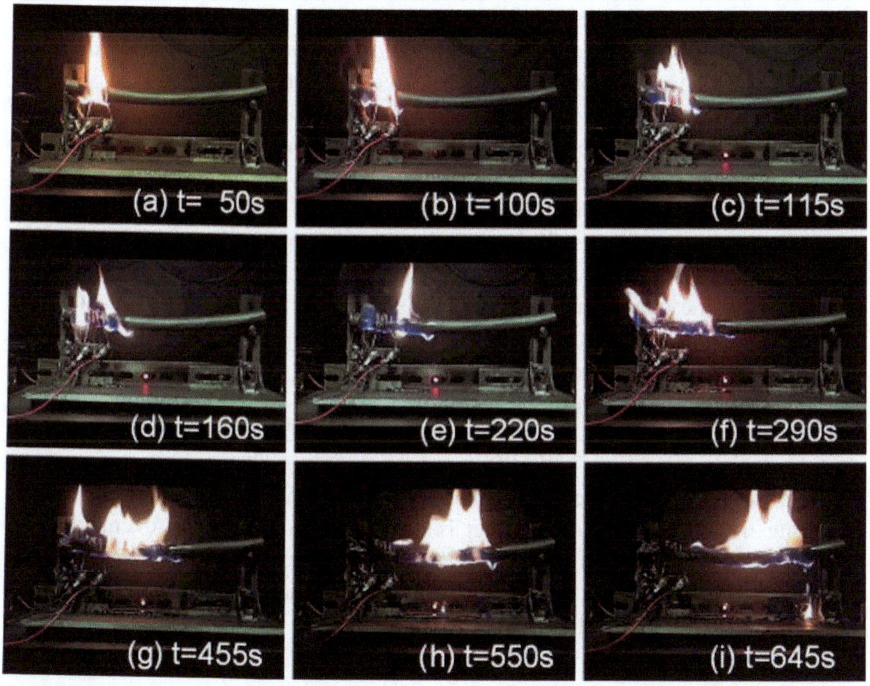

Fig. 10 Time sequence of ignition to flame spreading of 10C cable (ten-bundled 1Cs with sheath). Total duration of ignition stage is 130 s

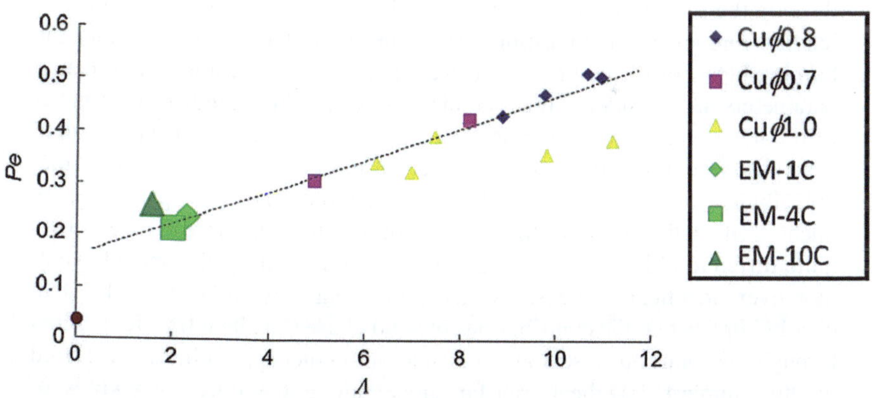

Fig. 11 Spread data summarized in *Pe-Λ* plane. Only selected data of copper wires (Cu ϕ0.7, Cu ϕ 0.8, Cu ϕ 1.0) are shown here with EM cable/wire data. EM-1C, EM-4C, and EM-10C denote 1C wire and 4C and 10C cables, respectively

further study would be necessary for validation. Nevertheless, it should be noted that the *same* linear correlation between *Pe* and *Λ* shown in Fig. 7 (for controlled wires) is adopted for practical cable/wires. In other words, the simple 1-D modeling approach is found to be valid to give the burning character of the wide range of cable/wire. In this way, we have a methodology to predict spread rate by flame width and residence time by the specification of the sample.

Although it seems that we have obtained the universal expression to describe the burning character of the electric wire, it should be noted that the combustible (coated insulator) considered in this study is a fixed material, i.e., polyethylene (PE). It is valuable to consider whether the correlation would be valid or not with other materials. As discussed already, the gradient of the line is related to the residence time, in other words, burning time. Of course the burning time is dependent on the character of the materials, so that it is natural to imagine that the gradient of the linear correlation line would be varied according to the kind of the burning material. Nevertheless, the methodology developed here should be consistent as long as the burning behavior is affected by the two major routes of heat input, namely, the one coming from the flame and the other coming through the core. Further burning tests with different burning materials would be necessary to upgrade and verify the present prediction formula (Eq. 5). Once the upgraded model including the burning material effect is developed in the future, eventually, we could have a way to simulate a real cable fire by using a scale model (simple, research-grade wire) based on the correlation so that the scale-down experiment of electric facility including the cable fire is possible.

Concluding Remarks

In this study, a series of combustion tests of controlled wire in a relatively wide range of parameters is summarized in the first place; then we attempt to develop the scale modeling methodology in order to apply the fundamentals learned through the combustion tests with controlled wires to predict the burning behavior of the practical ("scaled-up") cable. Various experimental parameters are considered (e.g., ambient pressure, three kinds of core metal material, various cable/wire scales) to study whether a universal expression to describe the burning behavior of the electric wire exists or not. Note that polyethylene is currently the burning material and this parameter is unchanged in the present study. It is notified that the spread rate shows quite different although the flame shape is similar during the spread event; moreover, identical flame shape gives different spreading rate. This fact clearly implies that the contributions in solid phase (i.e., heat transfer process through the solid) are essential to consider to predict spread behavior. Based on the simplest 1-D heat transfer model along the core, two kinds of non-dimensional groups [i.e., Peclet number (Pe) and one to describe the radial direction of heat transfer process: Λ] are introduced to find dynamic similarity. By introducing two length scales to represent the processes in gas and solid phases, all measured data are found to be collapsed into the single line in Pe-Λ plane, suggesting that the flame spread behavior would be

(continued)

(continued)

predictable based on their correlation. This correlation curve is justified with the spread date obtained by using the practical electric wire and cables (with/without sheath), confirming that scale modeling of flame spreading over electric wire is successfully proposed. Further study would be necessary to include the burning material effect into the correlation formula to propose an effective methodology to predict the burning behavior of the electric wire.

Acknowledgments This work is funded by JSPS (Grants-in-Aid for Young Scientists: #21681022, #17710131), JSF (Japan Space Forum), Nomura Foundation of Science and Technology, and the Association for Disaster Prevention Research. This work is also done as a part of the contract research with Japan Nuclear Energy Safety Organization (JNES). Backup support from the Combustion Society of Japan is quite helpful. YN (principal investigator) expresses his sincere thanks for their assistance and help.

References

1. Bakhman, N.N., Aldabaev, L.I., Kondrikov, B.N., Filippov, V.A.: Burning of polymeric coatings on copper wires and glass threads: I. flame propagation velocity. Combust. Flame **41**, 17–34 (1981)
2. Bakhman, N.N., Aldabaev, L.I., Kondrikov, B.N., Filippov, V.A.: Burning of polymeric coatings on copper wires and glass threads: II. critical conditions of burning. Combust. Flame **41**, 35–43 (1981)
3. Crescitelli, S., Pota, F., Santo, G., Tufano, V.: Influence of solid phase thermal properties on flame spread over polymers. Combust. Sci. Technol. **27**, 75–78 (1981)
4. Kikuchi, M., Fujita, O., Ito, K., Sato, A., Sakuraya, T.: Experimental study on flame spread over wire insulation in microgravity. Proc. Combust. Inst. **27**, 2507–2514 (1998)
5. Fujita, O., Kikuchi, M., Ito, K., Nishizawa, K.: Effective mechanisms to determine flame spread rate over ethylene-tetrafluoroethylene wire insulation: discussion on dilution gas effect based on temperature measurements. Proc. Combust. Inst. **28**, 2905–2911 (2000)
6. Umemura, A., Uchida, M., Hirata, T., Sato, J.: Physical model analysis of flame spreading along an electric wire in microgravity. Proc. Combust. Inst. **29**, 2535–2543 (2002)
7. Fujita, O., Nishizawa, K., Ito, K.: Effect of low external flow on flame spread over polyethylene-insulated wire in microgravity. Proc. Combust. Inst. **29**, 2545–2552 (2002)
8. Nakamura, Y., Yoshimura, N., Matsumura, T., Ito, H., Fujita, O.: Confined effect on flame spread over polymer-insulated electric wire in sub-atmospheric pressure. Proceedings of 6th Asia-Pacific Conference on Combustion, Nagoya, pp.198–201 (2007)
9. Matsumura, T.: Study on flame spread over polymer-insulated wire in reduced pressure with forced convection. M. Thesis, Hokkaido University (2007)
10. Nakamura, Y., Yoshimura, N., Matsumura, T., Ito, H., Fujita, O.: Flame spread over polymer-insulated wire in reduced pressure environments: similarity to microgravity phenomena. In: Progress in Scale Modeling. Springer, New York, NY, pp. 17–28 (2008)
11. Nakamura, Y., Yoshimura, N., Matsumura, T., Ito, H., Fujita, O.: Opposed-wind effect on flame spread of electric wire in sub-atmospheric pressure. J. Thermal Sci. Technol. **3**(3), 430–441 (2008)

12. Nakamura, Y., Kudo, T., Ito, H., Fujii, T., Kikuchi, M., Fujita, O.: Effects of radiation feedback on flame spread along polymer-insulated wire in microgravity. J. Combust. Soc. Jpn. **50**(153), 255–263 (2008) (in Japanese)
13. Nakamura, Y., Yoshimura, N., Ito, H., Azumaya, K., Fujita, O.: Flame spread over electric wire in sub-atmospheric pressure. Proc. Combust. Inst. **32**, 2559–2566 (2009)
14. Nakamura, Y., Azumaya, K., Ito, H., Fujita, O.: Flame spread over electric wire in space environment: steady or unsteady. Proceeding of 27th International Symposium on Space Technology and Science (27th ISTS), Tsukuba (2009), 2009-h-23 (on USB)
15. Nakamura, Y.: Fundamental study on fire in extraordinary environment. J. Combust. Soc. Jpn. **51–157**, 7–9 (2009) (in Japanese)
16. Nakamura, Y., Azumaya, K., Wakatsuki, K., Ito, H., and Fujita, O.: Experimental study of flame spread over electric cables at low pressure. Proceedings of Sixth International Seminar on Fire and Explosion Hazards, Leeds (2010.4), paper#38-2
17. Onishi, Y., Fujita, O., Agata, K., Takeuchi, H., Nakamura, Y., Ito, H., Kikuchi, M.: Observation of flame spreading over electric wire under reduced gravity condition given by parabolic flight and drop tower experiments. Trans. JSASS Space Tech. Japan 8: No.ists27 Ph_19-Ph_24 (2010)
18. Huang, X., Nakamura, Y., Williams, F.A.: Ignition-to-spread transition of externally-heated electrical wire. Proc. Combust. Inst. **34**(2), 2506–2512 (2013)
19. Takahashi, S., Takeuchi, H., Ito, H., Nakamura, Y., Fujita, O.: Study on unsteady molten insulation volume change during flame spreading over wire insulation in microgravity. Proc. Combust. Inst. **34**(2), 2657–2664 (2013)
20. Nakamura, Y., Wakatsuki, K., Hosogai, A.: Scale modeling of space fire. J. Jpn. Soc. Exp. Mech. (2013 Special Issue) **13**, s69–s74 (2013)
21. Nakamura, Y., Wakatsuki, K.: Strategy to diagnose a non-steadiness of flame spreading over electric wire. Proceedings of 50th Japanese-Section Symposium on Combustion, Nagoya, pp. 498–499 (2012)
22. Kim, Y. K., Hossain, A., Kim, S., Nakamura, Y.: A numerical study on time-dependent melting and deformation processes of phase change material (PCM) induced by localized thermal input (Chap. 23). In: Two Phase Flow, Phase Change and Numerical Modeling. In-Tech Open Access Publisher, Croatia, pp. 523–540 (2011)
23. Kim, Y.K., Hossain, A., Nakamura, Y.: Numerical study of melting of a phase change material (PCM) enhanced by deformation of a liquid–gas interface. Int. J. Heat Mass Transfer **63**, 101–112 (2013)
24. Kim, Y.K., Hossain, A., Nakamura, Y.: Numerical study of thermocapillary convection effect on melting process of phase change material subjected to local heating. J. Thermal Sci. Technol. **8**(1), 136–151 (2013)
25. http://www.jcma.jp/eco/em.htm (only Japanese is available on website)
26. Technical report of HS&T EM-CEE/F cables, Sumiden Hitachi Cable Ltd

Scale Modeling of Air-Dropped Water for Aerial Firefighting Against Urban Fire

Hironori Kikugawa, Tadashi Konishi, and Keita Hirano

Abstract In the event of a major earthquake, urban fires would spread unchecked for lack of firefighting water supply due to destruction of roads, hydrant damage, and so on. If aircraft, such as helicopters, could be used under specific conditions, it would be effective for firefighting in these conditions; aircraft can arrive at the fire site quickly and begin firefighting immediately. However, there is not enough research about the fire-extinguishing or fire-deterrent effect of aerial firefighting in urban fires. Until now, full-scale experiments using an actual helicopter have been carried out three times by the National Research Institute of Fire and Disaster. However, it is difficult to carry out full-scale experiments frequently. Therefore, in this research, a scale model experiment was proposed and similarity to a full-scale experiment was verified. Findings about air-spread phenomenon when water was dropped from a bucket and about the impact of water hitting the ground were reproduced.

Introduction

On January 17th in 1995, a major earthquake hit the southern part of Hyogo Prefecture in Japan. The earthquake would come to be known as the "Great Hanshin-Awaji Earthquake." In this great earthquake, many buildings, bridges, and roads were seriously damaged and heavy casualties were reported. Just after the earthquake, many fires broke out simultaneously, each fire spread, and damage from the disaster increased. During such a disaster, a firefighting team finds it difficult to arrive at the fire site due to road or bridge damage. In addition, hydrants may be inoperable because of water pipe breakage. Under such unique conditions, it

H. Kikugawa (✉) • T. Konishi
Department of Mechanical Engineering, National Institute of Technology, Oita College, Maki, 1666 Oita, Japan
e-mail: kikugawa@oita-ct.ac.jp

K. Hirano
Advanced Course Student, National Institute of Technology, Oita College, Maki, 1666 Oita, Japan

K. Saito et al. (eds.), *Progress in Scale Modeling, Volume II*,
DOI 10.1007/978-3-319-10308-2_23

has been hoped that aerial firefighting by helicopter would be a possibility in an urban area fire. However, the amount of water that can be conveyed by helicopter is restricted, and there is a limit to the fire-extinguishing effect. With an urban area fire, it is difficult to completely extinguish the fire over a large combustion area by aerial firefighting. Therefore, water was not used on the combustion zone directly, but on the surrounding zone, and the possibility of controlling the fire using a small amount of water was thereby examined. In addition, dropping water from the air is dangerous to human life in cases where inhabited houses collapse under the impact of water dropped from above.

From 1997 to 1999, the National Research Institute of Fire and Disaster (NRIFD) in Japan conducted full-scale experiments using an actual helicopter. In 1997, prevention effect on fire spread by dropping water on surrounding houses was examined [1]. In 1998, a case where aerial firefighting was started at an early stage for only one building fire was examined [2]. Since this experiment was carried out under strong wind conditions, the prevention effect against jumping sparks from the fire was also examined. In 1999, the fire-deterrent effect when the water was dropped continuously from many helicopters was examined [3,4]. In addition, the magnitude of the impact of dropped water was examined using a large helicopter. From 3 years' research work and the aerial firefighting experiments, the limits of aerial firefighting and the possibility of effective use were suggested. However, many problems were left unexplored; it became clear that the method of effective aerial firefighting needs to be inquired into more thoroughly. However, since a huge area and an enormous budget are needed, it is difficult to carry out full-scale experiments frequently.

Therefore, in this research work, a scale model experiment was proposed and similarity to a full-scale experiment was verified. As the first step, the scattering behavior in the air of the water dropped from the bucket was reproduced. Although there are many examples which have studied the collapse phenomenon and the physical phenomenon of liquid drops which fall through the air [e.g., 5–7], there are few examples which have studied the similarity of the spread of the water while scattering a huge amount of water. In this paper, the behavior of water dropped from a bucket into to the air was reproduced, and the similarity of the behavior to that of a full-scale experiment is discussed. In addition, the impact of water dropped on the ground is examined.

Experimental Method

Figure 1 is a photograph of a real water bucket for a medium-sized firefighting helicopter BK117 in Oita Prefecture, Japan. Water capacity of the bucket is 600 L; discharge orifice diameter is 200 mm. The bucket is hung under the helicopter, and in case of water scattering, the firefighter operated the opening and closing of the discharge hole manually. The small similarity tank model was produced, and we conducted two experiments, one for behavior of dropped water and the other is for impact of water dropped on the ground.

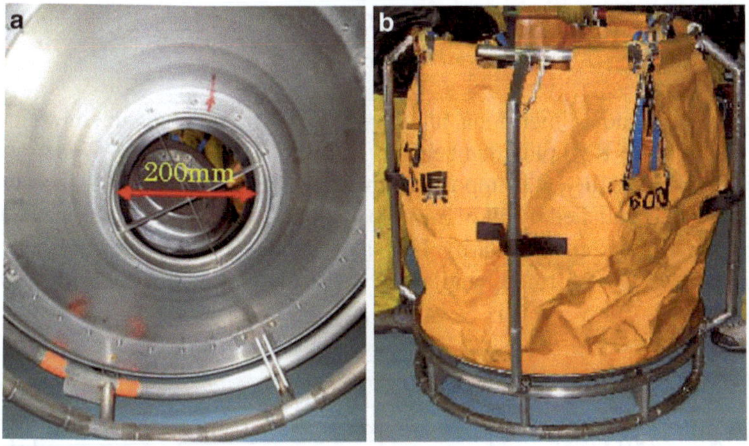

Fig. 1 Water bucket for firefighting helicopter. (**a**) Discharge hole (**b**) Exterior view

Scale Modeling

Scale modeling utilizes the laws of nature in a replica of the equipment which is smaller than the actual system but has the same key properties. The scaling-up or scaling-down process is carried out by similarity criteria which connect together the physical property, characteristics of the fluid flow, and the dimensions of the real system and the laboratory model. First of all, it is indispensable to observe and discern the phenomena of the real object carefully, in order to select the physical laws which govern the phenomena. The governing equations selected by key phenomena for the scale modeling are as follows [8]:

$$\text{Inertia force of water and air} \quad F_i = \rho l^2 u^2 \tag{1}$$

$$\text{Viscosity of water and air} \quad F_v = \mu l u \tag{2}$$

$$\text{Gravity force of water and air} \quad F_g = \rho g l^3 \tag{3}$$

$$\text{Surface tension of water} \quad F_s = \sigma l \tag{4}$$

Three non-dimensional numbers called π-numbers are derived by Eqs. (1), (2), (3), and (4) for momentum balance.

$$\pi_1 = \frac{F_i}{F_v} = \frac{\rho l^2 u^2}{\mu l u} = \frac{\rho l u}{\mu} \quad \text{(Reynolds number)} \tag{5}$$

$$\pi_2 = \frac{F_i}{F_g} = \frac{\rho l^2 u^2}{\rho g l^3} = \frac{u^2}{g l} \quad \text{(Froude number)} \tag{6}$$

$$\pi_3 = \frac{F_i}{F_s} = \frac{\rho l^2 u^2}{\sigma l} = \frac{\rho l u^2}{\sigma} \quad \text{(Weber number)} \tag{7}$$

The π_1 to π_3 are non-dimensional numbers for similarity laws of air and water flow, which shows the Reynolds number, the Froude number, and the Weber number, respectively. In reality, it is impossible to set up the law of similarity, because it is simultaneously satisfied by all of π-numbers in the scale model experiment. Therefore, relaxation of the similarity law, a procedure which eliminates the other physical laws, is performed, employing the analytical method and separation of phenomena.

$$\boxed{\pi_1 = \frac{F_i}{F_v} = \frac{\rho l u}{\mu}} \xrightarrow{\boxed{\rho = \rho'}\ \boxed{\mu = \mu'}} \boxed{\pi_1 = lu} \longrightarrow \boxed{\frac{u}{u'} = \frac{l'}{l}} \tag{8}$$

The Reynolds number of the full scale is in a turbulent flow range. In the turbulence range, since the effect of the inertia force is overriding, it is not necessary to take the effect of viscous force into consideration. Therefore, the equalization of the Reynolds number between a real model and a scale model is not required.

$$\pi_1 \neq \pi_1{}' \quad \text{(Relaxed)} \tag{9}$$

$$\boxed{\pi_2 = \frac{F_i}{F_g} = \frac{u^2}{gl}} \xrightarrow{\boxed{g = g'}} \boxed{\pi_2 = \frac{u^2}{l}} \longrightarrow \boxed{\frac{l}{l'} = \left(\frac{u}{u'}\right)^2} \tag{10}$$

$$\boxed{\pi_3 = \frac{F_i}{F_s} = \frac{\rho l u^2}{\sigma_w}} \xrightarrow{\boxed{\rho = \rho'}\ \boxed{\sigma = \sigma'}} \boxed{\pi_3 = lu^2} \longrightarrow \boxed{\frac{l}{l'} = \left(\frac{u'}{u}\right)^2} \tag{11}$$

The following relations are concluded if the flow in the real object and the model achieves similarity since the inertia and gravity forces are similar.

$$\pi_2 = \pi_2' \longrightarrow \boxed{\frac{l}{l'} = \left(\frac{u}{u'}\right)^2} \tag{12}$$

$$\pi_3 = \pi_3' \longrightarrow \boxed{\frac{l}{l'} = \left(\frac{u'}{u}\right)^2} \tag{13}$$

If the geometric similarity ratio and a velocity ratio are planned so that Eqs. (12) and (13) may be satisfied with the real object and the scale model. However, it is impossible to set up an experiment so that these conditions can be simultaneously satisfied as is shown above by Eqs. (12) and (13), because the velocity calculations are inversed. Therefore, two different experimental conditions were planned in order to satisfy Eqs. (12) and (13) separately.

Experiment for Behavior of Air-Dropped Water

For the setting of the condition to satisfy Eq. (12), the inertia and gravity forces are similar in a model situation, as in full scale. The experiment to satisfy Eq. (12) is the one concerned with behavior of air-dropped water discharged from bucket. The small pressure tank which served as the bucket in the scale model was set on the scaffold tower as shown in Fig. 2.

The plastic water bottle which was the scale model for the firefighting water bucket was attached in the upper part (2,030 mm height) of a scaffold tower. Water was put inside the plastic water bottle, the air pressure was introduced by the compressor, and the water was expelled from the discharge hole. The speed of the water which blows out from the tank was controlled by changing the air pressure (0.1–0.6 MPa) in the tank and the quantity of water (100–500 ml). The target flow velocity was set to satisfy Eq. (12), due to Froude number. In addition, a side wind was directed at the falling water from a wind tunnel, and the situation in which the helicopter was flying was reproduced. Side wind velocity was changed (5–16 m/s) for different flying conditions. The spread behavior of air-dropped water was captured with a high-speed video camera, and the velocity vector was analyzed using fluid image analysis software (Flow-vec32).

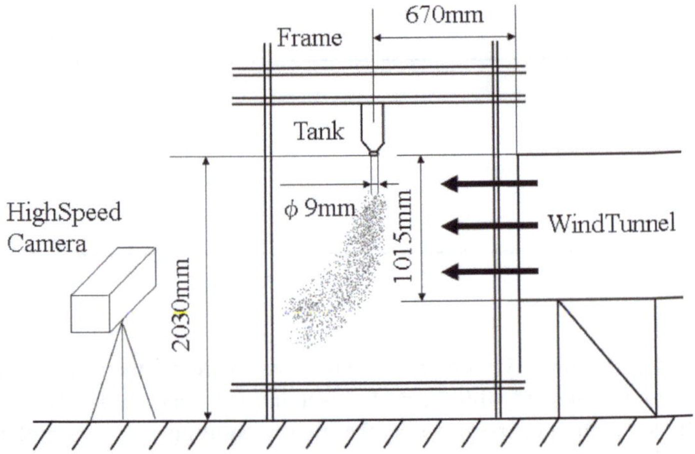

Fig. 2 Schematic of experimental apparatus for behavior of air-dropped water

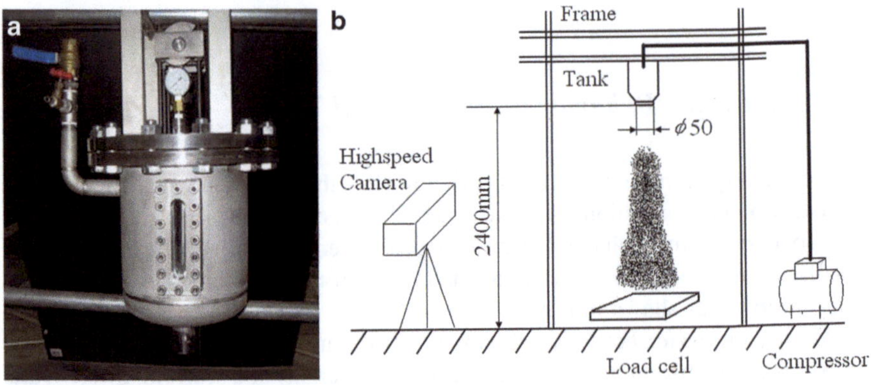

Fig. 3 Experimental apparatus for crashed water impact. (**a**) Pressure tank (**b**) Schematic of experimental apparatus

Experiment for Crashed Water Impact

To set the conditions to satisfy Eq. (13), the model achieves similarity between the inertia force and surface tension of the water. This is shown in the experiment by impact of water when the water hit the ground. The 1:4 scale small pressure tank which served as a scale model for the bucket was produced as shown in Fig. 3 and was attached to the upper part (2,400 mm height) of a scaffold tower. The tank was filled with water, air pressure was added by compressor, and water was expelled from the discharge hole. The speed of the water expelled from the tank was controlled by changing the air pressure (0.1–0.6 MPa) and the quantity of water (1–5 L). The target flow velocity was set to satisfy Eq. (13), due to Weber number.

The impact load when the water struck the ground was measured by the load cell. The receiver part of the load cell is 330 mm long and 330 mm wide. The measurement interval of the load cell was set to 10 ms.

Results and Discussion

Experiment for Behavior of Air-Dropped Water

As a result of the plastic water bottle pre-experiment, the air pressure at 0.3 MPa and the quantity of water at 300 mL would be the best conditions by which to observe the behavior of air-dropped water. Inject speed became quick as the pressure in a tank became high pressure. It came to be remarkably influenced by the air resistance while water dropped in the air, for water collapsing becomes a small water drop as the quantity of water is reduced. Figure 4 shows velocity vector distribution at crosswind velocity of 5 and 16 m/s, respectively. These behaviors of air-dropped water are quite similar to the full-scale experiment with the firefighting helicopter. Figure 5 shows velocity occurrence distribution at crosswind velocity of 5 and 16 m/s, respectively. In the case of crosswind velocity of 16 m/s, velocity occurrence distribution is predictable compared to the crosswind velocity of 5 m/s case. It seems to be successfully to reproduce for the behavior of air-dropped water from these results.

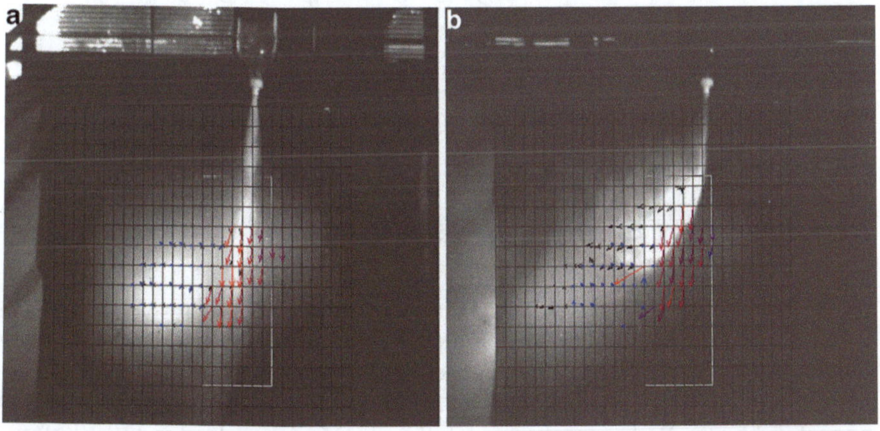

Fig. 4 Velocity vector distribution. (**a**) Crosswind speed 5 m/s (**b**) Crosswind speed 16 m/s

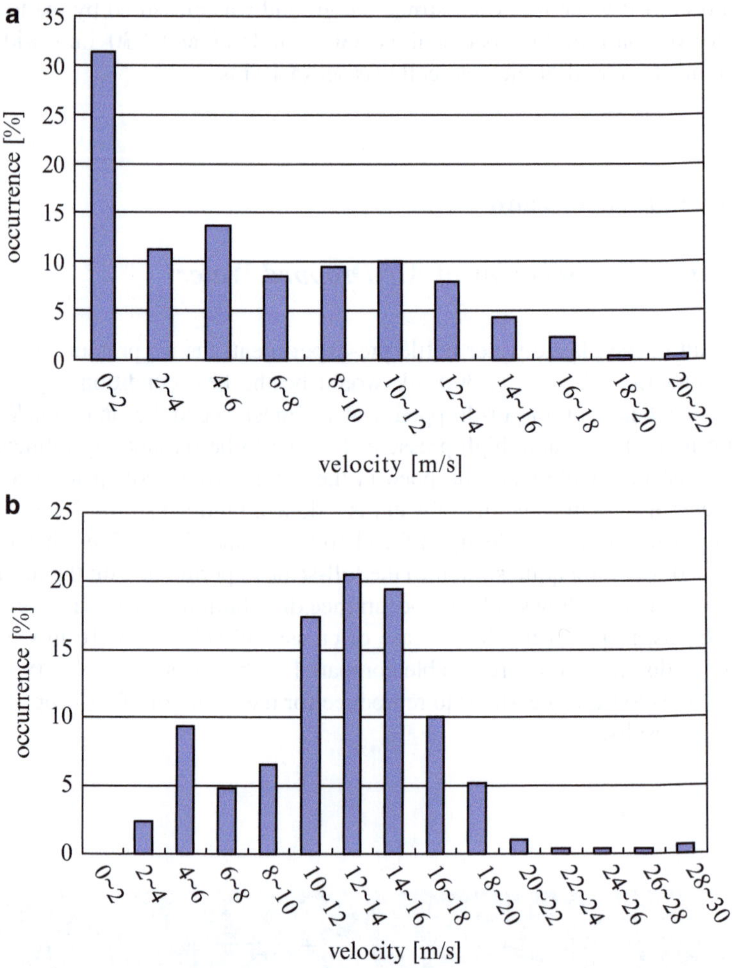

Fig. 5 Velocity occurrence distribution. (**a**) Crosswind speed 5 m/s (**b**) Crosswind speed 16 m/s

Experiment for Crashed Water Impact

As a result of the plastic water bottle pre-experiment, the air pressure which is 0.16 MPa and the quantity of water which is 5 L would be the best conditions to observe for experiment for water impact of air-dropped water. Figure 6 shows impact load of water striking the ground measured by the load cell, which is normalized by the Weber number on a full scale. Maximum load is 110.1 kgf/m^2. Since the maximum impact load in a full-scale experiment was reported to be about 100 kgf/m^2, almost the same impact load was obtained also in the experiment with a similarity model. In addition, impact time is normalized by water volume and tank

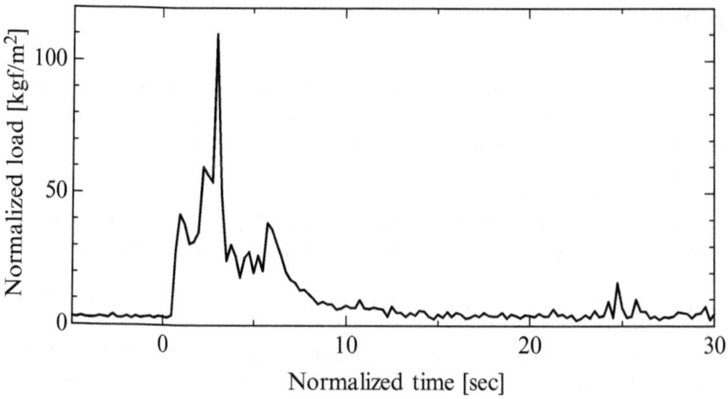

Fig. 6 Impact load of water crashed on the ground

scale ratio. The time water took to leave the tank until it was empty was calculated. It would be expected that similar impact time decay would be obtained under a full-scale experiment. For the next step, revised design of the discharge hole of the water bucket for reducing the impact has been planned.

Conclusions

In this report, an experiment which corresponds with a similarity rule paying attention to Froude number and Weber number was conducted. For water expelled from the pressure tank which served as a scale model for a full-scale bucket, the behavior of scattering water while the helicopter was flying was reproduced. In addition, for the similarity experiment about the impact of a load of water striking the ground, impact load was equivalent to that of a full-scale experiment. For future work, with an actual fire, we will have to take into consideration various influences, such as influence of the ascending current from a fire flame, and downwash from a helicopter blade, wind behavior due to particular geographical features, and so on. In order to propose an effective method of aerial firefighting, as a following step, a similarity experiment which will model a fire is under planning.

Acknowledgments This work was supported by Grants-in-Aid for Scientific Research (KAKENHI) on Innovative Areas "Optimization of air dropped water for fire-fighting in great urban fire disaster with active controlled wind tunnel" (19710158) of the Ministry of Education, Culture, Sports, Science and Technology (MEXT), Japan.

References

1. National Research Institute of Fire and Disaster: The research report on the spread-of-a-fire block effectiveness by aerial firefighting in case of an urban area fire. The National Research Institute of Fire and Disaster study materials No. 42 (1998) (in Japanese)
2. ibid. No. 45 (1999) (in Japanese)
3. ibid. No. 46 (1999) (in Japanese)
4. Konishi, T., Kikugawa, H., Iwata, Y., Koseki, H., Sagae, K., Ito, A., Kato, K.: Aerial firefighting against urban fire: mock-up house experiments of fire suppression by helicopters. Fire Saf. J. **43**, 363–375 (2008)
5. Merrington, A.C., Richardson, E.G.: The breakup of liquid jets. Proc. Phys. Soc. **59**(Part 1, No. 331), 1–13 (1947)
6. Lane, W.R.: Shatter of drops in streams of air. Ind. Eng. Chem. **43**(6), 1312–1317 (1951)
7. Demetrios, T.P.: On the breakup of viscous liquid threads. Phys. Fluids **7**(7), 1529–1545 (1995)
8. Emori, R.I., Saito, K., Sekimoto, K.: Scale models in engineering, 3rd edn. Giho-do, Tokyo (2000) (in Japanese)

Effect of Porosity on Flame Spread Along a Thin Combustible Solid with Randomly Distributed Pores

Yukito Watanabe, Akihiko Ito, and Hiroyuki Torikai

Abstract The flame spread route in a residential fire strongly depends on the arrangement of combustible materials like furniture, bedclothes, housewares, etc. In addition, the flame spread rate distinctly varies by individual heat transfer path. There are two situations in residential fires: one is the case in which the flame is extinguished while in progress, and the other is the case in which it burns out. The threshold of burnout or extinguishment may be determined by the quantity of combustible materials and their placement on the floor. Our objectives are to make clear the threshold of flame spread and to estimate the route of flame spread in residential fire. In this paper, we examined nonuniform flame spread along a thin combustible solid with randomly distributed pores. Experimental results show that the flame spread rate increases with increasing the porosity rate and reaches maximum value at around 20–30 % porosity, then decreases. The flame cannot spread and is completely extinguished at 55 % porosity in each pore diameter. In addition, we measured the characteristic length of thermal boundary layer ahead of the flame leading edge, L, using the shadowgraph method, and introduced the scale length ratio of pore diameter with the characteristic length, $S \equiv d/L$. The modified flame spread probability with the scale length ratio is unified as a function of the porosity rate.

Nomenclature

d	Pore diameter
F	Flame spread probability
F_j	Flame jumping probability
F_m	Modified flame spread probability
h	Slit width
L	Characteristic length of thermal boundary layer ahead of flame leading edge
N	Number of flame spread success
N_j	Number of flame jumping success

Y. Watanabe • A. Ito (✉) • H. Torikai
Department of Intelligent Machines and System Engineering, Hirosaki University, Hirosaki, Aomori 036-8561, Japan
e-mail: aito@cc.hirosaki-u.ac.jp

K. Saito et al. (eds.), *Progress in Scale Modeling, Volume II*,
DOI 10.1007/978-3-319-10308-2_24

N_s Average number of slits
P Porosity rate
R Load of combustible material
R_{th} Threshold of flame spread
S Scale ratio ($\equiv d/L$)
V Flame spread rate

Introduction

Generally, fires have been classified into six categories: residential fire, forest fire, vehicle fire, ship fire, aircraft fire, and other fires. Of these, residential fire has accounted for more than 50 % of all fires and a high death rate due to failing to escape from the fire [1]. The flame spread along a solid material in a residential fire is nonuniform due to different shapes and arrangements of combustible materials like furniture or bedclothes, housewares, etc. In addition, the flame spread distinctly varies by individual heat transfer path. The risk of death strongly depends on the speed and the route of flame spread. Therefore, to establish a safe escape route from a residential fire is important in order to protect people's lives.

Recently, a few papers have been published using numerical simulation based on percolation theory as a way to calculate safe escape routes in residential fires [2, 3]. Percolation theory shows a connection of particles which are randomly arranged on a grid. For example, assume that R is the rate of the existence of combustible materials and $(1-R)$ is the rate of noncombustible materials or space on a grid of room. When R is less than unity ($R < 1$), combustible materials randomly exist and are randomly distributed in the room. Thus, R has a threshold of flame spread (R_{th}) at the boundary of fire extinguished while in progress or fire that burns out in the situation of residential fire as shown in Fig. 1, [4, 5, 6]. The threshold of flame spread produces a different situation of damage. When combustible materials are distributed at $R < R_{th}$, flame cannot spread across intervening space between

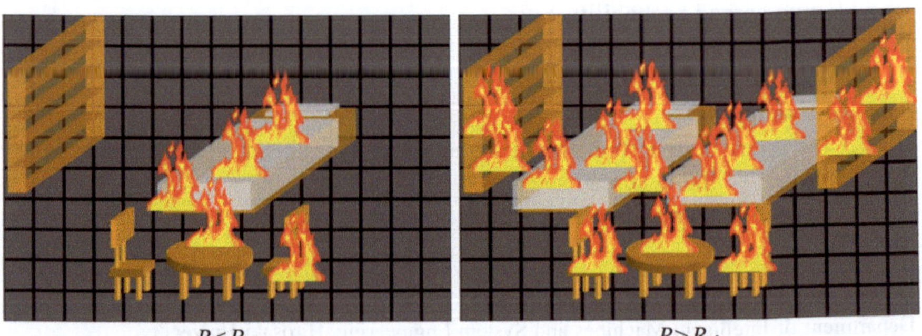

Fig. 1 Distribution of combustible materials which are run throughout a room on a grid on which are represented different situations of damage by flame spread

combustible materials, and the fire is extinguished because of lack of a viable flame spread route. When combustible materials are distributed at $R > R_{th}$, flame can spread across intervening space between combustible materials, and the fire burns out because of the existence of several viable flame spread routes. Thus, the different situations of damage are caused by various distributions and quantities of combustible materials in space. However, it has not been confirmed by experiment that percolation theory is useful to estimate whether there will be extinguishment or burnout. As a basic experiment for applying percolation theory to residential fire, we created an environment of nonuniform flame spread by randomly distributing pores on a thin combustible solid and experimented on variously changing porosity rate per total combustible solid area.

In this paper, we show threshold of flame spread for various pore diameters. In addition, we measure the characteristic length of thermal boundary ahead of the flame leading edge and introduce the scale ratio of pore diameter with the characteristic length. The flame spread probability is discussed related to the scale ratio and the porosity rate.

Experimental Apparatus and Procedure

Schematic experimental apparatus and a test sample are shown in Fig. 2. In this experiment, we used a sheet of filter paper having 200 mm length, 60 mm width, and 0.25 mm thickness as a test sample, on which many pores are opened to establish nonuniform flame spread. An approach section of 40 mm length is set up at the end of filter paper to establish uniform ignition and steady flame spread. The end of the filter paper is ignited by electrically heated nichrome wire. The filter paper is fixed on an aluminum frame tensioned for stabilizing the flame spread. A camera is set up above the test sample to record the flame spread from above. The experimental apparatus is covered by plastic sheets so that an ambient disturbance cannot affect the flame spread. We analyzed the flame spread rate by measuring the time interval from one end of the test piece to the other end using the captured images. The experiments were conducted in the atmosphere (oxygen concentration at 21 %).

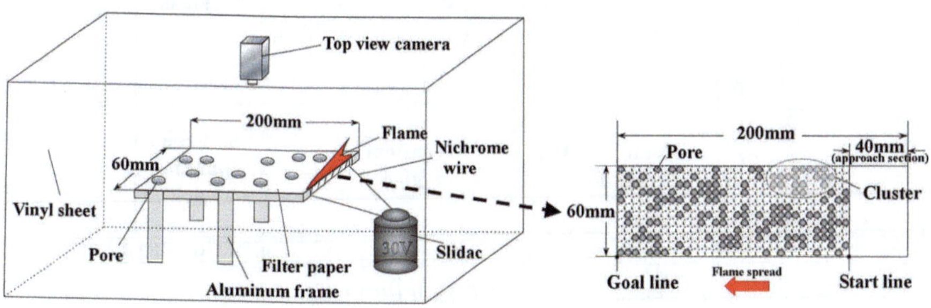

Fig. 2 Experimental apparatus and a test sample at 30 % porosity with 5 mm diameter pores

Size of Pore

The size of a pore was decided by the following method: a flame jumping test was conducted using the filter paper having a slit as shown in Fig. 3. The flame jumping probability is defined as:

$$\text{Flame jumping probability} = \frac{\text{Number of flame jumping success } (N_j)}{\text{10 types of filter paper with same slit width}} \times 100[\%] \quad (1)$$

The relationship between flame jumping probability and slit width is plotted in Fig. 3. When the slit width is less than 4 mm, the flame can always propagate. When the slit width is above 8 mm, flame cannot propagate. From these results, we selected 4 mm pore diameter for certainty of flame jumping, 5 mm pore diameter for half-and-half probability, and 8 and 15 mm pore diameter for no flame jumping.

Porosity Rate and Flame Spread Probability

The porosity rate is defined as Eq. (2), in which total porosity area is summation of pore area (each pore area × number of pores):

$$\text{Porosity rate } (P) = \frac{\text{Total porosity area}}{\text{Filter paper area (length} \times \text{width)}} \times 100[\%] \quad (2)$$

We examined ten types of pore distribution pattern per each porosity rate. To select a pore distribution pattern, pore marks are tightly arranged on a grid which was run

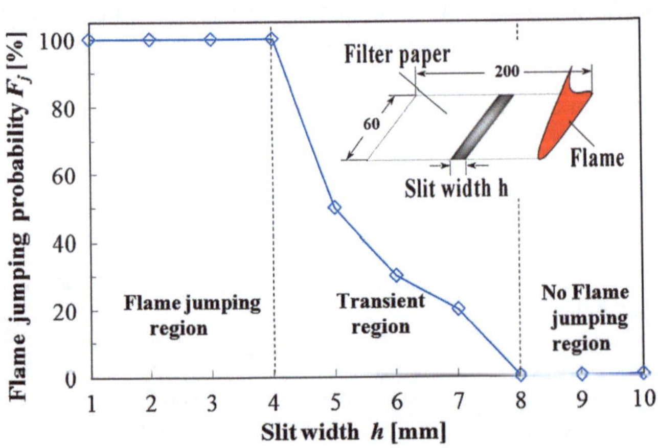

Fig. 3 Relationship between flame jumping probability and slit width

throughout a filter paper and numbered. The numbered pore marks were randomly selected according to Excel's "random numbers" function and punched. The flame spread probability is defined in Eq. (3) in which flame spread success means the flame can spread from start line to goal line:

$$\text{Flame spread probability } (F) = \frac{\text{Number of flame spread success } (N)}{\text{10 types of pores distribution pattern}} \times 100 \quad [\%] \tag{3}$$

Results and Discussion

Flame Spread Rate

The average flame spread rate was obtained by measuring the time interval between a flame spreading from start line to goal line. The relationship between the time-averaged flame spread rate, V, and the porosity rate, P, is shown in Fig. 4 for the four different pore diameters. In this figure, the flame spread rate, V_d, calculated by deRis model, is also plotted. In the deRis model [7, 8], the flame spread rate for a thermally thin solid material is expressed as follows:

$$V_d = \frac{\sqrt{2}\lambda}{\rho C_p d}\left(\frac{T_f - T_{\text{vap}}}{T_{\text{vap}} - T_\infty}\right) \tag{4}$$

where λ, ρ, C_p, d, T_f, and T_{vap} represent thermal conductivity, fuel-bed density, specific heat, solid of thickness, flame temperature, and vaporization temperature, respectively. The flame spread rate tends to increase with increasing the porosity rate from 0 % to around 30 % in each pore diameter. This tendency is caused by

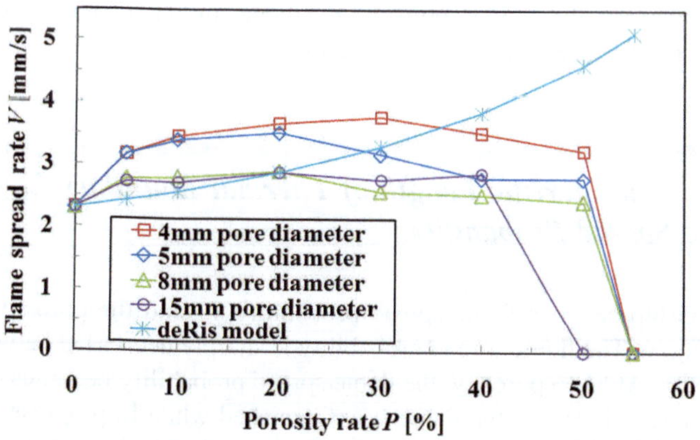

Fig. 4 Relationship between flame spread rate and porosity rate

decreasing the superficial density of filter paper in the deRis model. However, the flame spread rate is faster than that obtained by the deRis model. The difference between the experimental result and the deRis model is due to the oxygen diffusing through opening a pore on the filter paper [9]. In this experiment, we observed that a flame will jump a 4 and 5 mm pore diameter [10], while a flame cannot jump an 8 and a 15 mm pore diameter. This flame jump may be related to the preheated length ahead of the flame leading edge. The preheated length will be discussed later. The flame spread rate slightly decreases with increasing the porosity rate from 40 % to around 50 % in each pore diameter. Above 50 % porosity, the flame spread rate sharply decreases, and the flame extinguishes while in progress until about 55 % porosity. In reference to the sharply decreasing flame spread rate and flame extinction, two factors are to be considered. One factor is that clusters appear above 40 % porosity. When pores connect to each other and make a cluster, simple pore scale doubles. As a result, the flame cannot jump beyond a cluster. The other factor is that the heat required for flame spread decreases with increasing the porosity rate, and the lack of heat leads to flame extinguishment.

Cluster Distribution and Flame Spread Trajectory

The cluster distribution for four different porosities and the flame spread trajectory every 2 s are shown in Fig. 5. The term "cluster" means more than two pores are connected in both transverse and longitudinal directions. To take an example, at 5 mm pore diameter in Fig. 5, the cluster is distributed with various areas at 30 % porosity, and large clusters appear by the clusters are connected each other at 50 % porosity. Above 55 % porosity, the large cluster occupies the area from start line to goal line on the sample. The flame spread trajectory is not flat over 30 % porosity compared with that for no pore sample. At 50 % porosity, the flame spread trajectory is even more nonuniform, and the flame could spread to only one side of the goal line. Finally, at 55 % porosity, the flame could not spread from start line to goal line on the filter paper in this experiment.

Effect of Characteristic Length of Thermal Boundary Layer on Flame Spread Probability

The relationship between flame spread probability, F, and the porosity rate, P, is shown in Fig. 6. The flame spread probability is sharply decreased between 30 and 50 % porosity. At 55 % porosity, the flame spread probability becomes 0 % in each pore diameter. This means the flame is extinguished while in progress.

We observed the flame cannot jump up the slit on the filter paper on which pores are connected transversely to the direction of flame spread as shown in Fig. 7, and

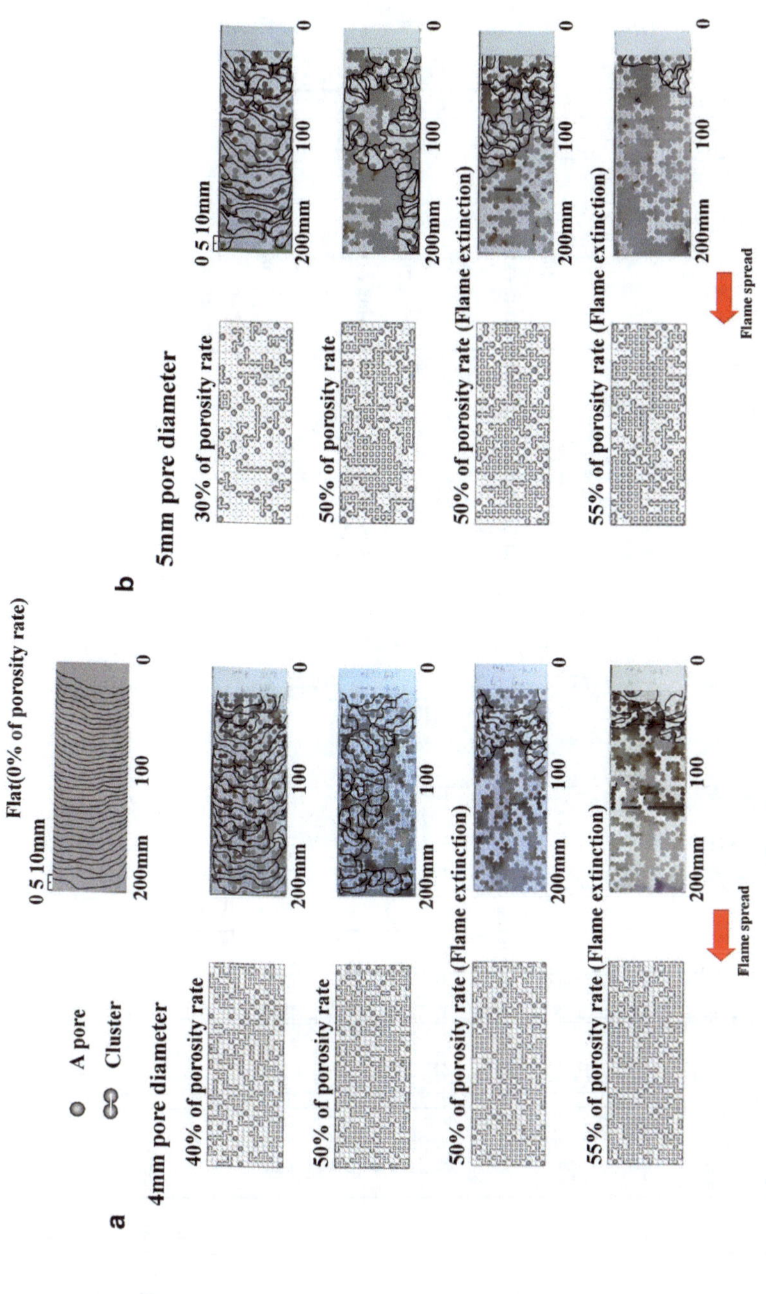

Fig. 5 Cluster distribution and flame spread trajectory for each porosity rate at (**a**) 4 mm, (**b**) 5 mm, (**c**) 8 mm, and (**d**) 15 mm pore diameter

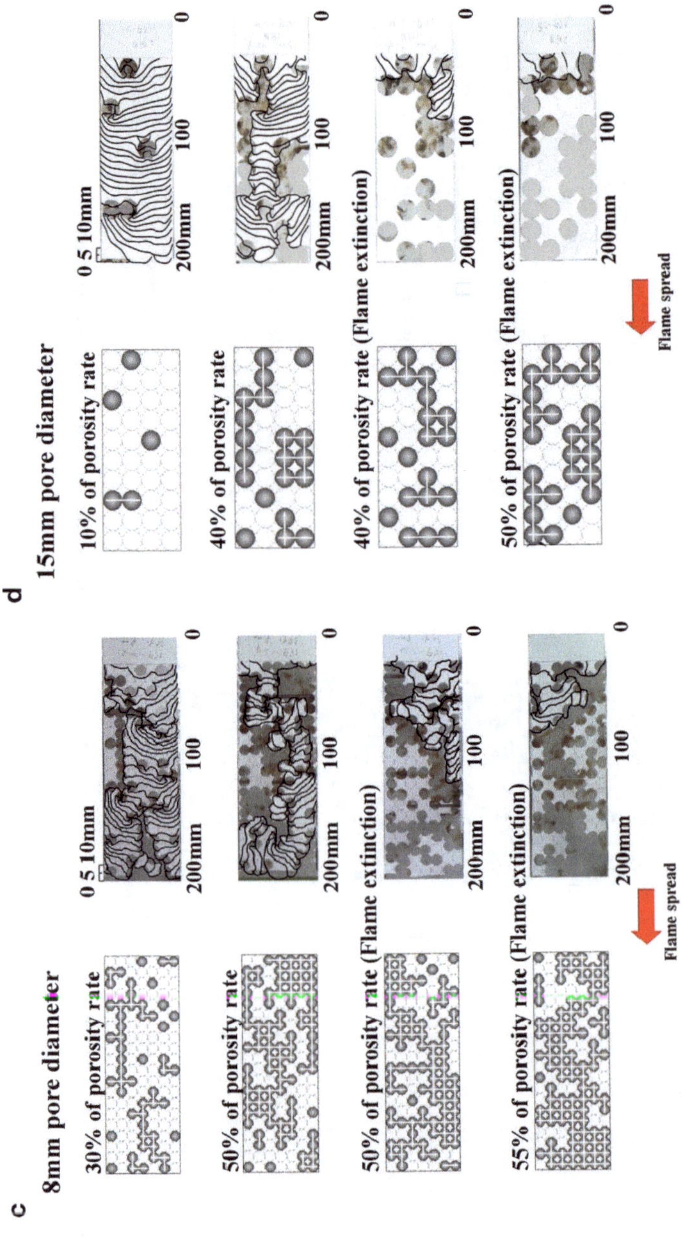

Fig. 5 (continued)

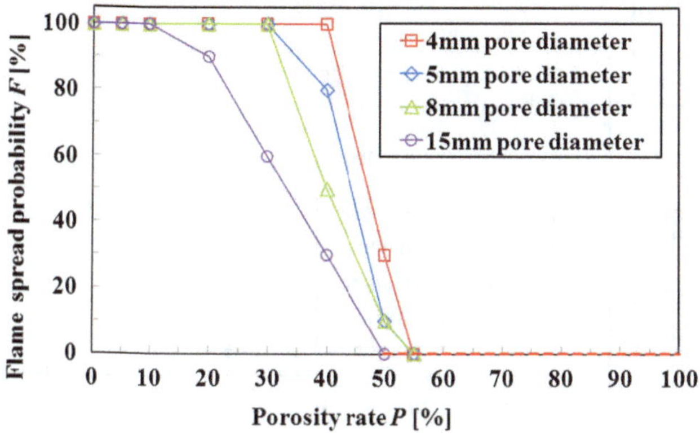

Fig. 6 Relationship between flame spread probability and porosity rate

the flame self-extinguishes while in progress. That is, flame extinguishment may be mainly attributed making the slit by connecting each pore. Therefore, we counted the average number of slits of 10 types of pore distribution patterns per porosity rate to investigate the effect of the slit on the flame spread probability. The relationship between the average number of slits, N_s, and the porosity rate, P, is shown in Fig. 7a. The average number of slits tends to sharply increase with increasing the porosity rate from 40–60 %. In addition, the slit width also affects flame extinction because the flame jumping probability changes owing to the slit width in the preliminary slit test. Therefore, we introduce the scale ratio for evaluating the effect of slits on flame spread and extinguishment. The heat transfer rate to the unburned portion just ahead of the flame leading edge is important for deciding if the flame can jump the slit or not. In fact, we measured the preheated length ahead of the flame leading edge to set the basis for the value of the slit. Shadowgraph method was employed for visualizing the thermal boundary layer, and the images were visualized in a transverse direction when the flame spread on the filter paper having 200 mm length, 60 mm width, and 0.25 mm thickness at flat (0 % porosity) (Fig. 8). The experimental apparatus and the typical visualization image are shown in Fig. 8. From the visualization image, we measured the preheated length ahead of the flame leading edge, L. Based on the preheated length without pores, we defined the scale ratio $S \equiv d/L$ as the ratio of pore scale, d, to the preheated length, L. The average number of slits is modified by multiplying the scale ratio. The modified average number of slits is shown in Fig. 7b. As a result, the increase in the average number of slits seems to be correlated with the decrease of flame spread probability and flame extinction as shown in Fig. 6. In addition, no flame spread region is divided by the contour above 5.5 of the average number of slits.

The relationship between the flame spread probability and the average number of slits is shown in Fig. 9a. From this result, the flame spread probability depends on both scale and area of pores. As the same manner, the relationship between the flame spread probability and the modified average number of slits by multiplying

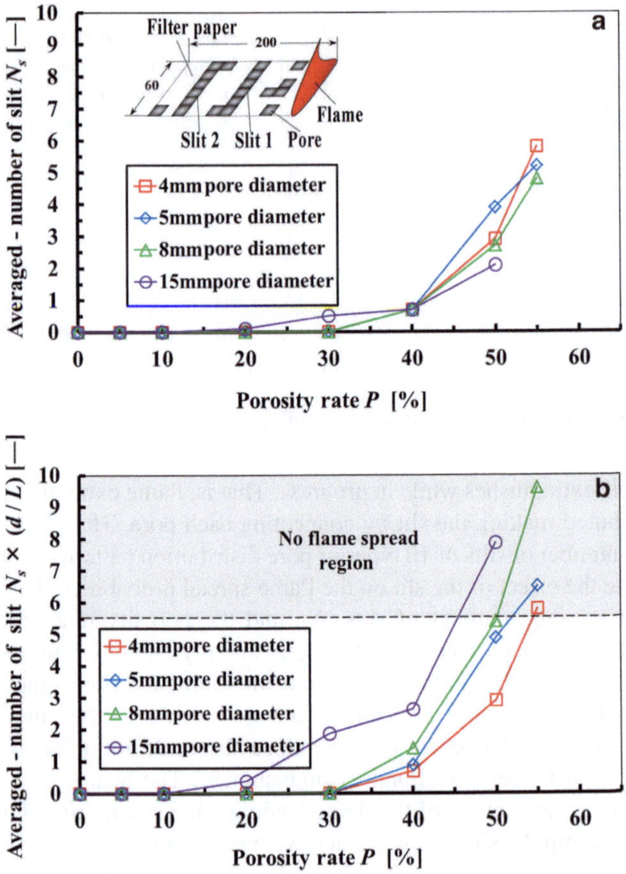

Fig. 7 (**a**) Relationship between average number of slits and porosity rate. (**b**) Relationship between modified average number of slits, scale ratio, and porosity rate

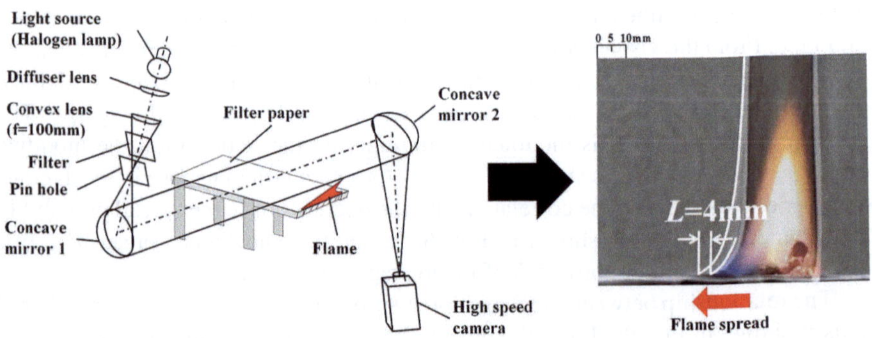

Fig. 8 Image visualized in transverse direction by shadowgraph method when flame spread on filter paper

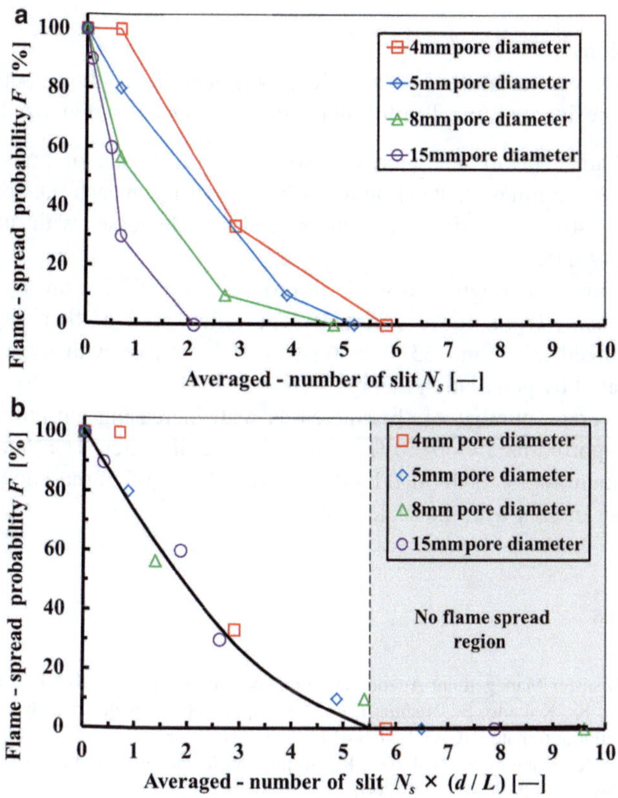

Fig. 9 (**a**) Relationship between flame spread probability and average number of slits. (**b**) Relationship between flame spread probability and modified average number of slits with scale ratio

the scale ratio is shown in Fig. 9b. As a result, the flame spread probability can be unified by the modified average number of slits with the scale ratio.

This experiment was conducted by using filter paper as combustible material when the slope of flame spread is horizontal, and then the value of L was measured. The situations of flame spread distinctly vary by each combustible material and slope of flame spread [11]; the L in each situation also may be changed. In an actual fire such as a residential fire, radiation heat transfer plays a major role in flame spread [12, 13]. Therefore, if the L can be measured in each situation, a radiation-dominant flame spread fire similarly would be discussed by using the scale ratio based on the L.

Conclusions

This study was conducted as a basic experiment for applying percolation theory to residential fire. The following conclusions are summarized here:

1. The flame spread rate increases with increasing rate of porosity and reaches maximum value around 20–30 % porosity in each pore diameter. At over 40 % porosity, the flame spread rate decreases with increasing porosity rate.
2. The flame is extinguished while in progress until 55 % porosity in each pore diameter in this experiment. Thus, threshold of flame spread is recognized at around 55 % porosity and is consist with the threshold estimated by percolation theory.
3. The average number of slits increases with increasing the porosity rate from approximately 40–60 % with each pore diameter. The flame extinguishment region is divided by the contour above 5.5 of modified average number of slits with the scale ratio.

References

1. Fire and Disaster Management Agency: White paper on fire service (2008) (in Japanese)
2. Yamamoto, K., Kokudo, S., Nishinari, K.: New approach for pedestrian dynamics by real-coded cellular automate (RCA). LNCS **4173**, 728–731 (2006)
3. Yamamoto, K., Kokudo, S., Nishinari, K.: Simulation for pedestrian dynamics by real-coded cellular. Physica A **379**, 654–660 (2007)
4. Stauffer, D., Aharony, A.: Introduction to Percolation Theory, 2nd edn. Taylor & Francis, London (1992)
5. Durrett, R.: Essentials of Stochastic Processes. Springer, New York (1999)
6. Grimmett, G.: Percolation. Springer, New York (1989)
7. deRis, J.N.: Spread of a laminar diffusion flame. Proc. Combust. Inst. **12**, 241–252 (1964)
8. Drysdale, D.: An Introduction to Fire Dynamics, 2nd edn. Wiley, New York (1998)
9. Sekimoto, T., Kudo, Y.: Flame spread over discontinuous flammable solid. Proceedings of JAFSE Annual Symposium, pp. 232–233 (2008) (in Japanese)
10. Nakamura, Y.: Irradiated ignition of solid combustibles: application to a fundamental fire research. J. Combust. Soc. Jpn. **48**(146), 328–334 (2006) (in Japanese)
11. Hirano, T., Sato, K., Tazawa, K.: Instability of downward flame spread over paper in an air stream. Combust. Flame. **26**, 191–200 (1976)
12. Ju, Y., et al.: Effect of radiative emission and absorption on the propagation and extinction of premixed gas flames. Proc. Combust. Inst. **27**, 2619–2626 (1998)
13. Ruan, J., et al.: Combined effects of nongray radiation and pressure on premixed $CH_4/O_2/CO_2$ flames. Combust. Flame. **124**, 225–230 (2001)

Epilogue: Scale Modeling and Meditation

Tadao Takeno

I first met Professor Kozo Saito, editor of this volume, during the Second International Symposium on Scale Modeling in Lexington KY in 1997. This lucky chance helped me to learn more about scale modeling and since then I have developed my personal and professional contact with him which has continued up to today. During one of my visits to his home in Lexington, I found Professor Saito was practicing daily meditation (40 min), and we had a series of discussions on meditation, whether or not it could help bring us enlightenment. Since then, somehow I felt that there may be some connection between scale modeling and meditation; both use a simple technique to capture the essence, governing physical laws in scale modeling and enlightenment in meditation. It is a challenge to reach the deepest level of achievements through a simple technique whose tradition can be traced back to centuries-old Japanese Zen practice method which only allowed a few to attain enlightenment. Scale modeling in a lesser degree shares a similar trend; the law approach in scale modeling seems simple, yet capturing the governing physics and developing scaling laws to achieve the R&D goal is not simple.

In relation to meditation, I have an interesting episode to share. During my sabbatical year in 1981 at Imperial College, London, with Professor Felix J. Weinberg, I found that he did daily meditation. His interest in meditation seems due to his unfortunate childhood experience in a Nazi concentration camp (Boy 30529: A Memoir. Verso, 2013). According to him, during his 40 min meditation, he received new ideas, jotted them down on a piece of paper, and reviewed them later. He discovered some interesting ideas among many ordinary ones. His style of meditating was unique due to his back problems caused during his time in the concentration camps, but he thought that the style should not matter as regards meditation.

I completed my sabbatical, returned to Tokyo, and a few years passed. Professor Weinberg was invited to Japan by the Japan Society for Promotion of Science, and I met him again in Tokyo. He was interested in Zen, so I accompanied him to a Zen temple in Ueno, Tokyo, where he met a young Zen monk who could speak English, and a discussion began on Zen. The Zen monk asked Professor Weinberg about his style of meditation. The professor replied: "I meditate in my own style due to my back problems." The monk replied: "Zen requires a particular style of meditation. If you can't follow that style, your meditation is fake, and any benefit from your meditation is also fake." The professor responded forcefully to justify his unique

T. Takeno
Emeritus Professor, Nagoya University, Nagoya, Japan

K. Saito et al. (eds.), *Progress in Scale Modeling, Volume II*,
DOI 10.1007/978-3-319-10308-2

style of meditation: "The style should not matter as long as you reach the ultimate goal." Unfortunately, they did not come to any agreement during this discussion.

Time passed and I met Professor Shunichi Tsuge and had a chance to share this story for his comment. Professor Tsuge immediately replied: "Professor Weinberg is wrong and the young Zen monk is right, because the absolute universal truths and laws exist beyond personal opinions and values; therefore people who are interested in learning Zen should follow the established Zen practice method to reach the ultimate."

I often wonder about that discussion, whether or not style matters when attempting to attain the ultimate, but I have not reached a conclusion yet. Therefore, I must conclude my epilogue by saying for the benefit of all scale modeling users: Enjoy scale modeling and hopefully learn how to reach the ultimate scale modeling world where knowledge and art coexist.

MIX
Papier aus verantwortungsvollen Quellen
Paper from responsible sources
FSC® C105338

If you have any concerns about our products,
you can contact us on
ProductSafety@springernature.com

In case Publisher is established outside the EU,
the EU authorized representative is:
Springer Nature Customer Service Center GmbH
Europaplatz 3, 69115 Heidelberg, Germany

Printed by Libri Plureos GmbH
in Hamburg, Germany